Distribution Centre:

Elsevier Science Publishers Ltd, Units 30–32,
Eastbury Road, The London Industrial Park,
Beckton, London E6 4LP

Sulphide deposits—their origin and processing

Sulphide deposits—their origin and processing

Editorial Committee: P.M.J. Gray, G.J. Bowyer,
J.F. Castle, D.J. Vaughan and N.A. Warner

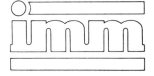 **The Institution of
Mining and Metallurgy**

Published at the office of
The Institution of Mining and Metallurgy
44 Portland Place London W1 England
© The Institution of Mining and Metallurgy 1990   ISBN 1 870706 23 4

Front cover is a false colour backscatter electron image, obtained from Cameca Camebax electron microprobe, showing colloform copper sulphides from Cattle Grid orebody, Mt. Gunson, South Australia. Photograph kindly supplied by A.R. Ramsden and D.H. French, CSIRO Division of Exploration Geoscience, Australia

Printed in England by Barnes Design + Print Group

Foreword

The sulphide deposits pose a challenge to every technology that the mineral exploitation industry uses. Their structure, composition and multi-element content could hardly be further from those of the end-products of highly refined single metals which must be won from them. The location of the deposits is rarely on the surface, close to end-product markets, and the disposal of the by-products of processing in environmentally acceptable ways is an unavoidable part of processing. The mining of sulphide deposits is probably more expensive per tonne of product than for any other of the tonnage metals since low unit cost mining methods cannot be used in deep veins or lodes or, if the sulphides are in disseminated form, the deposits are of low grade.

Research engineers and scientists are striving to relate, in usable scientific terms, the properties of these natural deposits to the physical and mechanical means of transforming a deposit into an engineering material.

The quantitative characterization of mineral textures has advanced considerably in recent years with the advent of more powerful equipment. It is now possible to identify all minerals present in any but the very finest of microstructures, measure the proportions of each present and obtain an accurate and full analysis of all the elements contained in them.

Even with these measurements we are still some way from being able to forecast the reaction of a natural combination of sulphide and gangue minerals to, for example, the application of mechanical forces such as are used in ore grinding, the rates of reaction in contact with an aqueous solution of well-defined and homogeneous characteristics or the behaviour of discrete particles of that texture when subjected to gravity, magnetic or surface forces in mineral separation.

This gap in rational understanding is not too comfortable for the process engineer to live with and, even if he has grown accustomed to it in dealing with mineral feed, it is, nevertheless, possibly still at its widest in dealing with the sulphides. Production engineers are accustomed to organizing their data, and thus their manufacturing technology, into rigid programs which can take command of day-to-day operation almost totally. It is a salutary experience for a mineral processing engineer to observe a modern motor-car or electronic component production line and realize how far he still is from understanding the basic parameters that still give him some unwelcome surprises — not infrequently at five o'clock on a Sunday morning!

This volume records some of the progress that has been achieved recently in understanding the processes by which sulphide minerals have been assembled by nature and the way in which the properties of those assemblages influence the metallurgical processes that have been derived by trial and error on a bulk scale to treat sulphides. There is still far to go.

Great advances in the technologies of medicine, electronics, telecommunications and data processing have followed from research into the fundamentals of the most basic molecular, atomic and electron units. Perhaps basic research into the fundamental forces that have formed the Earth's crust will, in time, provide metallurgists with the right tools for transforming minerals to metals on a controlled production line basis.

Meanwhile, the 'try-it-and-see' methods on the production or pilot plant and in the laboratory used by metallurgists for the sulphides are becoming more sophisticated and efficient.

The future for the major non-ferrous metals and all their many associated elements from the sulphide deposits is bright. The challenge to match the product quality and price demanded by the user is no less than it ever was.

Philip Gray
Technical Editor

July, 1990

Contents

Geology, petrology and mineralogy

Compositional and textural variations of the major iron and base-metal sulphide minerals

James R. Craig
Department of Geological Sciences, Virginia Polytechnic Institute and State University, Blacksburg, Virginia, U.S.A.
David J. Vaughan
Department of Geology, The University of Manchester, Manchester, England

ABSTRACT

The crystal structures, stoichiometries, electrical and magnetic properties, stabilities and mineral textures found in the metal sulphides are briefly reviewed. Eight of the major iron and base metal sulphide minerals, chosen because of their widespread occurrence (pyrite, pyrrhotite), role as the major ore mineral of a particular metal (chalcopyrite, sphalerite, galena, pentlandite), or importance as a carrier of rare or precious metals (arsenopyrite, tetrahedrite) are discussed in greater detail. The crystal structures and physical properties of these minerals are discussed, along with phase relations in the relevant sulphide systems. Particular emphasis is placed on the presentation of data on major and minor element compositional variations in these minerals and textural features commonly observed in ores containing them, both of which are of crucial importance in their metallurgical processing.

INTRODUCTION

The naturally occurring metal-sulphur compounds, collectively referred to as the sulphide ore minerals, serve both as actual metal sources and as the hosts for many of the world's precious, base and strategic metals. Distinct, named, sulphide species now number in the hundreds and have been variously classified on the basis of chemistry and crystal structure. Despite this large group, most completely listed in Fleischer (1987), the majority of sulphide-bearing deposits are dominantly composed of one or more of the small group of major sulphide minerals listed in Table 1. This listing of only eight minerals is admittedly arbitrary, but these minerals constitute more than 95% (and in many cases 99%) of the sulphide mineral volume in most sulphide-type

ore deposits. These minerals range from sulphides in which the principal metal extracted is a necessary, and usually dominant, constituent (e.g., galena or sphalerite) to sulphides in which the most valued metal is a minor to trace constituent (e.g., gold in pyrite or arsenopyrite). Sphalerite is an excellent example of a sulphide that serves in both of the ways noted above; it is mined for its zinc content, but it is today virtually the only source of cadmium, an element present as a minor to trace component. It also serves as a source of gallium, germanium and indium. It is important to remember that the sulphide minerals, especially pyrite, have also served as major sources of sulphur (and more rarely selenium and tellurium) as well as metals. A general listing of the major types of sulphide-rich ore deposits, as presented in Table 2, documents the abundance of these minerals in the ores. In addition, these major sulphide minerals, especially pyrite and pyrrhotite, occur as common accessory phases in a large variety of rock types.

These same minerals, so valuable as metal sources or as hosts to the minerals that contain the metals, are today also recognized as the potential sources of major environmental hazards, such as acid mine drainage and acid rain. The mineral of greatest concern in this regard is pyrite, also known as "fool's gold," because of its very superficial similarity to gold and its abundant widespread occurrence. Indeed, pyrite is overwhelmingly the most abundant metal sulphide in the crust of the earth. Its content in sulphide ores may range from only a few percent of the sulphide mass in Sudbury-type and some stratiform massive sulphide deposits, to virtually 100% of the sulphide minerals in coal beds.

The compositional variations of the major sulphide minerals are reasonably well characterized, both as a result of numerous analyses of natural samples from a wide variety

Table 1. The major iron and base-metal sulfide minerals.

Name	Ideal Formula	Principal Elements Derived (* = by-product)
pyrite (marcasite)	FeS_2	Co, Au, S
pyrrhotite	$Fe_{1-x}S$	Ni, S*
chalcopyrite	$CuFeS_2$	Cu
sphalerite	ZnS	Zn, Cd*, Ge*, Ga*, In*
galena	PbS	Pb, Ag*, Bi*, Sb*
arsenopyrite	$FeAsS$	As, Au
tetrahedrite	$Cu_{12}Sb_4S_{13}$	Cu, Sb, Ag, As
pentlandite	$(Fe,Ni)_9S_8$	Ni, Co, Pd*

1

Table 2. Abbreviated listing of the major types of sulfide ore deposits. This classification is modified and much simplified from that of Cox and Singer (1987).

Type	Major Minerals*	Metals Extracted	Examples
Ores related to mafic and ultramafic intrusions			
Sudbury nickel-copper	po, pn, py, cpy, viol	Ni, Cu, Co, PGM	Sudbury, Ontario
Merensky reef platinum	po, pn, cpy	Ni, Cu, PGM	Merensky Reef, S. Af.
			JM Reef, Montana
Ores related to felsic intrusive rocks			
Tin and tunsten skarns	py, cass, sph, cpy, wolf	Sn, W	Pine Creek, California
Zinc-lead skarns	py, sph, gn	Zn, Pb	Ban Ban, Australia
Copper skarns	py, cpy	Cu, Au	Carr Fork, Utah
Porphyry copper/ molybdenum	py, cpy, bn, mbd	Cu, Mo, Au	Bingham Canyon, Utah
			Climax, Colorado
Polymetallic veins	py, cpy, gn, sph, ttd		Camsell River, NWT
Ores related to marine mafic extrusive rocks			
Cyprus-type massive sulfides	py, cpy	Cu	Cyprus
Besshi-type massive sulfides	py, cpy, sph, gn	Cu, Pb, Zn	Japan
Ores related to subaerial felsic to mafic extrusive rocks			
Creede-type epithermal veins	py, sph, gn, cpy, ttd, asp	Cu, Pb, Zn, Ag, Au	Creede, Colorado
Almadén mercury type	py, cinn	Hg	Almadén, Spain
Ores related to marine felsic to mafic extrusive rocks			
Kuroko type	py, cpy, gn, sph, asp, ttd	Cu, Pb, Zn, Ag, Au	Japan
Ores in clastic sedimentary rocks			
Quartz pebble conglomerate gold-uranium	py, uran, Au	Au, U	Witwatersrand, S. Af.
Sandstone-hosted lead-zinc	py, sph, gn	Zn, Pb, Cd	Laisvall, Sweden
Sedimentary exhalative lead-zinc (Sedex)	py, sph, gn, cpy, asp, ttd, po	Cu, Pb, Zn, Au, Ag	Sullivan, BC
			Tynagh, Ireland
Ores in carbonate rocks			
Mississippi Valley type	py, gn, sph	Zn, Pb, Cd, Ga, Ge	SE Missouri

* Abbreviations used as follows: po = pyrrhotite, pn = pentlandite, py = pyrite, cpy = chalcopyrite, viol = violarite, cass = cassiterite, sph = sphalerite, wolf = wolframite, gn = galena, bn = bornite, mbd = molybdenite, ttd = tetrahedrite, asp = arsenopyrite, cinn = cinnabar, uran = uraninite.

of deposits, and as a result of systematic laboratory investigations of the phase equilibria. We can here only present a few relevant phase diagrams; for additional information the reader is referred to Barton and Skinner (1979) and Vaughan and Craig (1978, 1990). Table 3 contains a tabulation of the maximum concentrations of numerous elements in the common sulphide minerals. Analytical data for the common sulphide minerals is abundant but widely scattered and largely redundant in displaying minor amounts of a variety of elements. The large number of analyses results from these minerals being abundant and from the desire of investigators to ascertain the distribution of valued elements so that they can be effectively extracted. There have been relatively few extensive compilations of sulphide mineral compositional ranges. The largest (Fleischer, 1955) is now 35 years old and contains data derived primarily using analytical methods that indiscriminately included elements from mineral inclusions as well as from the mineral being studied. The development of the electron microprobe, which allows analysis of areas as small as a few micrometers, and more recently the ion probe and the proton probe (PIXE), have yielded new data that appear to give much more accurate measurements of the sulphide mineral compositions, especially for minor and trace elements. Table 3 presents a listing of the maximum contents of many elements in the major sulphides considered in this paper. Sources used were limited to those employing modern analytical techniques that should have largely avoided contamination by mineral inclusions.

STRUCTURES AND PROPERTIES OF THE MAJOR SULPHIDE MINERALS

Crystal structures

Several of the common sulphide minerals were among the first materials to be studied by X-ray crystallography, and since that time the structures of nearly all mineralogically significant sulphides have been determined. It is possible to categorize the mineral sulphides into a series of groups based on major structure types, or having key structural features in common, as shown in Table 4 (modified after Vaughan and Craig, 1978). In many cases, these are the classic structures of crystalline solids such as the rocksalt structure of the galena group (Fig. 1A), the sphalerite and wurtzite forms of ZnS (Fig. 1B,C), or the nickel arsenide structure (Fig. 1C). The disulphides are characterized by the presence of dianion (S-S, S-As, As-As, etc.) units in the structure; as well as the pyrite structure in which FeS_6 octahedral units share corners along the c-axis direction, there is the marcasite form of FeS_2 in which octahedra share edges to form chains of linked octahedra along the c-axis (Fig. 1D). The structures of $FeAs_2$ (loellingite) and FeAsS (arsenopyrite) are variants of the marcasite structure that have, respectively, shorter or alternate long and short metal-metal distances across the shared octahedral edge (see Fig. 1D). A few sulphides such as molybdenite or covellite (Fig. 1F) have layer structures, and a small number exhibit structures best characterized as containing rings or chains of linked atoms (e.g., realgar, AsS). A diverse group of sulphides, referred to by Vaughan and Craig (1978) as the metal-excess group, is composed of an unusual and diverse set of structures well illustrated by the important example of the mineral pentlandite ($(Ni,Fe)_9S_8$, see Fig. 1G).

As can be seen from Table 4, in many of these groups a number of minerals share the actual structure type, but there are also, commonly, other minerals that have structures based on these "parent" structures and that can be thought of as being "derived" from them. The relationship between a derivative structure and the parent structure may involve:

(1) Ordered substitution, e.g., the structure of chalcopyrite ($CuFeS_2$) is derived from sphalerite (ZnS) by alternate replacement of Zn atoms by Cu and Fe resulting in an enlarged (tetragonal) unit cell (see Fig. 2A). As also shown in Figure 2A, stannite (Cu_2FeSnS_4) results from further ordered substitution of half of the Fe atoms in $CuFeS_2$ by Sn.

(2) A stuffed derivative, e.g., talnakhite ($Cu_9Fe_8S_{16}$) is derived from chalcopyrite by the occupation of additional, normally empty cavities in the structure (Fig. 2B).

(3) Ordered omission, e.g., monoclinic pyrrhotite (Fe_7S_8) is derived from the NiAs structured FeS by removal of Fe atoms leaving holes (vacancies) that are ordered (Fig. 2C).

(4) Distortion, e.g., the troilite form of FeS is simply a distortion of the parent NiAs structure form (Fig. 2C).

Table 3. Maximum concentrations (in ppm unless otherwise indicated) of numerous elements in the eight major sulfide minerals discussed in the text. All data are from studies employing techniques such as electron microprobe or PIXE that are both sensitive and capable of avoiding contamination by mineral inclusions. References for the data are given in parentheses after the data: Full references are given at the end of text.

Element	pyrite	pyrrhotite	chalcopyrite	sphalerite	galena	arsenopyrite	pentlandite	tetrahedrite SS
V	32 (11)	-	-	-	-	-	-	-
Cr	11 (11)	-	-	-	-	-	-	5.7% (1)
Mn	-	-	-	11.69% (17)	158 (3)	essential	essential	13.6% (29)
Fe	essential	essential	essential	27.6% (17)	-	9.9% (33)	52.6% (20)	4.2% (8)
Co	major ss	415 (3)	-	-	-	4.3% (33)	essential	3.5% (8)
Ni	major ss	719 (3)	essential	0.21% (24)	-	-	essential	essential
Cu	40 (11)	-	essential	1.3% (26)	2000 (26)	-	-	12.7% (8)
Zn	3334 (5)	-	2570 (4)	essential	1.19% (26)	-	-	-
Ga	-	-	-	0.16% (14)	-	-	-	1.3% (28)
Ge	-	-	-	0.14% (14)	-	essential	-	30.1% (8)
As	8% (9)	-	-	-	-	-	-	41.1% (13)
Se	644 (3)	180 (3)	4383 (5)	396 (3)	3681 (3)	-	682 (4)	-
Zr	37(11)	-	-	-	-	-	-	-
Mo	-	-	-	-	7 (10)	-	80 (4)	0.8% (28)
Ru	-	-	-	-	-	-	86 (4)	-
Rh	-	5200 (4)	-	-	-	-	1.42% (4)	-
Pd	-	-	1.62% (19)	308 (5)	3.1% (10)	-	14.77% (27)	55.0% (22)
Ag	0.12% (5)	1685 (3)	-	2.84% (7)	899 (3)	-	-	11.9% (23)
Cd	-	-	1085 (5)	10.4% (31)	-	-	-	-
In	-	-	2.34% (15)	286 (3)	7 (10)	-	-	14% (8)
Sn	0.41% (5)	-	-	900 (24)	7900 (24)	-	-	37.1% (8)
Sb	-	-	-	-	200 (10)	-	-	26.4% (18)
Te	-	-	-	-	-	-	-	-
W	-	-	-	-	-	-	-	-
Pt	-	-	-	-	-	1.6% (25)	-	2.26% (32)
Au	110 (30)	1.8 (30)	7.7 (30)	-	3.4 (30)	-	-	24% (8)
Hg	17 (11)	-	-	-	-	-	-	2.6% (28)
Tl	-	-	-	-	400 (10)	-	-	6.3% (24)
Pb	51 (11)	-	-	0.38% (26)	essential	-	-	19.7% (2)
Bi	-	-	-	0.58% (24)	6.2% (10)	-	-	-

References
(1) Basu (1984)   (2) Boldryeva (1973)   (3) Brill (1989)   (4) Cabri (1984)   (5) Cabri (1985)
(6) Cabri (1989)   (7) Craig (1983)   (8) Doelter (1926)   (9) Fleet (1989)   (10) Foord (1989)
(11) Fralick (1989)   (12) Harris (1984)   (13) Johan (1982)   (14) Johan (1988)   (15) Kase (1987)
(16) Kieft (1990)   (17) Kissin (1986)   (18) Kovalenker (1986)   (19) Loucks (1988)   (20) Misra (1973)
(21) Nikitin (1929)   (22) Paar (1978)   (23) Pattrick (1985)   (24) Pearson (1988)   (25) Picot (1987)
(26) Scheubel (1988)   (27) Scott (1973)   (28) Spiridonov (1988)   (29) Godorikov (1973)   (30) Cook (1990)
(31) Burke (1980)   (32) Kovalenker (1980)   (33) Klemm (1965)

Table 4. Sulfide structural groups.

1) THE DISULFIDE GROUP

Pyrite Structure	Marcasite Structure	Arsenopyrite Structure	Loellingite Structure
FeS_2 pyrite	FeS_2 marcasite	FeAsS arsenopyrite	$FeAs_2$ loellingite
CoS_2 cattierite		FeSbS gudmundite	$CoAs_2$ safflorite
			$NiAs_2$ rammelsbergite

derived by As/S ordered substitution
(Co,Fe)AsS cobaltite
(Ni,Co,Fe)AsS gersdorffite (I)

2) THE GALENA GROUP
PbS galena
α-MnS alabandite

3) THE SPHALERITE GROUP

Sphalerite Structure $\longrightarrow$	derived by ordered substitution $\longrightarrow$	stuffed derivatives
ß-ZnS sphalerite	$CuFeS_2$ chalcopyrite	$Cu_9Fe_8S_{16}$ talnakhite
CdS hawleyite	Cu_2FeSnS_4 stannite	$Cu_9Fe_9S_{16}$ mooihoekite
Hg(S,Se) metacinnabar	Cu_2ZnSnS_4 kesterite	$Cu_4Fe_5S_8$ haycockite

4) THE WURTZITE GROUP

Wurtzite Structure $\longrightarrow$	composite structure derivatives $\longrightarrow$	?further derivatives
α-ZnS wurtzite	$CuFe_2S_3$ cubanite	$Cu_2Fe_2SnS_6$ hexastannite
CdS greenockite	$?AgFe_2S_3$ argentopyrite	

derived by ordered substitution
Cu_3AsS_4 enargite

5) THE NICKEL ARSENIDE GROUP

NiAs Structure $\longrightarrow$	distorted derivatives $\longrightarrow$	ordered ommission derivatives
NiAs niccolite	FeS troilite	Fe_7S_8 monoclinic pyrrhotite
NiSb breithauptite	CoAs modderite	Fe_9S_{10}, $Fe_{11}S_{12}$ hexagonal pyrrhotite, etc.?

6) THE THIOSPINEL GROUP
Co_3S_4 linnaeite
$FeNi_2S_4$ violarite
$CuCo_2S_4$ carrollite

7) THE LAYER SULFIDES GROUP

Molybdenite Structure	Tetragonal PbO Structure	Covellite Structure
MoS_2 molybdenite	$(Fe,Co,Ni,Cr,Cu)_{1+x}S$	CuS covellite
WS_2 tungstenite	mackinawite	~Cu_3FeS_4 idaite

8) METAL EXCESS GROUP

Pentlandite Structure	Argentite Structure	Chalcocite Structure
$(Ni,Fe)_9S_8$ pentlandite	Ag_2S argentite	Cu_2S chalcocite
Co_9S_8 cobalt pentlandite		↘?derivative
		$Cu_{1.96}S$ djurleite

Digenite Structure $\longrightarrow$	derived by ordered substitution	Nickel Sulfide Structures
Cu_9S_5 digenite	Cu_7S_4 anilite	NiS millerite
		Ni_3S_2 heazlewoodite

9) RING OR CHAIN STRUCTURE GROUP

Stibnite Structure	Realgar Structure	Cinnabar Structure
Sb_2S_3 stibnite	As_4S_4 realgar	HgS cinnabar
Bi_2S_3 bismuthinite		

In some cases, the relationships involved are more complex, as, for example, in certain of the sulphosalt minerals[*] where the resulting structure is composite and made up of slabs or units of a parent structure (or structures) arranged in some ordered fashion (an important example of a sulphosalt mineral, tetrahedrite, is further discussed below). It is also useful in certain cases (such as the stuffed derivatives or ex-amples of ordered omission mentioned above) to regard the mineral sulphide structures as a relatively "rigid" sulphur lattice framework from which metal atoms can be removed, or to which metal atoms may be added.

Stoichiometry

Many metal sulphides show evidence that the elements that comprise them are not combined in a simple whole number ratios, i.e., they exhibit non-stoichiometry.

In certain cases, the extent of deviation from a simple ratio is considerable. For example, the pyrrhotites are

---

[*] Defined as minerals with a general formula $AmTnXp$ in which common elements are A:Ag,Cu,Pb; T:As,Sb,Bi; X:S. They contain pyramidal TS_3 groups in the structure.

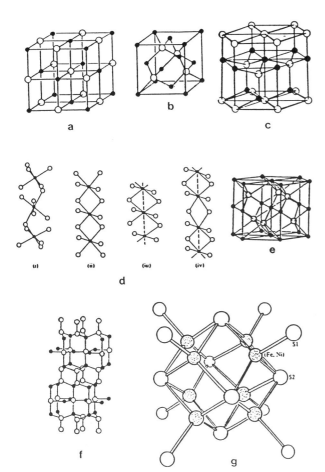

Figure 1. Crystal structures of the major sulphides: (A) galena (PbS) structure; (B) sphalerite (ZnS) structure; (C) wurtzite (ZnS) structure; (D) pyrite structure and the linkage of metal-sulphur octahedra along the c-axis direction in (i) pyrite FeS_2, (ii) marcasite FeS_2, (iii) loellingite $FeAs_2$, and (iv) arsenopyrite FeAsS; (E) niccolite (NiAs) structure; (F) covellite (CuS) structure; (G) cube cluster of tetrahedrally-coordinated metals in the pentlandite, $(Ni,Fe)_9S_8$, structure. In each case, metals are shown as the smaller or shaded spheres.

sometimes given the general formula $Fe_{1-x}S$ where $0 < x < 0.125$ and the varying compositions correspond to a varying concentration of vacancies "left" in sites that would otherwise be occupied by Fe atoms. However, in systems like these, ordering of the vacancies occurs at low temperatures, and the result may be a series of stoichiometric phases of slightly different compositions. Although Fe_7S_8 has a (monoclinic) superstructure (based on the NiAs cell) resulting from vacancy ordering (Tokonami et al., 1972), the situation in the "intermediate" or "hexagonal" pyrrhotites is more complex. Some of these pyrrhotites may represent ordered phases with clearly defined compositions (Fe_9S_{10}, $Fe_{11}S_{12}$, etc.), but more complex and partial ordering in these systems may occur. One problem is certainly that free energy differences between a series of phases resulting from vacancy ordering would be very small, making a successful investigation of relationships between synthetic products very difficult or impossible.

Other examples of non-stoichiometry involve deviations from the simple ratio that are so small as to have been ignored until fairly recently. For example, galena (PbS) exhibits a range of non-stoichiometry of 0.1 atomic %

(Bloem and Kroger, 1956). Galena is apparently stable over a wide range of values of aS_2, and at high aS_2 it has lead vacancies, whereas at low aS_2 there are sulphur vacancies. It has also been suggested that certain "polymorphic pairs" of minerals exhibit this type of non-stoichiometry, and these reports conflict with the rigid definition of polymorphism. For example, Scott and Barnes (1972) have suggested that wurtzite, formerly regarded as the high-temperature (>1020°C) polymorph of ZnS, is actually sulphur-deficient relative to sphalerite. Electrical measurements indicate that zinc vacancies occur in sphalerite and sulphur vacancies in wurtzite with a total range in non-stoichiometry of about 1 atomic %. Sphalerite-wurtzite equilibria would, therefore, be a function of aS_2 as well as temperature and pressure. However, another pair of minerals to which such reasoning might be applied are pyrite and marcasite, and here Tossell et al. (1981) have offered an interpretation that is based on the reaction mechanism by which marcasite is formed (commonly in acid solution) as a metastable species relative to pyrite. An important consequence of non-stoichiometry is the effect that it has on physical (electrical, optical, hardness) and chemical properties, although the latter have been less well studied. Preliminary studies (Vaughan et al., 1987) suggest that rates of surface alteration, that would in turn have a marked influence on such extraction methods as flotation, are strongly influenced by the stoichiometry of sulphides.

Electrical and magnetic properties

As well as exhibiting a richness and diversity in structural chemistry, the metal sulphides also show a tremendous range of electrical and magnetic behavior. As Table 5 indicates, whereas such non-transition metal sulphides as sphalerite and galena are diamagnetic, diamagnetism is also exhibited by pyrite and also marcasite (the other FeS_2 polymorph). Substitution of iron for zinc in sphalerite leads to paramagnetic behavior, and many transition metal sulphides show various forms of magnetic ordering at lower temperatures including antiferromagnetism (e.g., chalcopyrite), ferromagnetism (e.g., cattierite) and ferrimagnetism (e.g., monoclinic pyrrhotite, although "hexagonal" pyrrhotites are antiferromagnetic). Other transition metal sulphides, which are metallic conductors, exhibit the weak temperature-independent paramagnetism that is known as Pauli paramagnetism. As well as the metallic conductivity occurring in many sulphides of diverse magnetic character, numerous sulphides are semiconductors, and a smaller number (e.g., pure sphalerite) are insulators. Among semiconducting sulphides, both intrinsic and impurity (or extrinsic) conduction mechanisms occur, and conduction via electrons (n-type) or via holes (p-type) is common. Such variations are often a consequence of very minor impurities being present, or of slight non-stoichiometry (e.g., galena, PbS, exhibiting a total variation in composition of 0.1 atomic %, shows p-type conductivity in lead-deficient samples and n-type conductivity in sulphur-deficient samples).

Sulphide mineral stability

Much work has been done in the last 50 years to establish the stability relations of sulphide minerals in terms of variables that include temperature, pressure, composition and the activities of various components. Much of the data, whether derived from synthesis experiments or thermochemical measurements, are reviewed by Vaughan and Craig (1978), Mills (1974), Kostov and Mincheeva-Stefanova (1981) and sources therein. Although it is not the primary objective of the present paper to review these data, indeed that would not be possible in a relatively short contribution, certain aspects are worthy of discussion.

Table 5. Magnetic and electric properties of some major sulfide minerals
(after Vaughan and Craig, 1978).

Mineral Species	Magnetic Properties	Electrical Properties
Sphalerite (ZnS)	diamagnetic	insulator ($E_g \sim 3.7$ eV)
"Iron sphalerite" (Zn,Fe)S	paramagnetic	semiconductor ($E_g \sim 0.5$ eV for ~12 at % Fe)
Galena (PbS)	diamagnetic	semiconductor (n- and p-type, $E_g \sim 0.41$ eV)
Pyrite (FeS_2)	diamagnetic	semiconductor (n- and p-type, $E_g \sim 0.9$ eV)
Cattierite (CoS_2)	ferromagnetic ($T_c = 110$ K)	metallic
Chalcopyrite ($CuFeS_2$)	antiferromagnetic ($T_N = 823$ K)	semiconductor (n-type, $E_g \sim 0.5$ eV)
Covellite (CuS)	diamagnetic(?)	metallic
(Monoclinic) Pyrrhotite (Fe_7S_8)	ferrimagnetic ($T_c = 573$ K)	metallic
Carrollite ($CuCo_2S_4$)	Pauli paramagnetic	metallic
Pentlandite $(Ni,Fe)_9S_8$	Pauli paramagnetic	metallic

E_g = band or "energy" gap, T_c = Curie temperature; T_N = Néel temperature.

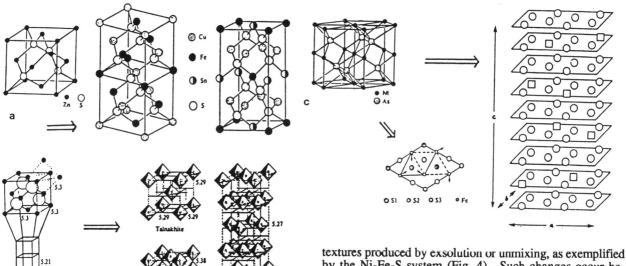

Figure 2. Parent and derivative crystal structures in the sulphide minerals: (a) the sphalerite (ZnS) structure with the chalcopyrite ($CuFeS_2$) and stannite Cu_2FeSnS_4 structures; (b) the sphalerite and chalcopyrite unit cells with an octahedron of metals outlined within which may be an additional metal ion in the minerals talnakhite ($Cu_9Fe_8S_{16}$), mooihoekite ($Cu_9Fe_9S_{16}$) and haycockite ($Cu_4Fe_5S_8$), the arrangement of additional occupied metal sites being as shown (also shown are the dimensions of the parent sphalerite cell in Å); (c) the niccolite unit cell of high temperature FeS that has vacancies in place of Fe atoms in monoclinic pyrrhotite (Fe_7S_8) that are ordered as shown in the diagram that has vacancies represented by squares (and only Fe atom layers shown); also shown in a projection onto the basal plane are the distortions that occur in the troilite modification of FeS.

In terms both of the interpretation of assemblages and textures in the context of understanding the genesis of the ores and from the point of view of mineral processing, temperature-composition relations in the subsolidus region are of the most value. This is because many sulphide minerals undergo changes in the solid state down to relatively low temperatures compared with silicates and oxides (Fig. 3). These changes give rise to a wealth of intergrowth textures produced by exsolution or unmixing, as exemplified by the Ni-Fe-S system (Fig. 4). Such changes occur because at elevated temperatures in many sulphide systems there are extensive fields of solid solution that shrink with falling temperature. In the case of the Ni-Fe-S system, it is the breakdown of the so-called monosulphide solid solution and segregation of pentlandite according to the reaction

$$(Fe,Ni)S = Fe_{1-x}S + (Fe,Ni)_9S_8$$

| (Fe,Ni)S | $Fe_{1-x}S$ | $(Fe,Ni)_9S_8$ |
| monosulphide solid solution | pyrrhotite | pentlandite |

that is central to understanding the assemblages and textures in the sulphide nickel ores. Such processes may lead to completely separate grains of the two phases being formed on exsolution, or the two phases may be intergrown as laths, blebs, etc., often with a clearly-defined crystallographic relationship. The crystal structure of the two phases will partly dictate the kind of texture that forms, but other crucial factors are related to the monosulphide solid solution bulk "starting" composition and the cooling history. The importance of kinetic factors is clear from such studies as have been undertaken (e.g., Kelly and Vaughan, 1983), although relatively little work has been done in this field. The majority of binary, ternary and many quaternary sulphide systems have been studied as regards temperature-composition relations, yielding evidence of the maximum stability temperatures of phases, solid solution limits, and coexistence of phases under the equilibrium conditions pertaining in such experiments.

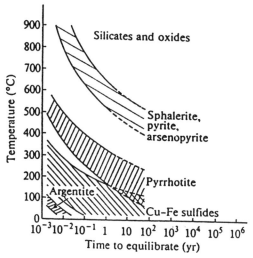

Figure 3. Relative rates of equilibration of various common sulphide minerals as a function of temperature. (After Barton, 1970.)

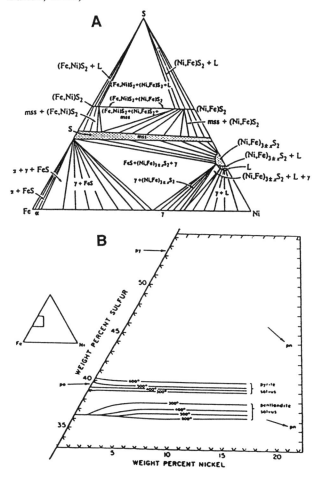

Figure 4. (A) The condensed phase relations in the Fe-Ni-S system at 650°C. Note the absence of pentlandite and the monosulphide solid solution (*mss*) that spans the system from Fe$_{1-x}$S to Ni$_{1-x}$S. (B) A portion of the Fe-Ni-S system showing the compositional limits of the monosulphide solid solution (*mss*) in (A) at 600, 500, 400, and 300°C. Pentlandite flames, as those shown in Fig. 17, exsolve from the *mss* as the sulphur-poor boundary retreats during cooling.

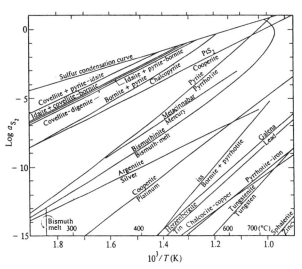

Figure 5. Sulphidation curves for several sulphides as a function of temperature.

The stable phase relations may also be defined in terms of the activities of components and, in regard to the sulphide minerals, it is the activity of sulphur (aS$_2$) that is of the most importance. Thus, a "petrogenetic grid" commonly used to present sulphide mineral relationships is that in which aS$_2$ is plotted against temperature (actually log aS$_2$ versus 1/T because sulphidation boundaries then plot as straight lines in many cases) as shown in Figure 5. Such diagrams show the progressive development of assemblages through sulphidation from metal to monosulphide to (where appropriate) disulphide.

Another important type of phase diagram is that in which the activity of sulphur is plotted against that of oxygen (at constant temperature). Such aS$_2$-aO$_2$ diagrams may be presented for one or several cations. In Figure 6 a plot of this type is shown for the iron sulphides and oxides at 25°C. In this case, of course, the value of the diagram lies in the interpretation of relationships and textures that may arise in the oxidative alteration of sulphide minerals during weathering or some analogous process.

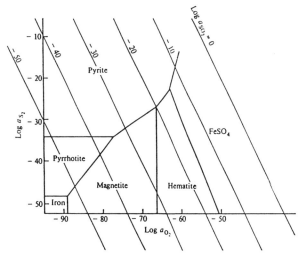

Figure 6. A log aS$_2$ versus log aO$_2$ plot of iron sulphides and oxides at 25°C contoured in terms of log aSO$_2$.

SULPHIDE ORE TEXTURES: PRIMARY AND SECONDARY DEVELOPMENT

The major sulphide ore minerals occur not only in a wide variety of deposits (Table 2) but also in an amazing array of textures. The textures, in effect the shapes and spatial distributions of mineral grains, are the products of the depositional and post-depositional histories of the minerals and the deposits in which they occur. The terminology customarily used to describe textures is summarized below:

Primary — textures reflecting the form and distribution of the minerals as they were originally deposited

Secondary — textures that have formed as the result of any post-depositional process

Hypogene — referring to the original minerals formed in a deposit

Supergene — referring to secondary minerals and textures that have formed as a result of meteoric weathering.

It is not possible to herein exhaustively present or discuss the many varieties of textures exhibited by the major sulphide minerals. We shall attempt instead to review some of the most common textures and refer the reader to Craig and Vaughan (1981) and Craig (1990) for additional discussion and illustrations.

Primary textures of the major sulphides range widely, depending upon the individual mineral phase or phases present (e.g., pyrite that typically forms crystals, Fig. 7A, versus pyrrhotite that is generally anhedral or interstitial to other phases), the environment of deposition (e.g., open space filling, Fig. 7B, versus replacement of preexisting phases, Fig. 7C), and chemical reactions (of multiple depositing solutions and between solutions and preexisting minerals).

Secondary textures of these minerals modify or replace primary textures by processes such as post-depositional deformation (e.g., brecciation, Fig. 8A, or flow), recrystallization (e.g., to form crystals or polycrystalline aggregates, Fig. 8B), exsolution (Fig. 9A), or supergene alteration (e.g., to form alteration rims, Fig. 9B, or to differentially corrode coexisting phases).

Analysis of the textures of the major sulphide minerals provides not only much insight into the origin and history of a deposit or occurrence, but also into the exploitability and potential problems of mineral processing. Thus, the very fine-grained nature of an ore (e.g., McArthur River, Australia, Fig. 10A) may result in its being very difficult to economically separate into clean concentrates by standard flotation techniques. In contrast, ores that are coarse grained as a result of primary deposition or metamorphic recrystallization (Fig. 10B) are relatively readily separated by flotation. The textures are also critical in determining which ores may readily yield precious metals during cyanide leaching (Fig. 11A) and those that are refractory (e.g., contain gold as fine inclusions within pyrite grains where it remains unreacted with cyanide solutions, Fig. 11B).

The differences in sulphide mineral structures (both crystal and electronic structures) result in a broad range in physical properties such as density, thermal stability, hardness, reflectance, electrical conductivity, etc. These variations result in significant differences in the rates and degrees of equilibration during cooling from original depositional conditions, and during post-depositional metamorphism or weathering. Thus, among mixtures of the common sulphide minerals, pyrite, arsenopyrite and sphalerite are the most refractory and the sulphides most likely to retain the evidence of their original deposition conditions, whereas chalcopyrite, galena, and pyrrhotite most readily adjust their compositions during cooling (Barton and Skinner, 1979). Similarly, under conditions of moderate dynamic metamorphism, pyrite

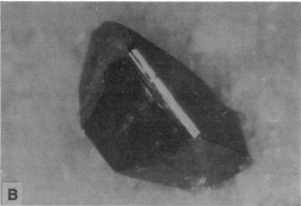

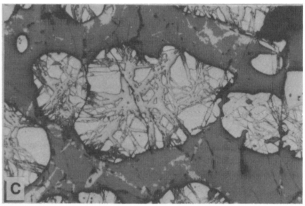

Figure 7. (A) Euhedral pyrite crystals in a matrix of interstitial anhedral pyrrhotite (field of view = 1.1 mm). (B) Euhedral 7 mm sphalerite crystal formed atop a mass of saddle dolomite crystals in an open vug in the central Tennessee ores. (C) Fractured pyrite crystal undergoing replacement by chalcopyrite as a result of reaction of the pyrite with copper-rich fluids.

remains a rigid phase retaining primary structure and composition, whereas phases such as pyrrhotite, chalcopyrite, and galena readily yield and deform plastically. This differential behavior under stress has been examined for several of the major sulphide minerals by Kelly and Clark (1975).

MAJOR IRON AND BASE METAL SULPHIDES

In the more detailed discussion that follows, the most important of the sulphides are considered in turn and aspects of their structures, chemistries and textural relationships in ores considered. A total of eight sulphides are considered in this way, chosen for their widespread occurrence (pyrite,

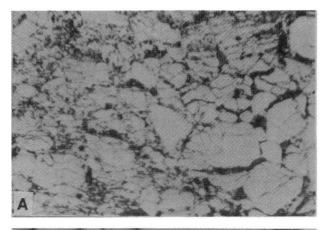

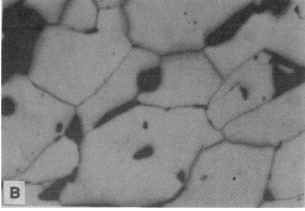

Figure 8. (A) Cataclastic texture of pyrite resulting from post-depositional deformation of sulphides (field of view = 0.6 mm). (B) Recrystallized pyrite exhibiting typical polygranular texture with 120° interstitial angles (field of view = 0.6 mm).

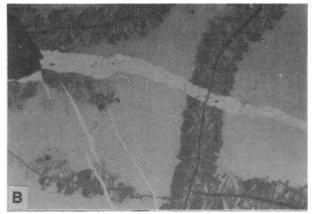

Figure 9. (A) Exsolution lamellae of chalcopyrite from bornite (field of view = 1.1 mm). (B) Veins of covellite formed on chalcopyrite as a result of supergene alteration (field of view = 1.1 mm).

pyrrhotite), role as the major ore mineral of a particular metal (chalcopyrite, sphalerite, galena, pentlandite), or importance as a carrier of rare or precious metals (arsenopyrite, tetrahedrite).

Pyrite

Pyrite (ideal formula FeS_2) is the most abundant of all sulphide minerals, occurring as a major phase in many sulphide ore deposits and as an accessory mineral in many other ores and rocks. Pyrite has rarely been mined for its iron content but has served as an ore of nickel and cobalt, both of which may substitute for iron in the structure. The pyrite of the Central African Copperbelt serves as a major source of the world's cobalt supply. Before the development of the Frasch technique for sulphur extraction, pyrite served as the major source of sulphur; today, pyrite from Iberian mines remains an important sulphur source for Europe. Because it is nearly ubiquitous and serves as a host for many other minerals containing valued metals, pyrite has been much analyzed. The maximum concentrations of many elements held within pyrite are given in Table 3.

The crystal structure of pyrite is illustrated in Figure 1D and, as already noted, is distinguished by the presence of S-S dianion units. Pyrite is a diamagnetic semiconductor (Table 4) that may exhibit p- or n-type conduction mechanisms. Marcasite (ideal formula also FeS_2) has generally been regarded as the dimorph of pyrite in which S-S dianion units again occur, and Fe is octahedrally coordinated to S. However, the mode of linkage of FeS_6 octahedral units is different to that found in pyrite (Fig. 1D).

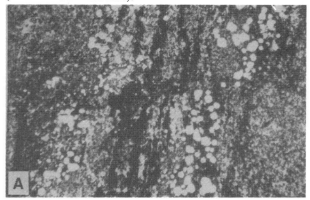

Figure 10. (A) Fine-grained, unmetamorphosed pyrite-sphalerite ore from McArthur River, Australia. (B) Coarsely recrystallized, metamorphosed sphalerite-pyrrhotite ore from Joma, Norway. Both samples have been photographed at the same scale (field of view = 0.6 mm).

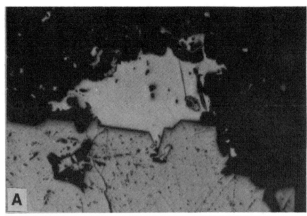

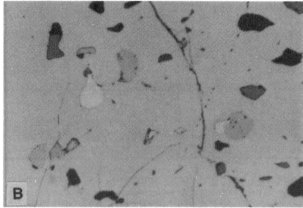

Figure 11.   (A) Gold grain along pyrite boundary where it would likely be dissolved by cyanide leach solutions. (B) A gold grain completely enclosed in pyrite where it would likely be protected from reaction from leaching solutions (field of view = 0.6 mm).

Pyrite, isostructural with cattierite (CoS_2) and vaesite (NiS_2) exhibits temperature-dependent extensive solid solutions toward both of these phases. Generally, however, the very low geochemical abundances of cobalt (25 ppm) and nickel (75 ppm) relative to that of iron (56,000 ppm) results in the formation of pyrites that contain no more than a few parts per million of cobalt and nickel. Pyrite is an important sink for cobalt and nickel and, occasionally, individual growth zones have concentrated the cobalt and nickel to form bravoites. Because pyrite is often the dominant sulphide in many ores and must be processed to extract other intimately intergrown ore minerals, much effort has been directed to determine what valuable elements the pyrite itself may contain (see Table 3). Many pyrites have been thought to contain gold within their lattices, but the highest documented value is 110 ppm (Cook and Chryssoulis, 1990).

Pyrite occurs in a very broad range of textures depending upon mode of origin, abundance relative to other minerals and post-depositional history. Although it can form in polycrystalline porous colloform bands at low temperatures, pyrite is best noted for its very common appearance as cubes (Fig. 7A) or pyritohedra. Its great tendency to form into euhedral crystals and its very great hardness (greater than all other sulphides and many silicates and oxides) results in it typically appearing as cubes, even in matrices that have undergone extensive deformation. Under conditions of thermal metamorphism, polycrystalline masses of pyrite recrystallize to typical annealed textures displaying 120° interstitial grain boundaries. Despite its hardness and refractory nature, pyrite, like all sulphides, is unstable in the oxidizing conditions of the earth's surface and most surface waters. The oxidation of pyrite, and sometimes coexisting

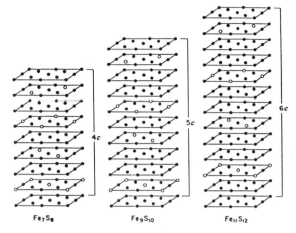

Figure 12.   Diagrammatic presentation of the structures of Fe_7S_8, Fe_9S_{10} and $Fe_{11}S_{12}$. These are the 4C, 5C and 6C pyrrhotites.

pyrrhotite or marcasite, leads to the generation of strongly acid waters that have been responsible for considerable damage to stream and vegetation. The solid phase resulting from pyrite oxidation commonly leads to the development of goethite (FeO-OH) pseudomorphous after pyrite cubes.

Pyrrhotite

Pyrrhotite (often simplistically referred to by the formula $Fe_{1-x}S$) is a major component in many important base metal deposits. It has been mined for its iron content. The structure of "pyrrhotite" is based upon the hexagonal NiAs (niccolite) structure (Fig. 1E), but this structure is only stable at higher temperatures (≳150°C). Stoichiometric FeS undergoes distortion at lower temperatures to give the troilite superstructure modification (see Fig. 2C). A very important feature of the pyrrhotites is the ability to omit iron atoms from the structure, leaving vacancies up to a composition of ~Fe_7S_8. At higher temperatures (~100-300°C), these vacancies are disordered, but at lower temperatures they undergo ordering so as to result in a series of superstructures based on the parent NiAs-type structure. The best known and characterized of these superstructures is that of "monoclinic pyrrhotite" (ideal composition Fe_7S_8) illustrated in Figure 2C. Here, the vacancies are ordered onto alternate iron atom layers parallel to the basal plane and alternate rows in these layers. The resulting structure has a unit cell with four times the c-axis dimension of the parent structure with some collapse around the vacancies causing a monoclinic distortion. The situation in the "intermediate pyrrhotites" at low temperatures is more uncertain, but it seems likely that various other vacancy-ordered superlattices are possible, centered on such compositions as Fe_9S_{10}, $Fe_{10}S_{11}$ and $Fe_{11}S_{12}$, and with 5C, 11C and 6C superstructures as illustrated in Figure 12 and further discussed by Vaughan and Craig (1990). An important aspect of the chemistry of these intermediate pyrrhotites arises from the kinetics of such vacancy ordering processes. Because the differences in free energies between various vacancy ordered structures is very small, a variety of metastable alternative structures and fine-scale intergrowths can be formed on cooling. This has led to considerable confusion in understanding equilibrium phase relations, although much progress has been made using Transmission Electron Microscopy.

The vacancy-ordered superstructures found in the pyrrhotites at low temperatures also have important consequences for their magnetic properties. Whereas FeS and the intermediate pyrrhotites are antiferromagnetically ordered at room temperature, the ordering of vacancies on to one sublattice in monoclinic pyrrhotite results in ferrimagnetism at room temperature. This property is of value in geophysical prospecting for pyrrhotite-containing ore bodies, and in the separation of monoclinic pyrrhotite by magnetic methods during mineral processing.

Although natural pyrrhotites rarely contain more than trace quantities of other transition metals, the high temperature hexagonal NiAs form of pyrrhotite exhibits complete solid solution with NiS (Fig. 4A) and CoS, and has been reported to accept up to ~6.8 wt % MnS in solid solution at 900°C (Skinner and Luce, 1971) Similarly, high temperature pyrrhotite can hold up to 4 wt % Cu in solid solution (Yund and Kullerud, 1966), but natural pyrrhotites contain only trace quantities of copper. Natural samples of the Fe-Ni-S system monosulphide solid solution (*mss*) have been reported from several localities, especially marine basalts, where rapid cooling has preserved intermediate *mss* compositions. Slow cooling apparently always results in exsolution of the nickel- and copper-bearing phases to leave a rather pure pyrrhotite.

Pyrrhotite occurs only very rarely in crystals; in ore deposits, it invariably is present as anhedral, commonly interstitial, masses. Although once considered a "high temperature" mineral, it is clear that pyrrhotite can form in deposits from magmatic temperatures above 1000°C to sea floor exhalative vents at 300°C and probably to much lower temperatures. It is now apparent that pyrrhotites in ores tend to continuously reequilibrate as temperatures drop. As noted in Figure 3, the pyrrhotites equilibrate much more readily than pyrite or sphalerite; hence, natural ores commonly actually preserve a variety of different temperatures when they are examined.

Pyrrhotites also reequilibrate rapidly in the weathering realm by suffering ready removal of iron from their lattices. As a consequence, meteoric weathering of pyrrhotite results in a progressive compositional change toward more sulphur-rich varieties (e.g., monoclinic pyrrhotite). Continued oxidation during weathering results in sufficient iron removal to form pyrite and/or marcasite.

Chalcopyrite

Chalcopyrite (ideal formula, $CuFeS_2$) is the major source of copper in almost all of the ore deposits worked for that metal. The structure of chalcopyrite (Fig. 2A) is derived from the sphalerite structure by ordered substitution of Cu and Fe for Zn. However, it is also possible to add or remove metal atoms from the structure, particularly at higher temperatures, such that extensive solid solutions exist in the Cu-Fe-S system at higher temperatures. Some (in particular, metal-rich) compositions near to that of chalcopyrite (such as talnakhite, $Cu_9Fe_8S_{16}$; mooihoekite, $Cu_9Fe_9S_{16}$; and haycockite, $Cu_4Fe_5S_8$) appear to be preserved in ores, although they may be metastable phases. Optically, they are almost impossible to distinguish from chalcopyrite and have very similar X-ray powder diffraction patterns. The structures of these stuffed derivatives of chalcopyrite are shown in Figure 2B and consist of ordered occupation of additional (normally empty) sites.

Chalcopyrite, like pyrrhotite, tends to occur as anhedral masses interstitial to other sulphides, although it has often been observed as euhedral tetrahedra lining vugs in ores and in fluid inclusions. It is known to accept significant quantities of other elements, especially zinc and tin (Table 3). The composition of chalcopyrite lies centrally within the Cu-Fe-S system; hence, it coexists with many of the common iron and copper sulphides (Fig. 13A). The weathering of chalcopyrite, as well as that of bornite and the copper sulphides, proceeds in a manner similar to that of pyrrhotite. The ready removal of iron first, and then copper, results in the driving of the copper-iron sulphide assemblages across the Cu-Fe-S system as shown by the arrow on Figure 13B.

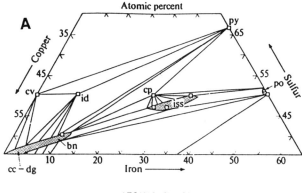

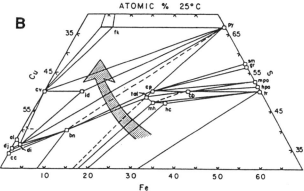

Figure 13. (A) Phase relations in the central portion of the Cu-Fe-S system; (a) schematic relations at 300°C. Abbreviations: cc, chalcopyrite; dj, djurleite; di, digenite; al, anilite; cv, covellite; bn, bornite; id, idaite; cp, chalcopyrite; tal, talnakhite; mh, mooihoekite; hc, haycockite; cb, cubanite; mpo, monoclinic pyrrhotite; hpo, hexagonal pyrrhotite; tr, troilite; py, pyrite; iss, intermediate solid solution. (B) Schematic low-temperature phase relations in the Cu-Fe-S system (after Craig and Scott, 1974). The shaded arrow indicates the general trend of copper-iron-sulphide minerals during the weathering process. The removal of iron, and subsequently copper, results in the formation of compositions richer in sulphur.

Sphalerite

Sphalerite (ideal formula, ZnS) is the only ore mineral of zinc, although it commonly contains other metals replacing Zn (most notably Fe, but also Cd, Co, Ni, Cu, Ga, Ge). The sphalerite or "zinc blende" structure is one of the fundamental structure-types found in inorganic solids (Fig. 1B). However, the other major structure type found in ZnS, the hexagonal structure of wurtzite, is also of fundamental importance (Fig. 1C). The structural relationship between sphalerite and wurtzite is such that, although both contain Zn and S in tetrahedral coordination, linking of the tetrahedra (via corners) in sphalerite results in a cubic unit cell, whereas in wurtzite the cell is hexagonal. It is also a relationship that can be viewed in terms of close-packing of the sulphur anions in layers parallel to the basal plane and the stacking of these layers to give cubic or hexagonal symmetry. Variations in stacking sequence can give rise to different polytypes and a submicroscopic interlayering of sphalerite and wurtzite structure types. As already noted, some authors have proposed that a compositional difference exists between sphalerite and wurtzite, although others would regard wurtzite formed at low temperatures as a metastable phase resulting from the mechanism of nucleation of ZnS in solution (Vaughan and Craig, 1990).

The sphalerite lattice and the similarity to the size of the Zn^{2+} ion permits the ready substitution of a number of divalent transition metal ions for the zinc. Indeed, naturally-occurring sphalerites always contain metals (dominantly iron) in addition to zinc. Iron, virtually always present, ranges from trace quantities to greater than 27 wt % and is the primary cause of color in sphalerite. Sphalerites from Mississippi Valley-type deposits nearly always contain less than 2% iron (commonly <1%) and are usually pale yellow in color, whereas those from vein and massive sulphide deposits may contain variable amounts of iron (5-10% are common) and are reddish to nearly black. The most iron-rich sphalerites occur in meteorites where values in the range of 18-27.6 wt % Fe are reported. The iron content of the sphalerite (generally quoted in terms of FeS) is controlled by the activity of sulphur. The definitive study of the Fe-Zn-S system (Barton and Toulmin, 1966) demonstrated that the FeS content of the sphalerite solid solution is buffered by coexisting iron sulphides. Using those data, Craig et al. (1984) demonstrated how knowledge of the dominant iron sulphides (pyrite and/or pyrrhotite) in massive ores can be used to predict the composition of the sphalerite and hence the purity of the zinc concentrate that is produced.

Barton and Toulmin (1966) defined, and Scott (1973) later calibrated, the sphalerite geobarometer. Application of the geobarometer is based upon the observation that increasing pressure reduces the FeS content of sphalerite equilibrated with pyrite and hexagonal pyrrhotite in a manner that is temperature-independent (Fig. 14A). Sphalerite is sufficiently refractory to preserve the original composition unless it remains in contact with pyrrhotite or chalcopyrite; low-temperature reequilibration of sphalerite generally reduces the original FeS content and results in pressure estimations that are too high.

The sphalerite structure easily accommodates a variety of elements substituting for zinc; hence, sphalerites serve as important sources of Cd, Ga, Ge, and In. Few data are available for the Ga, Ge, and In contents of commercially mined ores, but Table 3 presents values for the richest reported sphalerites.

Sphalerites from many types of ores contain disseminated to rod-like inclusions of chalcopyrite ($\pm$ pyrrhotite) in a texture (Fig. 14B) known as "chalcopyrite disease" (Barton and Berthke, 1987). The quantity of copper present in the chalcopyrite commonly far exceeds the solubility of copper in the sphalerite structure, and the disease is believed to result primarily from selective reaction of copper-bearing fluids along crystallographically-preferred directions within the sphalerite lattice.

Galena

Galena (PbS) is the only significant ore mineral of lead. The crystal structure of galena is the familiar "rock salt" structure, in which both metal and sulphur are in octahedral, six-fold coordination. This structure may be one of the building blocks found in some of the more complex sulphosalts. As already noted, galena has been shown to exhibit a total variation in Pb:S ratio of ~0.1 atomic % with both lead-deficiency and sulphur-deficiency being possible and hence both p-type and n-type semiconducting behavior.

Galena has been reported to contain small amounts of several elements (Table 3). Experimental studies have demonstrated that complete solid solution exists at high temperatures between galena (PbS) and clausthalite (PbSe) and between galena and matildite (AgBiS$_2$). Whereas the PbS-PbSe solid solution remains stable at low temperature, the PbS-AgBiS$_2$ solid solution decomposes on cooling below about 200°C. The breakdown of this solid solution commonly results in the development of very fine exsolution

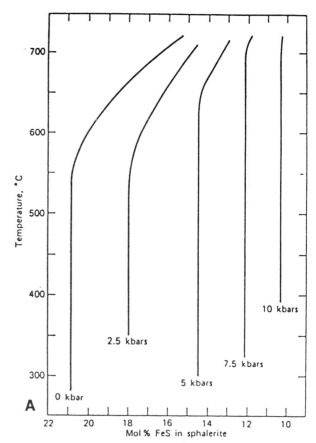

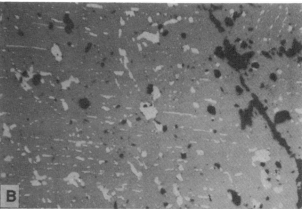

Figure 14. (A) Plot of the FeS content of sphalerite coexisting with pyrite and hexagonal pyrrhotite at 0, 2.5, 5, 7.5, and 10 kbar at temperatures from 300 to 700°C. (After Scott, 1976.) (B) "Chalcopyrite disease" consisting of blebs (actually rods) of chalcopyrite dispersed with sphalerite. See text; field of view = 0.6 mm.

lamellae of matildite in the host galena. Silver, commonly sought in galena-rich ores, is not taken into the galena structure except as a coupled substitution where the silver and bismuth atom are coupled to substitute for two lead atoms.

Arsenopyrite

Arsenopyrite (ideal formula, FeAsS) is worthy of inclusion in this review because of its importance as a carrier of gold, either as very fine particles or as so-called "invisible" gold. Arsenopyrite may also be of value to the ore mineralogist as an indicator of the temperature of formation of an assemblage (i.e., in geothermometry, Fig. 15), whereas to the mineral technologist it is a mineral more likely to cause problems during tailings disposal or smelting due to its arsenic content.

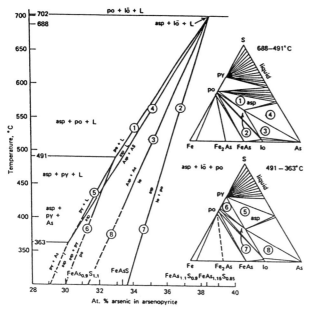

Figure 15. Pseudobinary temperature-composition plot showing arsenopyrite composition as a function of temperature and equilibrium mineral assemblage. The assemblages numbered 1-8 in the Fe-As-S phase diagrams on the right correspond to the labeled curves in the left-hand diagram. Abbreviations: asp, arsenopyrite; py, pyrite; po, pyrrhotite; lö, löllingite; L, liquid. (After Kretschman and Scott, 1976)

The crystal structure of arsenopyrite is closely related to that of marcasite. It is, of course, a "disulfide" in which Fe atoms are octahedrally coordinated to S-As dianion units. As in marcasite, the MX_6 octahedra share edges normal to the c-axis direction, but whereas in marcasite the Fe-Fe distances are of uniform length along the c-axis, in arsenopyrite there are alternate long and short Fe-Fe distances (Fig. 1D).

Arsenopyrite, even more than pyrite, exhibits a tendency to occur in well-developed subhedral to euhedral crystals. The correlation of arsenopyrite with gold has long been recognized but remains poorly understood. Most gold with arsenopyrite occurs as discrete electrum grains but Johan (1982) reported concentrations of up to 1.6 wt % gold in arsenopyrite, and Cabri et al. (1989) and Cook and Chryssoulis (1990) report 0.44% and 1.3%, respectively.

Kretschmar and Scott (1976) have demonstrated that the As content of arsenopyrite when equilibrated with pyrite and pyrrhotite varies as a function of temperature and hence can be used as a geothermometer (Fig. 15B). This has found considerable application but appears to often suffer from low-temperature reequilibration.

Tetrahedrite

Tetrahedrite is a complex sulphosalt mineral with an ideal composition of $Cu_{12}Sb_4S_{13}$ and exhibiting a complete solid solution in natural samples to tennantite ($Cu_{12}As_4S_{13}$). However, almost all natural tetrahedrites contain substantial amounts of other metals substituting for copper in the structure, notably Zn, Fe, Cd, Hg and Ag. A general formula for the tetrahedrites can be given as $M^I_{10}M^{II}_2M^{III}_4S_{13}$ (M^I = Cu, Ag; M^{II} = Fe, Zn, Cd, Mn, Cu; M^{III} = Sb, As, Bi). As a consequence, tetrahedrite may be an important ore mineral for a number of metals and is, in fact, the dominant ore mineral for extraction of silver. Thus, although volumetrically minor, the "tetrahedrites" are of considerable technological significance.

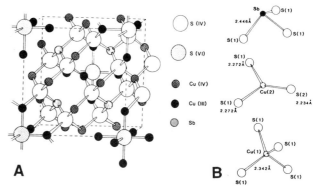

Figure 16. (A) A half-unit cell of tetrahedrite $Cu_{12}Sb_4S_{13}$. (B) Coordination polyhedra in tetrahedrite (Ref 21).

The crystal structure of $Cu_{12}Sb_4S_{13}$ tetrahedrite is shown in Figure 16. In each half-unit cell, ten M^I and two M^{III} atoms occupy six 4-fold and six 3-fold coordination sites, the M^{III} atoms occupy the equivalent of a tetrahedral site in sphalerite but are bonded to only three sulphur atoms resulting in a void in the structure; 12 S atoms are 4-coordinate and the other single S atom is 6-coordinate. Numerous studies have been undertaken of the nature and extent of the various substitutions that occur in the tetrahedrites. For example, using a variety of spectroscopic techniques, Charnock et al. (1989) were able to show that silver goes into trigonal rather than tetrahedral sites, cadmium into tetrahedral sites and iron mainly into tetrahedral sites, although in natural tetrahedrites of low silver content, it may enter the trigonal site.

The tetrahedrite-tennantite series exhibits a broad range of solid solutions (Table 3; Johnson et al., 1986). Most tetrahedrites are mined for their silver contents, and numerous studies have demonstrated that the silver contents correlate with antimony contents. It is clear that silver and arsenic have little tolerance for each other in the tetrahedrite structure. However, as antimony content rises, the capacity for silver substitution also rises.

Pentlandite

Pentlandite (($Ni,Fe)_9S_8$) is the major ore mineral of nickel. A complete solid solution occurs between pentlandite and cobalt pentlandite (Co_9S_8). The crystal structure of pentlandite is unique and contains "cube clusters" of tetrahedrally coordinated metals with very short metal-metal distances along the cube edges; the additional (ninth) metal atom in pentlandite occurs in octahedral coordination. The crystal chemistry and mineral chemistry of pentlandite exhibits a number of interesting features. One of these is that it undergoes substantial volume contraction on cooling, evidenced in natural samples by the presence of numerous fractures. It is also important to note that pentlandite almost always forms as a subsolidus phase in the Fe-Ni-S system (see Fig. 4).

Pentlandite formation occurs primarily as a result of exsolution in the form of so-called "flames" in nickeliferous pyrrhotite. Exsolution occurs such that pentlandite (111), (110) and (112) planes are parallel to the (001), (110) and (100) planes of the pyrrhotite (Francis et al., 1976). The exsolution occurs as a consequence of the shrinkage of the $(Fe,Ni)_{1-x}S$ monosulphide solid solution (mss) field during cooling. The mss phase that spans the Fe-Ni-S system at high temperatures (Fig. 4A) beings to thin (Fig. 4B) as temperature falls. Consequently, pentlandite exsolves as the

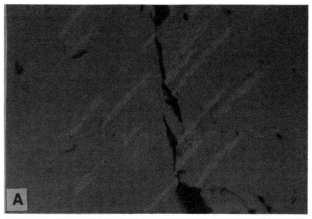

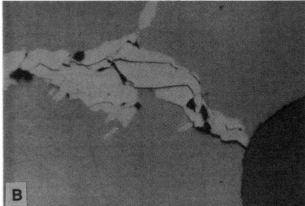

Figure 17. (A) Pentlandite "flames" formed as a result of crystallographically-oriented exsolution from *mss* on cooling (field of view = 0.6 mm). (B) Chain-like masses of pentlandite between grains of pyrrhotite. The characteristic fractures in the pentlandite are believed to result from the high coefficient of thermal expansion of the pentlandite during cooling.

sulphur-poor boundary of the *mss* retreats towards more sulphur-rich compositions.

Kelly and Vaughan (1983) have experimentally traced the development of the pentlandite textures that result from exsolution. Initial exsolution at the highest temperatures results in the formation of discrete blebs of pentlandite along the grain boundaries of the *mss*. During continued exsolution the blebs grow and coalesce to form continuous rims between the *mss* grains. At lower temperatures, where rates of diffusion and exsolution are far slower, rims no longer form, but rather there are developed the fine blades and flames shown in Figure 17A. No doubt, some of the earliest formed lamellae coalesce into the rims, but because the pentlandite exsolution continues during cooling, new flames continue to be formed as prior ones disappear. The pentlandite chains are characteristically highly fractured (Fig. 17B), apparently because pentlandite has a coefficient of thermal expansion two to 10 times greater than that of the sulphides (e.g., pyrrhotite and chalcopyrite) with which it is commonly associated (Rajamani and Prewitt, 1975). The greater shrinkage suffered by pentlandite results in greater stress and subsequent cracking.

Pentlandite has traditionally been considered an ore mineral of nickel that yields by-product cobalt. In the past 20 years, however, pentlandites have been found that also carry significant quantities of silver (Scott and Gasparrini, 1973) and ruthenium and rhodium (Cabri et al., 1984) as shown in Table 3.

ACKNOWLEDGMENTS

The authors acknowledge support from NERC Grant GR3/6823 for collaborative research on iron sulphides and the help of a large group of mineralogists and ore petrologists who have stimulated their curiosity toward, and educated them about, sulphides. JRC also acknowledges USBM allotment grant G1194151 to Virginia Tech that helps support his sulphide work, and DJV acknowledges SERC Grant GR/F/62841 that supports work on copper and copper-iron sulphides.

REFERENCES

Barton, P.B. and Berthke, P.M. (1987) Chalcopyrite disease in sphalerite: pathology and epidemiology. American Mineralogist, 72, 451-467.

Barton, P.B. and Skinner, B.J. (1979) Sulfide mineral stabilities. In Barnes, H.L. (ed.), Geochemistry of Hydrothermal Ore Deposits, 2nd edition. Wiley-Interscience, New York, p. 278-403.

Barton, P.B. and Toulmin, P. (1966) Phase relations involving sphalerite in the Fe-Zn-S system. Economic Geology, 61, 815-849.

Basu, K., Bortnikov, N.S., Mookherjee, A., Mozgova, N.N., Svitsov, A.V., Tsepin, A.I. and Vrubeskaja, Z.V. (1984) Rare minerals from Rajpura-Dariba, Rajasthan, India. V: The first recorded occurrence of a manganoan fahlore. Neues Jahrbuch Mineralogie, Abhandlungen, 149, 105-112.

Bloem, J. and Kroger, F.A. (1956) The P-T-X phase diagram of the lead-sulfur system. Z. Phys. Chem. NF, 7, 1-14.

Boldryeva, M.M. and Borodayev, Y.S. (1973) Zinc-bismuth tetrahedrite, a new variety of grey ore. Transactions USSR Academy of Science, Earth Science Sections, 212, 180-181.

Brill, B.A. (1989) Trace element contents and partitioning of elements in ore minerals from the CSA Cu-Pb-Zn deposit, Australia. Canadian Mineralogist, 27, 263-274.

Buerger, N.W. (1934) The unmixing of chalcopyrite from sphalerite. American Mineralogist, 19, 526-530.

Burke, E.A.J. and Kieft, C. (1980) Roquesite and Cu-In-bearing sphalerite from Långban, Bergslagen, Sweden. Canadian Mineralogist, 18, 361-363.

Cabri, L.J., Blank, H., El Goresy, A., LaFlamme, J.H.G., Nobiling, R., Sizgoric, M.B. and Traxel, K. (1984) Quantitative trace-element analyses of sulfides from Sudbury and Stillwater by proton microprobe. Canadian Mineralogist, 22, 521-542.

Cabri, L.J., Campbell, J.L., LaFlamme, J.H.G., Leigh, R.G., Maxwell, J.A. and Scott, J.D. (1985) Proton-microprobe analysis of trace elements in sulfides from some massive-sulfide deposits. Canadian Mineralogist, 23, 133-148.

Cabri, L.J., Chryssoulis, S.L., de Villiers, J.P.R., LaFlamme, J.H.G. and Buseck, P.R. (1989) The nature of "invisible" gold in arsenopyrite. Canadian Mineralogist, 27, 353-362.

Charnock, J.M., Garner, C.D., Pattrick, R.A.D. and Vaughan, D.J. (1989) Coordination sites of metals in tetrahedrite minerals determined by EXAFS. Journal of Solid State Chemistry, 82, 279-289.

Chryssoulis, S.L., Cabri, L.J. and Lennard, W. (1989) Calibration of the ion microprobe for quantitative trace precious metal analyses of ore minerals. Economic Geology, 84, 1684-1689.

Cook, N.J. and Chryssoulis, S.L. (1990) Concentrations of "invisible gold" in the common sulfides. Canadian Mineralogist, 28, 1-16,

Cox, D.P. and Singer, D.A. (1987) (eds.) Mineral Deposit Models. U.S. Geological Survey Professional Paper 1693.

Craig, J.R. (1990) Textures of the ore minerals. In Jambor, J.L. and Vaughan, D.J. (eds.), Advanced Microscopic Studies of Ore Minerals. Mineralogical Association of Canada Short Course Series, vol. 17, p. 213-262.

Craig, J.R. and Vaughan, D.J. (1981) Ore Microscopy and Ore Petrography. Wiley Interscience, New York.

Craig, J.R. and Vaughan, D.J. (1983) Growth characteristics of sphalerite in Appalachian zinc deposits. In G. Kisvarsanyi et al. (eds.), Proc. Int'l Conf. on Mississippi Valley Type Lead-Zinc Deposits. University of Missouri-Rolla, p. 317-327.

Craig, J.R. and Vaughan, D.J. (1986) Paragenetic studies of growth-banded sphalerites in Mississippi Valley-type zinc deposits of the Appalachians. In J.R. Craig et al. (eds.), Mineral Paragenesis. Theophrastus Publishers, S.A. Athens, Greece, p. 153-158.

Craig, J.R., Ljokjell, P. and Vokes, F.M. (1984) Sphalerite compositional variations in sulfide ores of the Norwegian Caledonides. Economic Geology, 79, 1727-1735.

Doelter, C. (1926) Handbook of Mineral Chemistry.

Eldridge, C.S., Boucier, W.L., Ohmoto, H. and Barnes, H.L. (1988) Hydrothermal inoculation and incubation of the chalcopyrite disease in sphalerite. Economic Geology, 83, 978-989.

Fleet, M.E., MacLean, P.J. and Barbier, J. (1989) Oscillatory-zoned As-bearing pyrite from stratabound and stratiform gold deposits: an indicator of ore-fluid evolution. Economic Geology Monograph 6, 356-362.

Fleischer, M. (1955) Minor elements in some silfide minerals. In Bateman, A.M. (ed.), Economic Geology Fiftieth Anniversary Volume, p. 970-1024.

Fleischer, M. (1987) Glossary of Mineral Species, 5th edition. The Mineral. Record, Inc., Tucson, 234 p.

Foord, E.E. and Shawe, D.R. (1989) The Pb-Bi-Ag-Cu-(Hg) chemistry of galena and some associated sulfosalts: a review and some new data from Colorado, California, and Pennsylvania. Canadian Mineralogist, 27, 363-382.

Fralick, P.W., Barrett, T.J., Jarvis, K.E., Jarvis, I., Schnieders, B.R. and Vandekemp, R. (1989) Sulfide-facies iron formation at the Archean Morley occurrence, Northwestern Ontario: contrasts with oceanic hydrothermal deposits. Canadian Mineralogist, 27, 601-616.

Francis, C.A., Fleet, M.E., Misra, K.C. and Craig, J.R. (1976) Orientation of exsolved pentlandite in natural and synthetic nickeliferous pyrrhotite. American Mineralogist, 61, 913-920.

Goble, R.J. (1980) Copper sulfides from Alberta: yarrowite, Cu_9S_8, and spionkopite, $Cu_{30}S_{28}$. Canadian Mineralogist, 18, 511-518.

Goble, R.J. and Robinson, G. (1980) Geerite, $Cu_{1.60}S$, a new copper sulfide from Dekalb Township, New York. Canadian Mineralogist, 18, 519-523.

Godorikov, A.A. and Il'yasheva, N.A. (1973) Chemical characteristics of fahlores. Int'l Geology Review, 15, 1413-1422.

Harris, D.C., Cabri, L.J. and Nobiling, R. (1984) Silver-bearing chalcopyrite, a principal source of silver in the Izok Lake massive-sulfide deposit: confirmation by electron- and proton-microprobe analyses. Canadian Mineralogist, 22, 493-498.

Hutchison, M.N. and Scott, S.D. (1981) Sphalerite geobarometry in the system Cu-Fe-Zn-S. Economic Geology, 76, 143-153.

Johan, Z. (1988) Indium and germanium in the structure of sphalerite: an example of coupled substitution with copper. Mineralogy and Petrology, 39, 211-229.

Johan, Z., Picot, P. and Ruhlmann, F. (1982) Paragenetic evolution of the uranium mineralization rich in selenium at Chaméane (Puy-de-Dôme), France: chaméanite, geffroyite and giraudite, three new selenides of Cu, Fe, Ag and As. Tscher. Mineral. Petrogr. Mitt., 29, 151-167.

Johnson, N.E., Craig, J.R. and Rimstidt, J.D. (1986) Compositional trends in tetrahedrite. Canadian Mineralogist, 24, 385-397.

Kase, K. (1987) Tin-bearing chalcopyrite from the Izumo vein, Toyoha Mine, Hokkaido, Japan. Canadian Mineralogist, 25, 9-13.

Kelly, D.P. and Vaughan, D.J. (1983) Pyrrhotite-pentlandite ore textures: a mechanistic approach. Mineralogical Magazine, 47, 453-463.

Kelly, W.C. and Clark, B.R. (1975) Sulfide deformation studies, III. Experimental deformation of chalcopyrite up to 2000 bars and 500°C. Economic Geology, 70, 431-453.

Kieft, K. and Damman, A.H. (1990) Indium-bearing chalcopyrite and sphalerite from the Gåsborn area, West Bergslagen, Central Sweden. Mineralogical Magazine, 54, 109-112.

Klemm, D.D. (1965) Synthesen und Analysen in den Dreiecks-diagrammen FeAsS-CoAsS-NiAsS und FeS_2-CoS_2-NiS_2. Neues Jahrbuch Mineralogie, Abhandlungen, 103, 205-255.

Kostov, I. and Mincheeva-Stefanova, J. (1981) Sulphide Minerals: Crystal Chemistry, Paragenesis and Systematics. Bulgarian Academy of Sciences.

Kovalenker, V.A. and Rusinov, V.L. (1986) Goldfieldites: peculiarities of the chemical composition, paragenesis and conditions of formation. Mineralogicheskiy Zhurnal, 8, 57-70.

Kovalenker, V.A. and Troneva, N.V. (1980) On the gold-bearing fahlore. Sulphosalts, platinum minerals and ore microscopy. Proceedings XI General Meeting of IMA Moscow, p. 75-83.

Kretschmar, U. and Scott, S.D. (1976) Phase relations involving arsenopyrite in the system Fe-As-S and their application. Canadian Mineralogist, 14, 364-386.

Loucks, R.R. and Petersen, U. (1988) Polymetallic epithermal fissure vein mineralization, Topia, Durango, Mexico: Part II. Silver mineral chemistry and high resolution patterns of chemical zoning in veins. Economic Geology, 83, 1529-1559.

Marignac, C. (1989) Sphalerite stars in chalcopyrite: are they always the result of an unmixing process? Mineralium Deposita, 24, 176-182.

McLimans, R.K., Barnes, H.L. and Ohmoto, H. (1980) Sphalerite stratigraphy of the Upper Mississippi Valley zinc-lead district, southwest Wisconsin. Economic Geology, 75, 351-361.

Mills, K.C. (1974) Thermodynamic Data for Inorganic Sulphides, Selenides and Tellurides. Butterworth, London.

Nikitin, W.W. (1929) Parallel intergrowths of fahlore and their chemical constitution. Zeitschrift für Kristallographie, 69, 482-502.

Paar, W.H., Chen, T.T. and Gunther, W. (1978) Extremely silver-rich freibergite in the Pb-Zn-Cu ores of the Bergbaues "Knappenstube," Hoctor, Salzberg. Carinthia 11, 168, 35-42.

Pattrick, R.A.D. (1985) Pb-Zn and minor U mineralization at Tyndrum, Scotland. Mineralogical Magazine, 49, 671-681.

Pearson, M.F., Clark, K.F. and Porter, E.W. (1988) Mineralogy, fluid characteristics, and silver distribution at Real de Angeles, Zacatecas, Mexico. Economic Geology, 83, 1737-1759.

Picot, P. and Marcoux, E. (1987) Nouvelles données sur la métallogénie de l'or. Academy of Science [Paris] Comptes Rendus, 304, 221-226.

Rajamani, V. and Prewitt, C.T. (1975) Thermal expansion of the pentlandite structure. American Mineralogist, 60, 39-48.

Scheubel, F.R., Clark, K.F. and Porter, E.W. (1988) Geology, tectonic environment, and structural controls in the San Martin de Bolaños District, Jalisco, Mexico. Economic Geology, 83, 1703-1720.

Scott, S.D. and Barnes, H.L. (1972) Sphalerite-wurtzite equilibria and stoichiometry. Geochimica et Cosmochimica Acta, 36, 1275-1295.

Scott, S.D. and Barnes, H.L. (1973) Experimental calibration of the sphalerite geobarometer. Economic Geology, 68, 466-474.

Scott, S.D. and Gasparrini, E. (1973) Argentian pentlandite, $(Fe,Ni)_8AgS_8$, from Bird River, Manitoba. Canadian Mineralogist, 12, 165-168.

Skinner, B.J. and Luce, F.D. (1971) Solid solutions of the type (Ca,Mg,Mn,Fe)S and their use as geothermometers for the enstatite chondrites. American Mineralogist, 56, 1269-1297.

Spiridonov, E.M., Kachalovskaya, V.M. and Chvileva, T.N. (1988) Thallium-bearing hakite, a new fahlore variety. Trans. USSR Academy of Science, Earth Science Sections, 290, 206-208.

Tokonami, M., Nishiguchi, K. and Morimoto, N. (1972) Crystal structure of a monoclinic pyrrhotite (Fe_7S_8). American Mineralogist, 57, 1066-1080.

Tossell, J.A., Vaughan, D.J. and Burdett, J.K. (1981) Disulfide minerals: crystal chemical and structural principles. Physics and Chemistry of Minerals, 7, 177-184.

Vaughan, D.J. and Craig, J.R. (1978) Mineral Chemistry of Metal Sulfides. Cambridge University Press.

Vaughan, D.J. and Craig, J.R. (1990) Sulfide ore mineral stabilities. In Barnes, H.L. and Ohmoto, H. (eds.), Hydrothermal Processes: Applications to Ore Genesis.

Vaughan, D.J., Tossell, J.A. and Stanley, C.J. (1987) The surface properties of bornite. Mineralogical Magazine, 51, 285-293.

Wiggins, L.B. and Craig, J.R. (1980) Reconnaissance of the Cu-Fe-Zn-S system: sphalerite phase relations. Economic Geology, 75, 742-751.

Wu, X. and Delbove, F. (1989) Hydrothermal synthesis of gold-bearing arsenopyrite. Economic Geology, 84, 2029-2932.

Yund, R.A. and Kullerud, G. (1966) Thermal stability of assemblages in the Cu-Fe-S system. Journal of Petrology, 7, 454-488.

Rio Tinto deposits — geology and geological models for their exploration and ore-reserve evaluation

F. García Palomero
Rio Tinto Minera, S.A. Minas de Riotinto, Huelva, Spain

SYNOPSIS

This paper describes the geology of the Rio Tinto area, and the morphological and genetic relationships between different mineralisation types (stockwork, massive sulphide and gossan) in each of the three deposits (San Dionisio, San Antonio and Cerro Colorado). The geological relationships and spatial distribution of the ore mineralisation, and associated alteration minerals, define a mineralisation model related to shallow submarine volcanic activity.

Mineralisation begins with precipitation of sulphides in fissures in the volcanic hostrocks at estimated depths of 400 metres below the seafloor, and temperatures of 400ºC. Sulphide precipitation from ascending hydrothermal fluids increases towards the seafloor, and leads to the development of mineralised stockworks as the temperatures decrease to 100ºC at the top of the volcanic pile. The remaining sulphur and metals are deposited on the seafloor in massive sulphide lenses. The distribution of minerals in haloes to both types of mineralisation are related to cooling of the orefluids as they move outwards from the source area.

After the cessation of volcanic activity in the area flysch sedimentation continued followed by tectonism associated with the Hercynian orogeny. Subsequent erosion and strong weathering of the deposits, during the late Tertiary period, led to the formation of extensive gossan deposits.

The geological model of metal distribution in the sulphide and gossan deposits has been successfully used as a guide in exploration for new reserves, and also as a control on subsequent reserve evaluation using geostatistical methods.

INTRODUCTION

The Riotinto sulphide deposits are located in the Huelva province of southern Spain, 90 km northwest of Seville and 70 km northeast of Huelva (Figure 1).

Mining activity in the area dates from pre-Roman times, with a peak of activity between the third and first centuries BC (Rothenberg et al, 1987)[1], as illustrated by the large amount of slags (6 million tonnes) derived from treatment of gold and silver ore.

After a long interval during which only sporadic mining occurred, production of copper from the zones of secondary sulphide enrichment, with grades of 3-5% Cu, began in middle of the last century. Towards the end of the last century, as this material was worked out, activity switched to the exploitation of the primary massive sulphides for sulphur production. With the drop in pyrite prices, at the end of the 1960's, new mining operations were started for the production of copper from the low grade sulphide stockworks (enriched by supergene activity near the surface), and production of gold and silver from the surfical oxide cap (Gossan) formed by the weathering of underlying sulphides.

GENERAL GEOLOGY

On a regional scale, the Riotinto deposits lie in the Iberian Pyrite Belt, an upper Palaeozoic unit containing volcanics (andesites and rhyolites) at the base of the Carboniferous, with which the abundant sulphide deposits are associeted both spatially and genetically (Figure 1).

Riotinto anticline where it displays intense chloritic alteration associated with the mineralisation process. It shows a fairly uniform thickness of 400-500 metres in the Riotinto area.

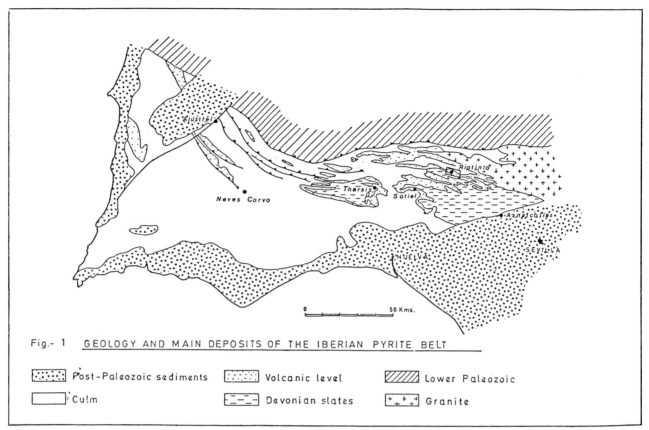

Fig.- 1   GEOLOGY AND MAIN DEPOSITS OF THE IBERIAN PYRITE BELT

[:::::] Post-Paleozoic sediments   [:::::] Volcanic level   [////] Lower Paleozoic

[___] Culm   [=-=] Devonian slates   [+ + +] Granite

Figure 2 shows the geology of the Riotinto area, where sulphide mineralisation is associated with the volcanic rocks of the Riotinto anticline (García Palomero, 1974)[2], being overlain by barren rocks of Culm facies. The main lithostratigraphic units are described briefly below.

Devonian

This unit outcrops to the south of the Riotinto area, in the core of a mainly volcanic anticlinal structure. It consists mainly of grey argillaceous slates with some quartzitic levels near the top. The first evidence for Devono-Carboniferous volcanic activity, dated as Tournaisian (Schermerhorn, 1971)[3], occurs near the top of this unit, .

Basic Complex

This unit conformably overlies the Devonian, and consists of alternations of basic pyroclastics, vesicular andesites, slates, fine grained acid pyroclastics and a variety of breccias.

   It outcrops to the south of the area, as well as in the core of the main

Acid Complex

This unit consists of very homogeneous acid volcanics (dacitic to rhyolitic in composition), with predominant pyroclastics, and locally abundant breccias, autobreccias and lavas. These rocks have a porphyritic texture with phenocrysts of quartz and feldspar in a microcrystalline matrix of the same composition, which shows a clear schistosity and frequent alteration (chlorite, haematite, and sericite).

   The thickness of the acid volcanic horizon is generally fairly uniform and similar to that of the Basic Complex (400-500m), although in the area of the mineralisation it diminishes considerably thinning to only 30-50 metres in the centre of Corta Atalaya (Figure 2).

Transition Series

This is typically a heterogeneous pyroclastic level with abundant lateral and vertical changes in both grain size and composition of the constituent

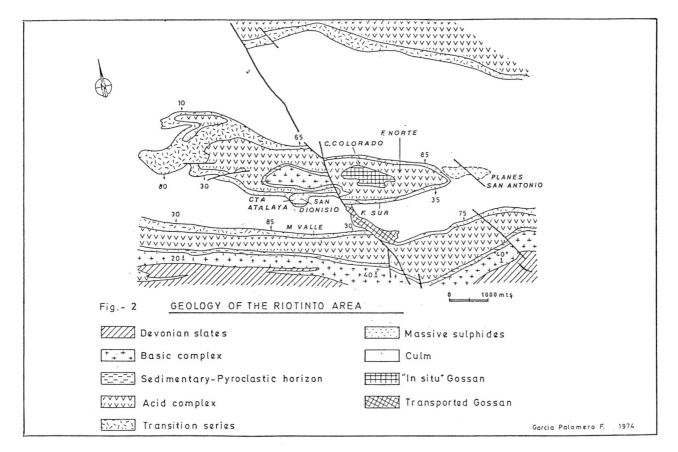

Fig.- 2   GEOLOGY OF THE RIOTINTO AREA

▨ Devonian slates	▦ Massive sulphides	
⊹ Basic complex	☐ Culm	
⊟ Sedimentary-Pyroclastic horizon	▦ "In situ" Gossan	
⋁ Acid complex	▨ Transported Gossan	
⋋ Transition series		

Garcia Palomero F.   1974

lithologies (polygenic breccias, litho-crystalline tuffs, finely crystalline tuffs, chloritised-haematised-serici-tised tuffs, massive sulphides, etc.). These represent the lithologies formed at the end of a phase of volcanic activity with the transition to the flysch sedimentation of the Culm. The thickness of this unit is very variable (5-50 metres).

The most notable feature of this series is that mineralised massive sulphide lenses, of probable syn-sedimentary origin, occur within it.

Culm

This unit consists of a monotonous series of turbiditic sediments, although within it different units can sometimes be distinguished on the basis of fossiliferous horizons, and differences in the relative proportions of greywacke and shale.

Tertiary

This is represented by subhorizontal deposits, 1-10 metres in thickness, of a limonite cemented mixture of fragments of volcanic rocks, slates, and iron oxides. These are laterally fairly restricted, and occur at specific

elevations (400 metres, 350 metres, etc) coinciding with possible terraces associated with the evolution of the river systems.

All the units described, except for the Tertiary, are affected by the Hercynian orogeny. This resulted in various phases of folding which formed the main east-west folds, and an associated schistosity (dipping north at 75-85 degrees), as well as faulting and thrusting. A weak metamorphism of greenschist facies is also developed.

Geological studies of the units described above have identified some features which indicate the palaeogeographic conditions prevailing at the time of mineralisation. Some of these phenomena are specific to the Rio Tinto area, and may be related to the presence of mineralisation, including the following:

1. The predominance of pyroclastic rocks with respect to lavas, in the Basic Complex of the Rio Tinto area. The associated lavas contain abundant vesicles of quartz and sulphides. These features disappear away from the Rio Tinto area.
2. The decrease in thickness of the Acid Volcanic unit close to the mineralisation (from 500 to 100 metres, with a minimum of 30 metres). A parallel

19

decrease in the thickness of the
pyroclastic elements in the transition
series is also noted.
3. At the eastern and western limits
of the area, coincident with strong
accumulations of pyroclastic rocks in
the acid volcanic level, abundant
fragments of oxidised lapilli tuff
and coarser pyroclastics are seen,
suggesting very shallow water volcanism.
4. At the base of the Culm frequent
slump features are seen, and the basal
unit of the Culm is absent close to
the mineralised zone.

These features seem to indicate a
very shallow marine environment
coinciding with the present-day
anticline, with marginal shallow
volcanic centres. This difference in
relief was eliminated during deposition
of the Culm.

MINERALISATION

Mineralisation in the Riotinto area
occurs in three separate zones, situated
in different parts of the Riotinto
anticline. Although locally inter-
connected, each zone belongs to a
separate genetic unit, sharing a common
genetic model.

The three mineralised zones are:
San Dionisio, Planes/San Antonio,
Cerro Colorado/Filon Norte/Filon Sur.
Each of these zones is made up of
three different types of mineralisation,
namely;
- Sulphide stockworks in the volcanic
  rocks,
- Stratiform massive sulphide lenses
  situated on top of the stockwork
  mineralisation, and intercalated
  tith other rocks of the transition
  series,
- Gossan (iron oxide cap), formed
  by weathering of outcropping massive
  sulphide and stockwork mineralisation,
  and restricted to within 70 metres
  of the surface.

The total amount of sulphide mineralisation in the Riotinto area suggests
that the area was the locus of one of
the most intense mineralising episodes
documented to date.

The importance of the mineralising
event is indicated by the following
statistics:

Mineralised area	4 km^2
Vertical extent	400-500 m.

Massive Sulphides:
Amount	500-600 Mt
Grades	S 50% Cu 1%,
	Zn 2%, Pb 1%,
	Ag 30ppm, Au 0.3ppm

Stockwork sulphides:
Amount	1900-2000 Mt
Grades	S 6%, Cu 0.15%,
	Zn 0.15%, Pb 0.06%,
	Ag 7ppm, Au 0.07ppm

Of the total sulphide mineralisation
formed, at least 400 million tonnes of
massive sulphides and 400 million tonnes
of stockwork mineralisation have been
affected by weathering and erosion.
Part of this remains in the form of
oxide caps (Gossan) totalling in excess
of 100Mt within which the concentration
of less soluble metals (Au, Ag, Pb, Ba,
Sn, etc.) has been enriched by the
weathering process.
The three different mineralised zones
are described separately below.

San Dionisio

This zone is located on the south flank
of the Riotinto anticline, associated
with a secondary synclinal structure
which plunges east at 30-35 degrees
(Figure 2).

Figure 3 shows a transverse N-S
section through the centre of this
deposit, illustrating the interrela-
tionship between the different types of
mineralisation, and lithologies present.
The different types of mineralisation
are described below.

Stockwork

Stockwork mineralisation in San Dionisio
consists of irregular veins containing
pyrite, chalcopyrite, galena, sphalerite,
magnetite, quartz, chlorite, calcite and
barite, cutting the volcanic host rocks.
The top of the stockwork marks the
base of the massive sulphides, and it
extends for up to 400 metres stratigra-
phically downwards from this contact,
down to the Basic Volcanic complex.
The mineralisation takes the form of
an originally semivertical chimney with
a semicircular cross section of about
600-700 metres in diameter. Folding
subsequent to formation of the deposit
has tilted this chimney through 90
degrees, so that it now lies in a
sub-horizontal position, elongated
north-south.

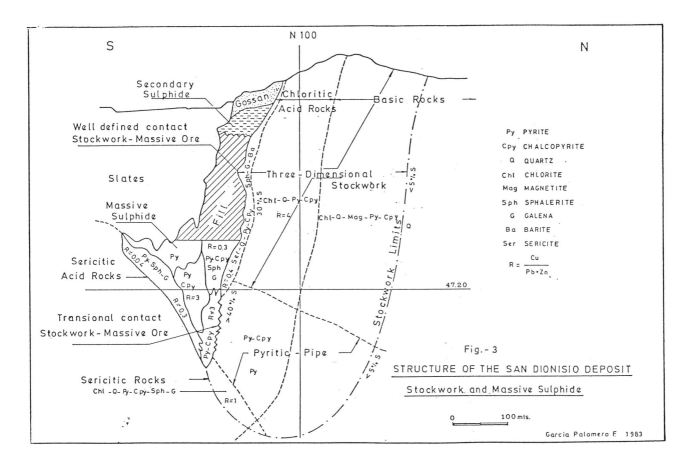

S   N 100   N

Secondary Sulphide

Gossan / Chloritic / Acid Rocks

Basic Rocks

Well defined contact Stockwork-Massive Ore

Slates

Three - Dimensional Stockwork

Massive Sulphide

Fill

Chl-Q-Py-Cpy
R=4

Chl-Q-Mag-Py-Cpy

Sericitic Acid Rocks

R=0.3
Py
R=0.01
Py+Sph-G
Py-Cpy
Sph
G
Ser-Q-Py-Cpy 30%S
Sph-G-Ba
R=0.4 Ser-Q-
40%S

R=3
R=0.3
Py-Cpy
R=3

Transional contact Stockwork-Massive Ore

Py-Cpy

Py-Cpy

Pyritic - Pipe

Stockwork Limits

<5%S

47.20

<5%S

Py PYRITE
Cpy CHALCOPYRITE
Q QUARTZ
Chl CHLORITE
Mag MAGNETITE
Sph SPHALERITE
G GALENA
Ba BARITE
Ser SERICITE

$R = \dfrac{Cu}{Pb+Zn}$

Fig.- 3

STRUCTURE OF THE SAN DIONISIO DEPOSIT

Stockwork and Massive Sulphide

Sericitic Rocks
Chl -Q- Py-Cpy-Sph-G
Py
R=1

0   100 mts.

Garcia Palomero F. 1983

Variations in vein density, thickness, texture, orientation, mineralogical composition, wallrock alteration, and relationship with the massive sulphides, allow the definition of a spatial zonation within the stockwork mineralisation. This reflects the physio-chemical conditions affecting the hydrothermal solutions as they ascended towards the seafloor. The zonation is made up of the following three units:

- Pyrite Chimney (stockwork nucleus), consisting of very intense pyrite mineralisation, with local chalcopyrite, in which original wallrock lithologies are almost indistinguishable. No clear contact with the massive sulphides can be determined (Figure 3).
- Three Dimensional Stockwork (typical stockwork), which is widely developed spatially, and surrounds the pyrite chimney. It is made up of a three dimensional network of sulphide veins which do not cut the overlying massive sulphides, with a clear contact between the two units.

Within this type of stockwork a varied mineral paragenesis is seen, along with intense alteration (quartz, chlorite, and sericite) of the hostrocks.

The mineral paragenesis present is indicated in Figures 3 and 4.

- Sericite Envelope, consisting of weak mineralisation in the outer fringes of the stockwork zone. It is restricted to the acid volcanic rocks within which strong sericitisation occurs contiguous with the top of the three dimensional stockwork.

Massive Sulphides

Massive sulphide mineralisation in San Dionisio forms a large lens of sulphides sitting directly on top of the stockwork. This has a more irregular shape and greater lateral extent than the underlying stockwork.

Figure 3 shows the folded nature of this lens, and its stratigraphic location between the acid volcanic rocks below, and the Culm above. Its spatial relationship with respect to the stockwork is also shown.

Within the massive sulphide body, the accumulation of sulphides is fairly monotonous with very few other minerals, apart from some thin pyroclastic horizons near the footwall and hangingwall contacts, and a weak interstitial siliceous matrix which

21

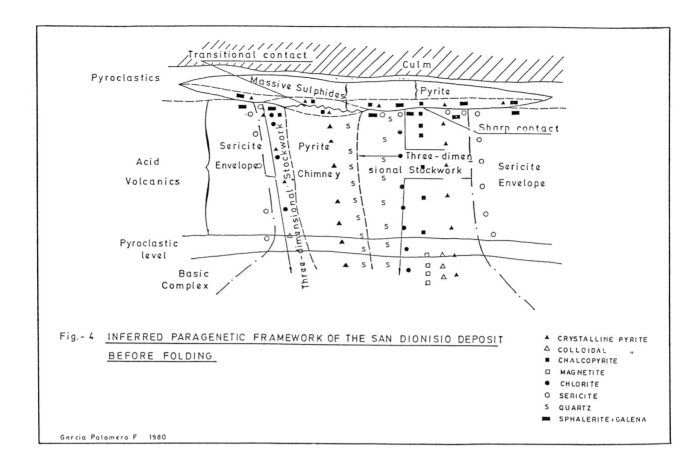

Garcia Palomero F   1980

Fig.- 4   INFERRED PARAGENETIC FRAMEWORK OF THE SAN DIONISIO DEPOSIT
BEFORE FOLDING

▲ CRYSTALLINE PYRITE
△ COLLOIDAL   "
■ CHALCOPYRITE
◻ MAGNETITE
● CHLORITE
○ SERICITE
S QUARTZ
▬ SPHALERITE,GALENA

never exceeds 2% by weight. The sulphide
minerals are totally recrystallised and
show some relict textures of a primary
sedimentary character (colloform,
framboidal, etc.) (Read, 1967)[4],
particularly within the pyrite. Less
stable sulphide minerals, such as
chalcopyrite, galena, tetrahedrite,
stannite, and bornite, are recrystallised in intergranular spaces between the
pyrite and sphalerite crystals, or in
microfractures within them. Despite
the monotonous character of the sulphide
body, two well defined units can be
distinguished on the basis of degree
of recrystallation, and mineral paragenesis. These are as follows:
a) Basal unit. This shows occasional
sulphide banding (pyrite-sphalerite-
galena), is medium-fine grained, and
contains the bulk of the base metals in
the massive sulphides. Within this
unit, metal distribution appears to be
related to structure and/or the underlying stockwork types.
    Variations in paragenesis are
illustrated in Figure 3, with the
preferential occurrence of pyrite-
chalcopyrite immediately above the
pyrite chimney, and pyrite-chalcopyrite-

sphalerite-galena above the three
dimensional stockwork. Beyond the
limits of the stockwork mineralisation
pyrite with sphalerite-galena occurs.
b) Upper unit. This is made up of
very fine grained pyrite, with a low
base metal content, and occasional
disseminated silica especially towards
the top. Local irregular concentrations
of galena and chalcopyrite are related
to remobilisation associated with
Hercynian folding, which also caused
strong remobilisation of quartz.
    The massive sulphides are overlain
by shales of the Culm series, as
indicated in Figure 3. At the lateral
limits of the massive sulphide mineralisation interfingering with the
Transition Series occurs. This series
thins over the centre of the sulphide
lens where it is represented by a 1-3
metre thick pyroclastic level.

Gossan

As indicated in Figure 3, the near
surface part of the stockwork and
massive sulphide mineralisation shows
intense weathering. Oxidation of the

22

sulphide minerals and leaching of the volcanic rocks has led to the formation of an Oxide Cap (Gossan), which extends down to a maximum depth of 70 metres below the surface in the most permeable parts of the orebody (within the massive sulphide lens).

Planes/San Antonio

This deposit is located at the eastern end of the Riotinto anticline, where the anticlinal axis closes and plunges eastwards at 30-35 degrees beneath the Culm (Figure 2).

The same types of mineralisation, as in San Dionisio, occur although some differences in overall size and mineral paragenesis patterns are seen.

Stockwork

The Planes stockwork, down to the limits of evaluation by drilling (Figure 5), is of reduced size and somewhat different characteristics to that of San Dionisio. Its structure in terms of mineral paragenesis, and relationship with the massive sulphides, is shown in Figure 5.

As in San Dionisio, this stockwork is again made up of three different units, namely:
- Pyrite Chimney, formed by semimassive sulphide mineralisation, generally without base metals, and lacking a clear contact with the overlying massive sulphides. In this transition zone into stratiform massive sulphides, at the top of the pyrite chimney, a strong concentration of copper occurs.
- Three Dimensional Stockwork is poorly developed here, and although lead-zinc sulphides are dominant over copper, overall base metal contents are low. The volcanic host rocks show less alteration than in San Dionisio.
- Sericite Envelope, is fairly well developed with several small minera- lised zones containing sphalerite and galena.

As in San Dionisio, both the three dimensional stockwork, and the sericite envelope, show a clear contact with the overlying massive sulphides with no evidence for stockwork veins cutting the massive sulphides.

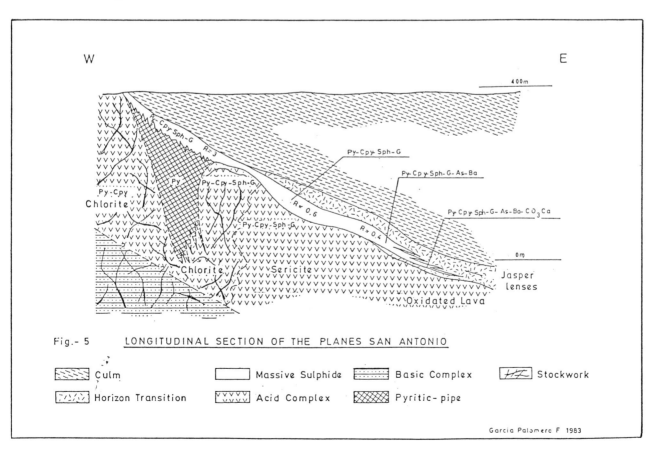

Fig.- 5   LONGITUDINAL SECTION OF THE PLANES SAN ANTONIO

Culm   Massive Sulphide   Basic Complex   Stockwork
Horizon Transition   Acid Complex   Pyritic- pipe

Garcia Palomero F 1983

23

Massive Sulphides

The massive sulphides in this deposit are planar in form, with average thicknesses of 10-15 metres. They contain enhanced grades of copper, as well as several minor metals such as Ag, Au, As, and Sn. They appear to form two separate lenses with some isolated links between the two (Figure 5). The lens directly overlying the stockwork, known as the Planes deposit, has been almost totally worked out. The other lens, known as the San Antonio deposit, occurs downdip, further away from the stockwork. It was discovered in the 1960's and remains unexploited.

The Planes lens, occurs directly above the pyrite chimney and shows no clear contact with it. It consists mainly of pyrite with abundant chalcopyrite. Apart from the massive sulphides the transition series is only represented by only 2-5 metres of tuffs.

The San Antonio lens occurs outside the limits of the stockwork in the underlying rocks, with a gradual development of a thick Transition Series between it and the Culm eastwards. An increase in typical slumping textures, new mineral parageneses, and additional lithologies within the pyroclastic rocks, are also seen going eastwards.

Although the average lens thickness is lower than usual at 10-15 metres, two parallel eastwest zones show sulphide thicknesses of up to 50 metres. Within these zones a great variety of separate sulphide lenses, as well as intercalated pyroclastic horizons, are recognised. These include compact polymetallic massive sulphides, banded-brecciated lenses, pyritic lenses, and rhyolite-sulphide breccia horizons. In these two zones the lenses and mineral parageneses show different lateral and vertical distributions, suggesting that they occupy two troughs, or parallel depressions, predating the sulphide formation. These depressions were filled with sylphide and pyroclastic material, displaced from higher elevations, during movements of the sea floor caused by seismic or volcanic activity.

The distribution of the parageneses shown in Figure 5, together with the changes in the nature of the sulphides and pyroclastics going eastwards,

suggest that the sulphides originated from the pyrite chimney located, to the west, upslope from the San Antonio lens (García Palomero F, 1980)[5].

Cerro Colorado/Filon Norte/Filon Sur

This deposit, located at the centre of the Riotinto anticline, to the east of the Eduardo fault (Figure 2), is the most extensive mineralised zone in the area. It consists of small remnants of massive sulphide (Filon Norte and Filon Sur) connected by a large oxide cap and extensive stockwork mineralisation in the underlying volcanics. Figure 6 shows a NE-SW section through the deposit, indicating the relationship between mineralisation types and their hostrocks.

The main geological features of this zone are described below, although it should be noted that the widespread effects of weathering, and mining activity. developed in this area make it difficult to study some of the most important geological aspects.

Stockwork

As shown in Figure 6, the stockwork mineralisation is very extensive, extending over 1000m from north to south, and 2000m from east to west. It is developed over a vertical thickness of 400-500m, affecting the top of the Basic volcanics and all the Acid Complex.

A geological reconstruction of the uneroded remnants defines a similar structure to the stockworks already described, with the presence of a pyrite chimney, three dimensional stockwork in chloritised volcanics, and a sericitic envelope. A general description of the stockwork mineralisation is not repeated, but some characteristics particular to this zone are listed below:
- wide spatial extent,
- low intensity of mineralisation in the three dimensional stockwork (4-6%),
- localisation at the edge of a possible rhyolite dome, inferred from the thinning of the acid volcanics from north to south,
- widespread disseminated copper mineralisation,
- widespread remobilised copper mineralisation due to its structural position at an anticlinal crest,

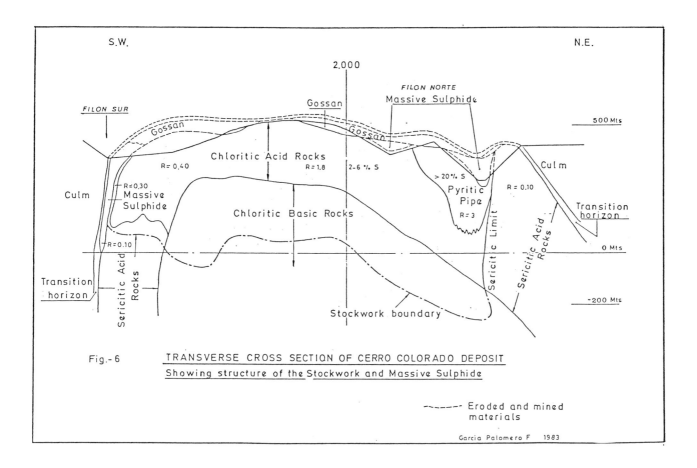

Garcia Palomero F 1983

Fig.- 6   TRANSVERSE CROSS SECTION OF CERRO COLORADO DEPOSIT
Showing structure of the Stockwork and Massive Sulphide

------- Eroded and mined
materials

- weathering which postdates the deposition of the overlying massive sulphides.

The absence of visible magnetite at depth, and the abundance of rhyolite breccias and autobreccias confirms the proximity to an active volcanic centre.

Massive Sulphides

As mentioned above, almost all (70-90%) of the massive sulphides in this zone have either been eroded away, or altered by weathering to form a very large iron oxide cap (Figure 6). Remnant unweathered massive sulphides are preserved on the south flank of the main anticline (Filon Sur), and in the axes of subsidiary synclines on the north flank (Filon Norte).

Only a small amount of information is available from the remaining massive sulphides since they have almost entirely been removed by mining. Available information shows several parallels with the other massive sulphide deposits, for example massive copper-rich lead/zinc-poor massive pyrite occuring on top of a pyrite chimney in the old Filon Norte pit.

Based on the abundance of barium and lead in the massive iron oxide cap, the four sulphide minerals (pyrite, chalcopyrite, sphalerite and galena), probably occurred in the massive sulphide overlying the three dimensional stockwork, which is now totally oxidised or eroded away,.

Filon Sur shows a lens of lead/zinc-rich copper-poor massive sulphide near the limits of stockwork mineralisation, overlying the sericite envelope, similar to the other massive sulphide lenses in the area.

Despite the lack of information in some areas, the geological characteristics of Cerro Colorado, Planes/San Antonio, and San Dionisio, show a distribution of sulphide mineralisation, and alteration patterns, etc. which indicate similar genetic processes and conditions of formation. These genetic factors can be combined to form a single genetic model for all the Riotinto sulphide deposits, as described in the next section.

Gossan (Iron Oxide Cap)

Surface oxidation of both the massive sulphides and stockwork sulphide mineralisation, in all three mineralised zones, has led to the formation of an iron oxide cap up to 70 metres thick. This iron oxide cap material is normally referred to as the Gossan orebody. San Dionisio and Planes / San Antonio are not affected to a great extent because of their location on the flanks of the main anticline. By contrast, in the widespread near surface sulphide mineralisation in Cerro Colorado, an estimated 70-90% of the original stratiform massive sulphides have been oxidised. A total of between 100 and 150 million tonnes of oxidised material, has been formed from the weathering of 400-500 million tonnes of sulphide bearing material.

The proportion of stockwork sulphides affected by oxidation is much less, and the processes involved in their weathering are less well know. The presence within the stockwork zone of an extensive layer of secondary sulphide enrichment, thought to be derived from the weathering of the overlying massive sulphides, indicate the importance of these weathering processes in enhancing the economic value of the original mineralisation.

Figure 2 indicates the wide extent of the Gossan derived from the weathering of the massive sulphides. The affects of weathering on the Cerro Colorado deposit, are shown in the transverse section presented in Figure 7.

Some of the main affects of weathering on the sulphide deposits are listed below:
- oxidation of sulphide minerals to form iron oxides and hydroxides
- total loss of sulphur, zinc, arsenic, etc.
- loss of copper, part of which is precipitated in a zone of secondary enrichment, the rest being lost in surface meteoric water
- partial loss of silver (80%), some of which is deposited in the secondary enrichment zone, the rest being lost in surface meteoric water
- residual enrichment of the least soluble elements, including gold, barium, lead, and tin, some of which are enriched up to 5 times their levels in the massive sulphides

The end product of sulphide weathering is a Gossan deposit consisting of an upper zone of massive iron oxides and hydroxides containing enriched levels of the least soluble metals (in particular gold), and a lower zone of paler weathering rocks in which the more soluble metals are enriched (silver and copper)

GENETIC MODEL

The historic development of ideas on the genesis of sulphide deposits associated with acid volcanic rocks is not discussed in detail here. Williams (1934)[6], one of the earliest workers to publish ideas on the genesis of the deposits, proposed an epigenetic origin for the deposits which was subsequently revised in later publications (Williams et al, 1975)[7]. According to this genetic theory the stockwork was formed by the crystallisation of sulphides in the channelways of a fumarolic system.

Given the close spatial and genetic relationship between volcanism and sulphide mineralisation in the area, a detailed genetic model can be defined. This genetic model relates the different mineralisation types in the three deposits described above, and takes into account the following factors (Garcia Palomero F, 1980 and 1982)[5,8]:
- the geological characteristics of the three mineralised deposits,
- the relationship between the mineralisation types and their host rocks,
- the relationship between different mineralisation types in each deposit,
- the distribution of the mineral parageneses,
- the relationships between the mineralisation and alteration in the volcanic hostrocks

Several aspects of this genetic model have been confirmed by geochemical studies. The genetic processes which formed the Riotinto deposits are thought to have evolved through the following sequence:
1) Submarine volcanic activity leading to the development of rhyolitic volcanic domes, and associated submarine topographic variations. The abundance of oxidation products suggests that this volcanic activity occurred at shallow depth. Formation of troughs parallel to the Hercynian fold axes were probably related to early tectonic events (Strauss, 1970)[9].
2) Waning of volcanic activity was accompanied by strong fumarolic activity in areas on the edges of the volanic centres, where the acid volcanic rocks

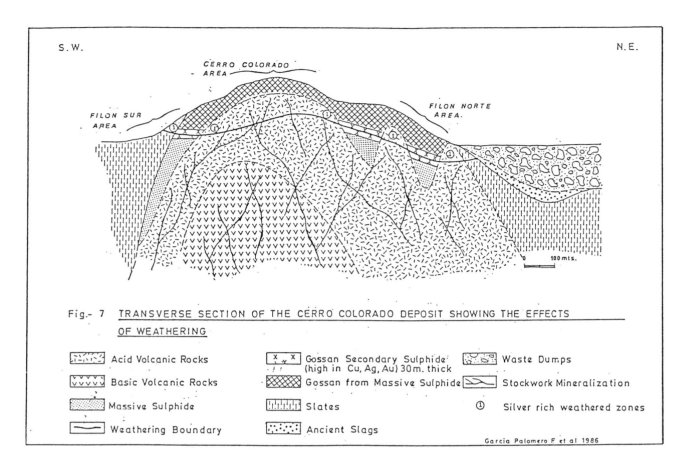

Fig.- 7 TRANSVERSE SECTION OF THE CERRO COLORADO DEPOSIT SHOWING THE EFFECTS OF WEATHERING

Acid Volcanic Rocks	Gossan Secondary Sulphide (high in Cu, Ag, Au) 30 m. thick	Waste Dumps
Basic Volcanic Rocks	Gossan from Massive Sulphide	Stockwork Mineralization
Massive Sulphide	Slates	① Silver rich weathered zones
Weathering Boundary	Ancient Slags	

Garcia Palomero F et al 1986

are thinnest. This activity led to concentration of sulphur and metallic elements which precipitated in favourable conditions giving rise to the sulphide mineralisation. Conditions necessary for precipitation were created by a hydrothermal convention cell driven by heat generated in magmatic bodies underlying the volcanic centres. The fluids involved were mainly phreatic seawater, with the possible addition of some fluids of magmatic origin. The large amount of sulphides accumulated in the area suggest that seawater was the main fluid involved in the convection cells, given that volcanic activity at this point was very weak and shortlived. Similar convection cells have been suggested as the formation mechanism for some Canadian deposits (Ridler 1973)[10].

The sulphur and metals are thought to have a magmatic origin, since the large quantity of metals present are unlikely to have been derived from the thin volcanic sequence. This is supported by studies of the isotopic composition of the sulphur.

3) As the metal bearing fluids ascended towards the seafloor, temperatures dropped causing some of the metals to be precipitated in fractures in the volcanic rocks giving rise to the stockwork mineralisation. This occurred under the following conditions: Precipitation started 400-500m depth below the seafloor within the volcanic pile. The amount and intensity of mineralisation increases upwards from this point.

The three different types of stockwork mineralisation (pyrite chimney, three dimensional stockwork and sericitic envelope) appear to evolve outwards from cylindrical cores. These units, defined by changes in intensity of mineralisation, mineral paragenesis, and wallrock alteration, are thought to be related to the cooling of fluids as they moved upwards and outwards from the hydrothermal centres. The pyrite chimneys are thought to represent centres of maximum fluid movement.

As well as the overall zoning of the mineral paragenesis, a sequence of formation of different minerals can also be recognised (chlorite, magnetite, pyrite-chalcopyrite, pyrite-chalcopyrite-sphalerite-galena-sericite, barite).

27

These features suggest that at the start of hydrothermal activity the stockwork was an open system, with sulphur and metals deposited on the seafloor. In the later stages parts of the system continued to be open to ascending fluids (the pyrite chimneys), the remainder consisting of tight fractures which were filled with late stage mineralisation (sphalerite-galena-chalcopyrite-barite in the three dimensional stockwork). The stockwork seen in Planes/San Antonio was mainly an open system, with most of the metals reaching the seafloor to form sedimentary massive sulphides, rather than being precipitated in stockwork veins.

From the distribution of mineral paragenesis shown in Figure 4, formation temperatures of 350-400°C at the roots of the stockwork (quartz-chlorite-magnetite), and around 100°C near the seafloor (sphalerite-galena-barite-sericite-etc). are indicated (Garcia Palomero F, 1980)[5]. This is supported by geochemical studies which indicate a similar range of formation temperatures, dependent on the location of the samples used (Eastoe CJ, et al, 1986)[11]. Oxygen isotope studies indicate formation temperatures of 210-230°C in the Cerro Colorado stockwork (Munha H, et al 1986)[12].

4) Sulphide deposition occurred on the seafloor in depressions close to the stockwork zones, with sulphur derived from sulphates in the seawater, temperature gradients, and thermal zonation, existing in these depressions at the start of sulphide precipitation, led to mineral and metal zonations as shown in Figures 3 and 5. Taking the pyrite chimneys as the centre of maximum temperature, the paragenesis changes outwards in the following sequence: pyrite-chalcopyrite, pyrite-chalcopyrite-sphalerite-galena, pyrite-chalcopyrite-shpalerite-galena-barite, pyrite-chalcopyrite-sphalerite-galena-barite-calcium carbonate, jasper. At a later stage an accumulation of pyrite with low contents of other metals was deposited, forming a cap on top of the other zones.

An average isotopic composition of the stratiform sulphides of between -6‰ and +8‰ ^{34}S was determined by Eastoe et al (1986)[11], and -14.1‰ to +8.0‰ by Misuno et al (1986)[13]. Eastoe et al explain several isolated negative

values of around -15‰, both in the stockwork and at the extremities of the stratiform sulphides, as possibly relating to biogenic sulphur.

The values of ^{34}S found in the stratiform sulphides contrast strongly with those found in the stockqork, where typical values in the range +6‰ to +12‰ were found by Eastoe et al[11], and +4.9‰ to +12.4‰ by Misuno et al[13]. These results suggest derivation from a mixing of cold sea water and ascending hydrothermal fluids, resulting in an isotopically heavier product.

During the deposition of the stratiform sulphides, some movement of the seafloor occurred, associated with volcanic activity. This caused slumping of sulphides and pyroclastic sediments downslope, giving rise to mixed sedimentary breccias, as seen in some parts of the San Antonio deposit.

The spatial distribution of the stockwork and the massive sulphides in the area shows an asymmetry in each deposit, suggesting the presence of a well defined topography, with deposition of massive downslope from areas close to the volcanic centres.
6) Deposition of the Culm sediments, followed by the Hercynian tectonism.
7) Uplift and erosion during Tertiary times, leading to the exposure of the volcanic rocks. Subsequent weathering of the sulphide mineralisation led to the formation of the extensive gossan mineralisation.

During weathering oxidation of sulphides occurred in both the massive sulphides and stockwork mineralisation. In both cases the more soluble elements such as copper, zinc and silver were leached out and redeposited at depth in a zone of secondary enrichment (70-100 metres below the surface), or removed in surface water runoff. Less soluble elements such as tin, lead, barium, and gold remained in the oxide cap where residual enrichment occurred. In the case of gold the resulting enrichment reaches economically exploitable grades.

ECONOMIC GEOLOGY AND MINING

Although the economic geology of mineral deposits is closely linked to mining and mineral processing activity, because mining activity at Riotinto commenced

before there was much geological
knowledge of the deposits, these two
aspects are discussed separately below.

Mining

The earliest known mining activity on
these deposits occurred in pre-Roman
times, with a peak of activity in the
period 300-100 BC (Rothenberg et al,
1986)[1]. At this stage mining activity
concentrated on zones of secondary
mineralisation, at the base of the
Gossan cap, which contained high grades
of silver, copper and gold. The extensive nature of these workings is
evidenced by the large quantity of
slags produced by treatment of the ores
(6-8 million tonnes in total). Known
workings from this period are concentrated in the Filon Sur and Filon Norte
deposits.

The next main phase of mining
activity commenced during the last
century, with the exploitation of the
main stratiform sulphide bodies
containing high copper grades. Copper
was produced either by directly leaching
the sulphides, or after pretreatment by
calcination.

At the turn of the century, the drop
in copper grades of the massive
sulphides, and their high sulphur content
(49-51%), led to the initiation of their
exploitation for sulphuric acid production
was initiated, with subsequent extraction
of copper from roaster slags outside the
mine area. At the same time working of
copper-rich stockwork zones was possible
due to the development of sulphide
flotation methods of ore treatment, with
copper being produced from the concentrates at a smelter in the mine area.

These two types of production, namely
pyrite for sulphric acid production, and
copper from sulphide stockwork zones,
continue to the present-day, despite
the ups and downs of the market for the
two end products. Overall the production
of copper from stockwork zones has
increased with time, as new reserves
have been discovered, and ore with
lower head-grades has been worked. The
amount of ore treated rose from 350 000
tonnes per annum in the 1960's to
8 000 000 tonnes per year in 1985. The
amount treated has dropped again in
more recent times, due to changing
economic circumstances. Production of
pyrite for sulphuric acid production
has been at a more regular rate of
about 1 000 000 tonnes per year for the
last 110 years!

In 1970 the gossan derived from the
massive sulphides was again worked,
this time for the extraction of gold
and silver. The initial production rate
of 1 million tonnes per annum has
gradually increased, due to increases
in precious metal prices, and the
definition of extensive additional
reserves. At present about 6 million
tonnes per year is treated, averaging
about 1g/t gold and 56g/t silver.

Most recent mining activity has
been by open pit mining methods. The
increase in reserves, along with
improved mine desing and production
control methods, required to support
these large operations is due largely
to the development of geological
knowledge of the deposits, as well as
to new exploration techniques and
evaluation methods.

Economic Geology

Since the start of mining activity
in the area, geological criteria used
in the exploration for new reserves
concentrated on the search for
continuations of the massive sulphides.
This focussed attention on the horizon
marked by the contact between the
slaty rocks of the Culm and underlying
porphyritic textured rocks (volcanics),
without being aware of their volcano-
sedimentary origin.

In the 1960's, with the acceptance
of a volcano-sedimentary origin for
these sulphides (Rambaud F, 1969)[14],
and an increased awareness of their
regional stratigraphic setting, the
first regional geological/geophysical
exploration programmes were commenced
in the area. Such exploration work
concentrated on the know host horizon
of the sulphides which had been defined
by earlier regional geological studies.

During the 1970's and 1980's, as
a result of the detailed studies of
the paragenesis of the sulphides and
the weathering processes which had
affected them, exploration was
concentrated on the full potencial of
all the economic metals present (Cu,
Pb, Zn, Au, Ag), resulting in a large
increase of know mineral reserves. At
the same time new methods of reserve
evaluation were implemented, based on
an integrated system of computerised
reserve estimation and mine planning
programs, allowing the selection of

the most appropriate mine design and scheduling creteria for maximising the economic benefit from exploiting these reserves.

The computerised reserve evaluation techniques used are based on a knowledge of the paragenesis and grade distribution of each element studied, in order to define the geological/statistical divisions used as a basis for subsequent geostatistical analysis. These techniques allow all the parameters or assay grades available for each sample, in this case mostly derived from diamond drillhole core (sampled on standard 2 metre lengths) to be integrated into the final reserve models.

The capacity of computer processing methods for handling large quantities of data, and considering different evaluation alternatives, is a fundamental improvement on previously used techniques. Totally manual approaches would not allow detailed study of all the available geological and geochemical information derived from a total of about 160 000 metres of drilling carried out at Rio Tinto, from which about 80 000 samples have been taken, and analysed for six different elements in most cases (Cu, S, Pb, Zn, Au, Ag, etc.).

Geological studies carried out on the different deposits for reserve evaluation purposes are described below:

Stockwork

Within the three stockwork deposits at Riotinto, attention has concentrated on the search for zones of copper enrichment within the overall stockwork mineralisation. Only in the cases of San Dionisio and Cerro Colorado have areas with economically viable copper grades been encountered.

Initial exploration for such mineralisation was carried out in the absence of an overall geological framework, being based on know mineral occurrences intersected by development carried out for mine infrastructure. During the mid sixties systematic studies were started, both in the exploration for further reserves, and also mine planning studies. This work was based on the hypothesis that the copper mineralisation was localised in the form of chimneys, or vertical structures, which cut across the geological structure, and which alternated with other barren structures

(Pryor RN, Rhoden N, and Villalon M, 1972)[15].

In the mid seventies, as a result of the synthesis of the available geological studies into an overall interpretation of the form and origin of the mineralisation, a new hypothesis was established. The copper concentrations within the stockwork were thus thought to be preferentially located in zones semiparallel with the volcanic stratigraphy, with lesser amounts in vertical zones which link the others, as shown in Figure 8 (García Palomero F, 1983)[8]. Subsequent exploration work on the San Dionisio and Cerro Colorado stockwork mineralisation was based on this hypothesis, and resulted in a spectacular increase in known reserves between 1977 and 1980. Reserve evaluation work was based on similar geological thinking, which was incorporated into the geoestatistical estimation methods implemented on a Perkin-Elmer 3210 computer on site in 1983.

The geoestatistically bases reserve estimation work carried out on Cerro Colorado and San Dionisio followed a similar approach, as summarised below (Garcia Palonero F et al, 1987, Sides, in prep.) [16,17].
1. Database preparation.
Drillhole datafiles are prepared for available drillholes, containing the spatial location of each sample, along with their geological characteristics and available assay data for S, Cu, Pb, Zn, etc.
2. Geological interpretation.
The spatial distribution of the main lithological units, alteration types, and mineralised zones are defined on sections and plans.

As part of the geological modelling, different populations of sample grades are defined for each element studied, on the basis of statistical studies. In the case of the Cerro Colorado, three different copper grade zones were defined as indicated in Figure 9. In the case of the San Dionisio stockwork a similar approach is used, but the basis for defining the zones is slightly different.
3. Experimental variograms.
Samples are classified according to the zonal divisions defined spatially by geological interpretations on sections and plans. Experimental variograms are then calculated for the three principal co-ordinate axes,

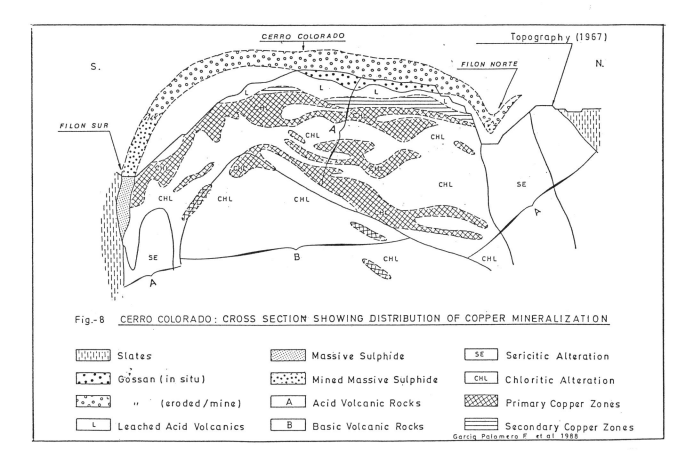

Fig.-8 CERRO COLORADO: CROSS SECTION SHOWING DISTRIBUTION OF COPPER MINERALIZATION

Slates		Massive Sulphide		SE	Sericitic Alteration
Gossan (in situ)		Mined Massive Sulphide		CHL	Chloritic Alteration
" (eroded/mine)		A	Acid Volcanic Rocks		Primary Copper Zones
L	Leached Acid Volcanics	B	Basic Volcanic Rocks		Secondary Copper Zones

Garcia Palomero F. et al 1988

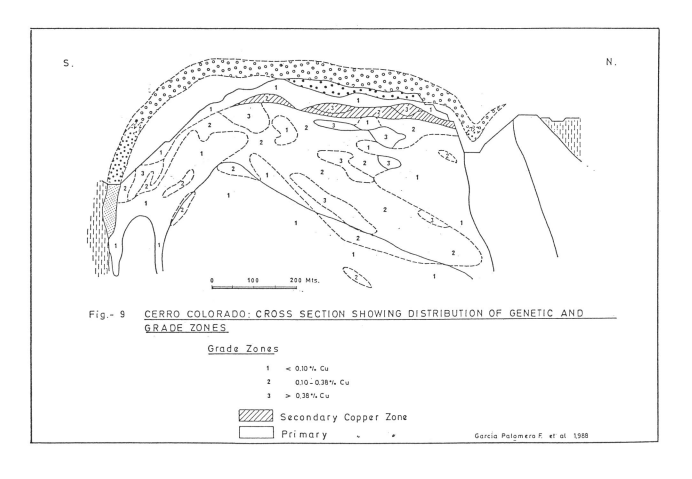

0   100   200 Mts.

Fig.- 9   CERRO COLORADO: CROSS SECTION SHOWING DISTRIBUTION OF GENETIC AND
GRADE ZONES

Grade Zones

1   < 0,10 °/. Cu

2   0,10 - 0,38 °/. Cu

3   > 0,38 °/. Cu

Secondary Copper Zone

Primary   "   "

Garcia Palomero F. et al 1988

separately for each zone, in order to study the autocorrelation of the metal being studied (copper grade) in each axis direction, within each zone.

4. Block model. Parameters of the reserve block models, or matrices, are selected based on the nature of the deposit being studied, and the mining methods to be investigated. In the case of Cerro Colorado block sizes of 25m east-west, 20m north-south, and 10-12m vertically, are used. The co-ordinates of each block are defined from its position within an array of blocks, and each block is assigned codes defining the main material types of interest (rock type, alteration, grade zone, etc.), by digitising in the geological interpretations made on section or plan, so as to build up a complete three dimensional model of the deposit.

5. Variogram models. Variogram models are fitted to the experimental variograms, defined by parameters such as the range, nugget variance, and sill variance. Separate models are used for each different grade zone, and models are tested statistically by kriging the grades of known sample points using surrounding samples (cross-validation point kriging), and carrying out statistical comparison between the estimated and known grades, as illustrated in Fig. 10.

6. Kriging of block grades. The selected variogram models are used to estimate grades for each block in the matrix by kriging. Only samples from the same zone are used in estimating the grade of a block within it, the number of samples used being restricted to those falling inside a search prism centred on the block.

7. Planning. The block models containing all the variables estimated, or assigned to them, form the basic for studies of the design of future mining units, as well as production control within existing operations.

The procedure of reserve evaluation used also allows the reserve models to be updated using results obtained during exploitation of the deposits. Reconciliation studies are carried out to determine the predictive accuracy of the methods used, both in terms of comparison of predicted and actual production figures (grades and tonnages), as well as the prediction of ore/waste boundaries, allowing the reserve estimation to be revised and improved. This is illustrated in the case of Cerro Colorado where block grades estimated from original exploration holes on a grid of 50m by 50m are used for long range mine planning (Primary Reserves). For short to medium range planning the reserve models are updated using information obtained from in-pit percussion drilling carried out on a closer spaced grid of 20m by 25m, in zones of interest (Investigation Reserves). The estimation models used in this instance are selected after detail study of the relationship between the estimated reserves and actual production in adjacent areas.

After 16 years of operation from 1970 to 1986, the remaining reserves for the Cerro Colorado open-pit operation are 180 million tonnes averaging 0.43% copper. The reserves of stockwork mineralisation in San Dionisio are currently being re-evaluated.

Gossan (Iron Oxide Cap)

The use of geological criteria in the exploration and evaluation of reserves of precious metals associated with this type of mineralisation, are described in detail elsewhere (Garcia Palomero et al, 1986)[18]. A brief description of the evaluation of the different types of gold-silver bearing gossanised rocks is given below.

A recent exploration campaign for Gossan reserves, based on the geological reconstruction of the sulphide mineralisation before weathering commenced, proved very successful. This was based on comparing the reconstructed pre-erosion geology with the situation when minign activity commenced in the area (Figure 7) and the present-day situation where most of the zones of weathered mineralisation had been affected by extensive mining activity. The gold-silver content of the weathered zone was estimated by extrapolation from the gold-silver contents of the known sulphides, taking into account the amount of ore already extracted. On this basic, an overall unexploited potential of 100-200 tonnes of gold and 10000 - 15000 tonnes of silver was estimated in 1984. This potencial formed the basis for the exploration campaign commenced at that date, which concentrated on looking for both in-situ areas of

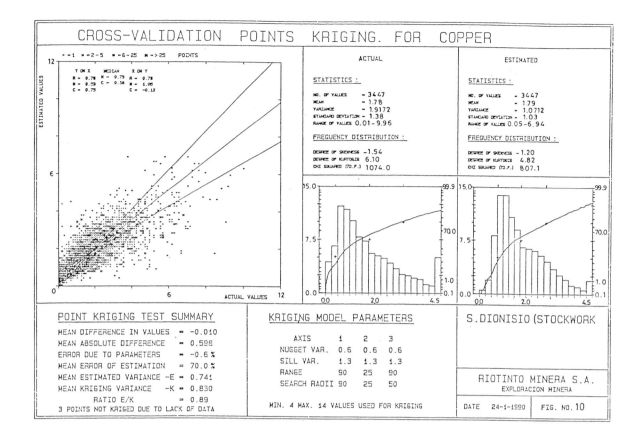

CROSS-VALIDATION POINTS KRIGING. FOR COPPER

Y ON X   MEDIAN   X ON Y
R = 0.78   M = 0.79   R = 0.78
M = 0.93   C = 0.58   M = 1.05
C = 0.75   C = -0.13

. = 1   × = 2-5   ■ = 6-25   ■ = >25   POINTS

ESTIMATED VALUES / ACTUAL VALUES

ACTUAL

STATISTICS :

NO. OF VALUES   = 3447
MEAN   = 1.78
VARIANCE   = 1.9172
STANDARD DEVIATION = 1.38
RANGE OF VALUES 0.01-9.96

FREQUENCY DISTRIBUTION :

DEGREE OF SKEWNESS -1.54
DEGREE OF KURTOSIS 6.10
CHI SQUARED (73.F.) 1074.0

ESTIMATED

STATISTICS :

NO. OF VALUES   = 3447
MEAN   = 1.79
VARIANCE   = 1.0712
STANDARD DEVIATION = 1.03
RANGE OF VALUES 0.05-6.94

FREQUENCY DISTRIBUTION :

DEGREE OF SKEWNESS -1.20
DEGREE OF KURTOSIS 4.82
CHI SQUARED (73.F.) 807.1

POINT KRIGING TEST SUMMARY

MEAN DIFFERENCE IN VALUES   = -0.010
MEAN ABSOLUTE DIFFERENCE   = 0.598
ERROR DUE TO PARAMETERS   = -0.6 %
MEAN ERROR OF ESTIMATION   = 70.0 %
MEAN ESTIMATED VARIANCE -E = 0.741
MEAN KRIGING VARIANCE   -K = 0.830
RATIO E/K   = 0.89
3 POINTS NOT KRIGED DUE TO LACK OF DATA

KRIGING MODEL PARAMETERS

	AXIS	1	2	3
NUGGET VAR.		0.6	0.6	0.6
SILL VAR.		1.3	1.3	1.3
RANGE		90	25	90
SEARCH RADII		90	25	50

MIN. 4 MAX. 14 VALUES USED FOR KRIGING

S.DIONISIO (STOCKWORK

RIOTINTO MINERA S.A.
EXPLORACION MINERA

DATE   24-1-1990   FIG. NO. 10

Gossan that had not yet been mined, and also for potential reserves in waste dumps from earlier mining activity.

The evaluation campaign was based on geological mapping, drilling and pitting aimed at determining the types of gossanised material, their in-situ location in relation to sulphide mineralisation, dump areas with gossan potential, gold-silver grades, material densities, metallurgical recovery, etc.. This work resulted in a database similar to that used for the Cerro Colorado copper evaluation, as described above.

The reserve evaluation was carried out using similar procedures to those followed for Cerro Colorado, except that in this case two economic metals, gold and silver, were studied. The result of the statistical and geostatistical analysis showed that whilst gold behaved differently in each in-situ zone, silver showed a similar pattern in all types of in-situ gossan (gossan derived from massive sulphides and from stockwork, gossanised shales and gossanised pyroclastics). These geostatistical models of the gold and silver distribution support the

geological ideas on the behaviour of these metals during weathering; gold, due to its lower mobility, being concentrated in-situ being more affected by local geological controls; silver, being more mobile, is deposited by cementation from groundwaters both in rocks which previously contained sulphides and also in barren wallrocks.

The computerised reserve models generated consisted of a block model, covering all the areas affected by gossanisation and with gossan-bearing dumps. Evaluation parameters were stored for each block (material type, grade zone classification, gold and silver grades, density, etc.) Using the same system as for Cerro Colorado, this allowed the plotting of plans and sections, showing these parameters, for use in mine planning and further geological studies.

The initial results of this exploration campaign were to outline identified resources (proven, probable and possible reserves), at the start of 1984, of 100 million tonnes averaging about 1g/t Au, and 56g/t Ag. The contained precious metal content of these reserves, 100t of gold, and 5,600t of silver

33

compared favourably with the initial exploration target (100t Au, and 10000-15000t Ag). The close match of the gold reflects its lesser mobility, whereas the deficit in silver reflects its greater mobility and probable losses in a subgrade dispersion halo as well as through removal in surface water runoff.

After several years of mining using the new geostatistically based reserve models as a basis for planning, a very good match has been found between estimated and actual production figures. The close match in predicted gold and silver content, both in tonnages and grades, as well as accurate spatial prediction, reflects in part the homogenisation of the metal distributions as a result of weathering processes.

Massive Sulphides

Historically, the massive sulphide deposits exploited for sulphur, require little or no geological control during mining, once their overall limits, and the nature of their contacts with hostrocks, have been defined. However, these deposits do also contain significant contents of base metals, 1% Cu, 1% Pb, 2% Zn in San Dionisio, and 1.6% Cu, 1% Pb and 2% Zn in San Antonio. A reconsideration of these deposits in the light of the recent studies of the paragenetic distribution of these metals suggested the possibility that zones of metal enrichment, of potencial economic interest for their base metal contents, might occur within them.

Based on geological ideas of the origin and stratigraphic control of metal zonation patterns, drilling campaigns were carried out of San Dionisio and San Antonio to test their potential base metal reserves. As a result of these campaigns zones, or lenses, within the massive sulphides, containing preferential concentrations of one or more base metal sulphides, were defined (see Figure 3).

These metal-rich zones have been evaluated in the San Dionisio deposit using the geostatistical techniques already described above. In this case each of the metals of potencial economic interest (copper, lead, and zinc) were studied with separate grade zones being defined for each. It is of interest that the stratigraphic origin

of these metal concentrations is supported by the experimental variograms and geostatistical models used. Variograms show similar ranges in the vertical and east-west directions, corresponding with the plane of stratification, but have much shorter ranges in the north-south direction, which corresponds to the stratigraphic-normal sense.

The in-situ geological reserves of the San Dionisio massive sulphide deposit were evaluated using a block model. This reserve model is being used as a basis for continuing studies necessary to define appropriate mining and mineral beneficiation methods for exploiting these massive sulphides. The project is currently in a phase of detailed metallurgical study which will determine whether or not these sulphides can be economically exploited for their base metal contents.

The total geological reserves of the San Dionisio deposit in 1986 were:

45Mt averaging:
    0.83% Cu, 50% S,
    0.65% Pb, 2.14% Zn

The current geological reserves of the San Antonio deposit, calculated using classical geological methods, are:

9.5Mt averaging:
    1.6% Cu, 1% Pb, 2% Zn,
    60ppm Ag, 0.6ppm Au

In order to improve the reserve estimation methods used in the stratiform sulphides, which are generally folded such that the main co-ordinate axes do not correspond to the stratigraphic axes at the time of deposition, the possibility of studying variograms in the plane of stratification and within different structural zones is being investigated. In more complicated folded situations "unfolding" procedures can be adopted, whereby the present-day co-ordinates are transformed into stratigraphic co-ordinates which are used during the grade estimation process (Sides E, 1984)[19].

In general the use of computerised methods in the exploration and evaluation of mineral deposits provides an invaluable tool. It is important to realise, however, that to obtain acceptable results, rigorous geological creteria must be applied. A maximum amount of geological detail, should also be used, since even in apparently homogeneous geological zones, stratigraphic, petrographic and structural discontinuities may divide an area into different spatial domains which are best studied separately.

ACKNOWLEDGEMENTS

The assistance of all the geological team at Rio Tinto Minera SA is acknowledged, along with the support received at all times from the company management. Special thanks are due to Prof. Antonio Arribas for his advice, and Edmund Sides for his technical and linguistic assistance.

REFERENCES
1. Rothenberg B, Garcia Palomero F, (1986) The "Rio Tinto Enigma - no more". IAMS No. 8. London.
2. Garcia Palomero F (1974) Caracteres estratigráficos del anticlinal de Riotinto. Studia Geologica VIII. Salamanca.
3. Schermerhorn LJG (1971) An outline stratigraphy of the Iberian Pyrite Belt. Bol. Geol. y Min. LXXXII. Madrid.
4. Read RA (1967) Deformation and metamorphism of the Pyritic San Dionisio orebody, Rio Tinto, Spain. PhD thesis, Royal School of Mines, Imperial College, London.
5. Garcia Palomero F (1980) Caracteres estratigráficos y relaciones morfologicas y genéticas de las mineralizaciones del anticlinal de Riotinto. Exma. Diputación Provincial de Huelva.
6. Williams D (1934) The geology of the Rio Tinto Mines, Spain. Trans. Inst. Min. and Metal. Vol. 71, London.
7. Williams D, Stanton RL, Rambaud F (1975) The Planes-San Antonio pyritic deposit of Rio Tinto, Spain: its nature, environment and genesis. Inst. of Min. and Metal. Vol. 84, London.
8. Garcia Palomero F (1983) Geología de las mineralizaciones de Riotinto y su modelo genético. Libro Homenaje a Carlos Felgueroso, Madrid.
9. Strauss G (1970) Sobre la geología de la provincia piritífera del Suroeste de la península Iberica y de sus yacimientos, en especial sobre mina de pirita de Lousal (Portugal). Mem. IGME -77, Madrid.
10. Ridler RH (Nov. 1973) Exhalite concept a new tool for exploration. The Northern Miner, Canada.
11. Eastoe CJ, Salomon M, & Garcia Palomero F (1986). sulphur isotope study of massive and stockwork pyrite deposits at Rio Tinto, Spain. Trans. Inst. Min. Metall. Sect B. Vol 95, London.
12. Munha J and Barriga FJAS (1986). High O^{18} ore forming fluids in volcanic-hosted base metal massive sulphide deposits. Econ. Geol. Vol 81.
13. Mitsuno Ch and coworkers (1986). Geological studies of the Iberian Pyrite Belt. Okayama Univ., Japan.
14. Rambaud F (1969). El sinclinal Carbonífero de Rio Tinto y sus mineralizaciones asociadas. Mem. IGME T.LXXI, Madrid.
15. Pryor RN, Rhoden HN and Villalon M (1972). Sampling of Cerro Colorado, Riotinto, Spain. Trans. Inst. Min. Metal.
16. Garcia Palomero F, Malave J, Marariño M, Sobol F, Sides E (1988) Modelos geológicos para la exploración y cálculo de reservas a largo y corto plazo, en el yacimiento de Cobre de Cerro Colorado (Minas de Riotinto). Bol. Geol. y Min. T.XCIX-I, Madrid.
17. Sides EJ (in prep). Computerised reserve estimation at Santiago and Rio Tinto, Spain, and Neves-Corvo, Portugal. PhD thesis, Imperial College, London.
18. Garcia Palomero F, Bedia Fernandez JL, Garcia Magariño M, Sides EJ (1986). Nuevas investigaciones y trabajos de evaluacion de reservas de gossan en Minas de Riotinto. Bol. Geol. y Min. T.XCVII-V, Madrid.
19. Sides EJ (1984). Internal company report, Rio Tinto Minera.

The massive sulphide deposit of Aznalcóllar, Spain, Iberian Pyrite Belt: review of geology and mineralogy

Josefina Sierra
Departamento de Cristalografía y Mineralogía, Facultad de CC. Geológicas, Universidad Complutense, Madrid, Spain

SYNOPSIS

Aznalcóllar is a strabound deposit located within the so-called Iberian Pyrite Belt. The deposit has reserves of 44×10^6 t of pyrite ore grading 0.44 % Cu, 1.76 % Pb, 3.33 % Zn, 67 g/t Ag, 1 g/t Au, and 34×10^6 t of sulphides grading 0.58 % Cu, 0.40 % Zn and 10 g/t Ag, the latter belonging to the so-called "cupriferous pyroclastic". The pyrite-Cu-Zn-Pb orebodies strike E-W and dip 45° towards the north. The pit is 1,500 m long, 700 m wide and has a depth of 300 m.

The orebodies occur within a lower Carboniferous volcano-sedimentary unit named "Complejo Volcánico Sedimentario", in which the volcanic rocks account for most of the sequence. Three episodes of acid volcanism (V_1, V_2, V_3) and one of basic volcanism (Vb, which ended the first acid volcanic episode) are recognized within this unit. The massive sulphide mineralization is genetically related to V_1; V_2 is thought to be responsible for some minor jasperoid-Mn mineralizations, and V_3 is "barren". The most important rock types at Aznalcóllar are tuffs, ashes, rhyolites, volcanic breccias and shales. These rocks were affected by the Hercynian compressions, which resulted in folding, penetrative flow schistosity and partial obliteration of S_0 planes.

Alteration is characterized by an early stage of silicification as a consequence of submarine fumarolic activity. This was followed in time by the formation of new minerals (quartz, plagioclase, chlorite, sericite) due to hydrothermal alteration and devitrification. The massive mineralization is mostly formed by pyrite with minor sphalerite, galena, chalcopyrite, and lead sulfosalts. A 3D statistical analysis of 576 samples taken from the mine showed no zonality of metal distribution and a reverse behaviour of Cu versus the pair Pb-Zn.

The origin of Aznalcóllar is time-related to the initial stage of the first acid volcanic episode (V_1), during which exhalative activity occurred, thus leading to metal leaching from the volcanosedimentary complex, expulsion of brines from vents, and eventually, to precipitation of the metal load onto the seafloor. Metal transport was probably in the form of halide complexes in weakly acid brines, which shortly after expulsion reacted with H_2S to form colloidal sulphide precipitates. Either gravitational instability or earthquake activity ultimately led to the generation of turbidite flows with rhythmic layering of black shales and sulphide layers.

The acid volcanics and sulphide mineralization are here thought to be a consequence of the emplacement of a subvolcanic plug locally known as the "felsite". Since the ore is spatially related to this subvolcanic body, Aznalcóllar is here considered to be a "proximal", although a non-rooted, volcanosedimentary deposit.

INTRODUCTION

The ore deposit of Aznalcóllar is located within the so-called "Iberian Pyrite Belt" (Fig. 1), one of the world's most outstanding metallogenic provinces of massive sulphide deposits. This roughly 30 Km wide belt of Dinantian age stretches from southern Portugal to Spain. The belt trends NW in Portugal and becomes progressively E-W in Spain, along some 230 Km, and includes important stratabound massive sulphide deposits, such as those of Neves Corvo, Rio Tinto, Tharsis and Aznalcóllar. Its northern boundary is defined by the Variscan antiform of Beja; southward, the Pyrite Belt is bounded by Carboniferous formations and Miocene marine sedimentary sequences.

Aznalcóllar constitutes a rather large E-W/45° mineralized sequence comprising the old mining works of Cuchillón, Silillos and Higueretas. As were most of the deposits in the Pyrite belt, Aznalcóllar was first mined by the Romans. Its modern exploitation began in 1876 by the Seville Sulphur and Copper Co. Ltd., and lasted until 1942. Further exploration and assesment of reserves were carried out by Peñarroya España in 1956. In 1960, a holding headed by the Banco Central (Spain) took over and a new company was formed (Andaluza de Piritas S.A.). Finally, in 1988, the Swedish company Boliden adquired the mine properties, and is its current owner.

Ore reserves amount to (a) 44×10^{6} t of massive sulphides grading 0.44 % Cu, 1.76 % Pb, 3.33 % Zn and 0.67 g/t Ag; and (b) 34×10^{6} t of the so-called cupriferous pyroclastic ("piroclasto cuprifero"), which grades 0.58 % Cu, 0.40 % Zn and 10 g/t Ag.

Fig. 2 View of the Aznalcóllar open pit works (taken from the Felsita hill).

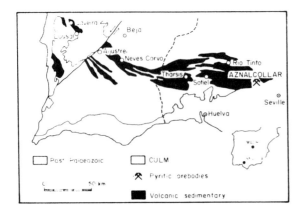

Fig. 1 Location map of the Iberian Pyrite Belt and Aznalcóllar

The geometrical characteristics of the mineralized body, the ore grade

distribution and geotechnical characteristics of the host-rock made it advisable for Andaluza de Piritas to exploit Aznalcóllar as an open pit mine (production started in 1979) (Fig. 2). The pit is 1500 m long, 700 m wide (maximus) and 300 m deep. At the current extraction rate (3.5 Mt/y, with a 3.1:1 stripping ratio), ore reserves will allow continuous exploitation till 1992. Boliden is currently carrying out a vigorous exploration programme aimed at finding new reserves.

The mineral processing flowsheet of Aznalcóllar is shown in Fig. 3.

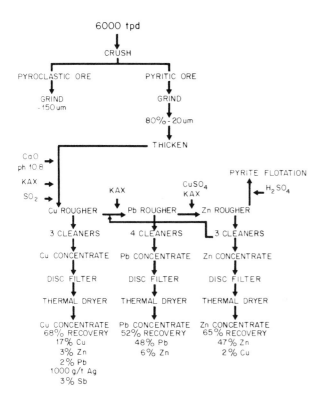

Fig. 3 Mineral processing flowsheet of Aznalcóllar (Fox, 1990).

According to Fox[1] the plant has a production capacity of: a) (pyrite circuit) 25,000 t/y of 21.5 % Cu, 1,500 g/t Ag copper concentrates; 38,000 t/y of 47.5 % Pb, 500 g/t Ag lead concentrates; and 100,000 t/y of 47.5 % Zn concentrates; and b)

(pyroclastic circuit) 40,000 t/y of 21 % Cu, 200 g/t Ag copper concentrates, and 7,000 t/y of 40% Zn concentrates. Sulphuric acid is produced as a byproduct at the current rate of 250,000-270,000 t/y.

Current research on bio-leaching is now being done at the University of Madrid (Dept. of Cristalografía y Mineralogía) in order to test wheten metals could be selectively leached from crude ores. Flask agitation, column tests and mineralogical studies are now in progress.

GENERAL CHARACTERISTICS OF THE STRATIGRAPHIC UNITS

Three well differentiated units account for most of the Pyrite Belt[2]: unit 1) A footwall sequence formed by probably Devonian slates and quartzites of some 500 m thick; unit 2) The Volcano-Sedimentary Complex (CVS), which overlies conformably the footwall sequence. This Dinantian complex hosts the mineralized bodies, and has a maximus thickness of 800 m. At the Pyrite Belt scale, the CVS consiste of a sequence of acid and basic volcanics, and interbedded shales/slates. Three episodes of acid volcanism (V_1, V_2, V_3) and one of basic volcanics (V_b; which ended the first acid volcanic episode) are recognized within the Iberian Pyrite Belt. The massive sulphides are genetically related to the first volcanic episode (V_1). V_2 is thought to be responsible for some minor jasperoid-Mn mineralizations and V_3 is "barren". Unit 3) conformably overlying the CVS, a Culm sequence completes the Paleozoic stratigraphy. These Culm facies are formed by slates and greywackes of variable thicknesses (several thousands metres in the west and 500 m in the east). The Culm is eroded at Aznalcóllar and the CVS is unconformably covered by Miocene marine sediments.

The Dinantian volcanism that formed the CVS is to be regarded as bimodal, and can be related to an important extensional episode of continental rifting in the southern realm of the Iberian Peninsula[3]. This episode belongs to the initial stages of the Variscan Cycle.

The local stratigraphy of the CVS at Aznalcóllar (from the base to the top) includes the following facies (Fig. 4):

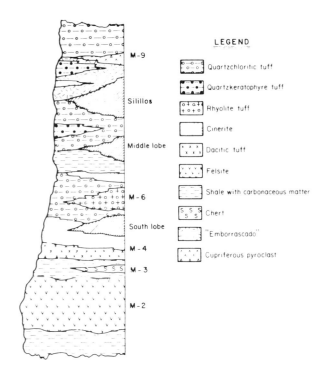

Fig. 4 Aznalcóllar stratigraphic column (Sierra, 1984 a).

1) a thick sequence of rhyolites; 2) carbon-rich slates and cherts; 3) dacitic tuffs; 4) massive sulphides (the so-called "Filón Sur" body, hosted by ashes); 5) an alternating sequence of tuffs and slates; 6) a second body of massive pyrite (the so-called "Filón medio"); 7) pyroclastics; 8) a third body of massive sulphides (the so-called "Filón Silillos"); and 9) quartz-chloritic tuffs and ashes

forming the so-called "piroclasto cuprífero", a zone locally enriched in chalcopyrite.

The acid volcanics underwent a series of complex processes of mineral recrystallization (Table 1). Up to five generations of quartz are recognized (Q_1-Q_5): the former two were related to magmatic (Q_1) (Fig. 5) and early

Fig. 5 Hexagonal clasts of igneous quartz (Q_1). Crossed polars x 10.

fumarolic alteration (Q_2) processes, respectively. The last three stages of quartz crystallization were related to later Variscan tectonometamorphic processes (Fig. 6). Plagioclase was formed during two episodes: a magmatic (time-related to Q_1) and a metamorphic (time-related to Q_4). Chlorite is mostly a product of devitrification and alteration of magmatic biotite. Three generations of this mineral are recognized:[4] the former is of uncertain relative age, but it might to related to earlier alteration

Fig. 6 Quartz-chlorite devitrified lava. Pyrite euhedra surrounded by Q_3 quartz. Crossed polars x 16.

processes (time-related to Q_2?); the second and third episode of chlorite formation are of metamorphic origin and are time-related to Q_3 and Q_4, respectively. Sericite formed in relation to devitrification-type processes during early metamorphic processes. K-feldspar is the most important mineral within the rhyolitic tuffs, and formed during a late magmatic stage. In the other rock types, K-feldspar is of metamorphic origin and time-related to Q_4.

Minerals of minor importance are: sphene, zircon, apatite, rutile, carbonates, tourmaline, kaolin and barite (the last three are rare).

TECTONIC DEFORMATIONS

Similar to the other ore deposits of the Iberian Pyrite Belt, Aznalcóllar is pre-orogenic mineralization, which

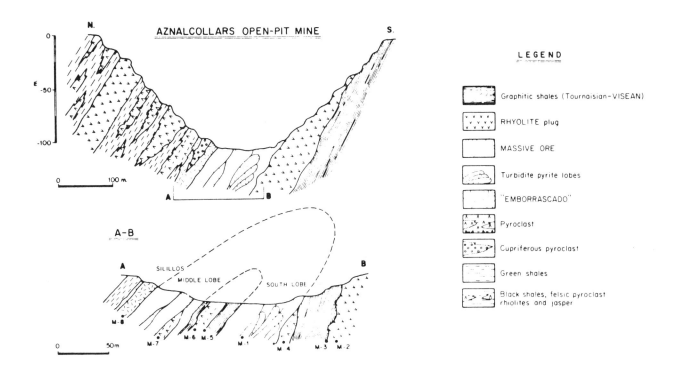

Fig. 7 Schematic cross-section of Aznalcóllar (Sierra et al., 1985).

41

underwent a series of deformation phases during Variscan time. The general structural trend of the Pyrite Belt follows a NW (Portugal) to E-W (Spain) arc in which folding was extensive and intensive. Four Variscan phases of deformation are recognized within the belt. A) Phase 1 (mid Westphalian[5]): is responsible for the major structures of Aznalcóllar[4], and generated tight isoclinal recumbent folds vergent to the South with E-W axes (Fig. 7). This phase is characterized by reverse faulting and overthrusting leading to the complete obliteration of the isoclinal folds. The S_1 schistosity developed during this phase. B) Phase 2 (Stephanian[5]): it generated N90ª-120ªE coaxial assymetric folds and the S_2 foliation (a non-penetrative fracture cleavage locally observed in the metapelitic rocks). C) Phase 3: this phase produced E-W orientated kink-bands in relation to a late episode of thrust tectonics[6]; and D) Phase 4: this phase is of late Variscan[6] age and is characterized by the development of E-W and NE brittle faulting.

THE OREBODIES

General

As mentioned earlier, four main mineralized zones are recognized at Aznalcóllar: Filón Sur, Filón Medio and Filón Sililos (massive sulphides and the Piroclasto Cuprífero, a mostly disseminated mineralization. Stockwork-type mineralization has been also suggested for Aznalcóllar[7], but a series of features makes that very improbable: 1) absence of zoned related alteration processes[8]; 2) the minor importance and small size of the mineralized veinlets (as compared to "true" stockworks e.g. Rio Tinto); and 3) the supposed "stockwork-like" structure occurs towards the top of one

of the mineralized zones, which completely discard it as a possible feeder zone[9].

Mineralogy

The ore mineralogy is mostly formed by massive recrystallized pyrite. The content of pyrite decreases towards the orebody contacts with the barren host-rocks. A wide variety of ore textures are recognized in pyrite[10]: a) annealed, b) euhedral, c) "caries" (Fig. 8), d) atolls in sphalerite,

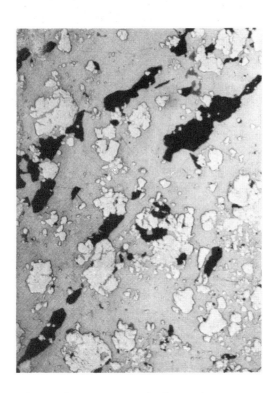

Fig. 8 Caries texture of pyrite cemented by chalcopyrite. Plain polars x 200.

e) zoned, f) brecciated, g) framboidal (Fig. 9) and h) colloform. The rest of the minerals infilled spaces left by the uncomplete annealling of pyrite during the metamorphism. Other textural features include the enveloping of pyrite by sphalerite, galena, chalcopyrite or bournonite.

Sphalerite is second in volumetric importance within the ore overwhelmingly dominated by pyrite. This mineral occurs

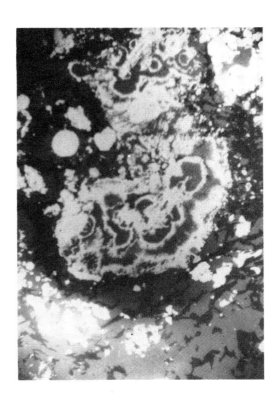

Fig. 9 Framboids and colloforms of
pyrite in sphalerite. Plain polars x 200

as thin film coatings, colloforms,
mineralized fractures, and bands.
Chalcopyrite infills cavities and
fractures in pyrite. Lead occurs as
galena, bournonite and/or menenghinite.
All these minerals infill cavities in
pyrite or occur in close association to
sphalerite. Other mineral phases of very
minor importance are: melnicovite,
arsenopyrite, pyrrhotite, tennantite-
tetrahedrite, stannite, gudmundite,
marcasite, nagyagite, silver and
bismuth. The paragenetic study[10]
indicates that the different mineral
phases must have been precipitated
simultaneously (first stage of mineral
formation). Mineral precipitation must
have been colloidal, and later, due to
diagenetic and metamorphic processes,
the minerals underwent remobilizations
and recrystallizations.

Different from other examples of
massive sulphide mineralization,

Aznalcóllar does not display any
features reflecting ore zoning at
orebody scale. In fact, minerals such
as chalcopyrite, sphalerite and galena
follow a rather random distribution[4.10].

Alteration

As stated earlier, the orebodies occur
within a volcanogenic sequence which
has undergone a variety of alteration
related processes. These phenomena can
be regarded as either hydrothermal or
metamorphic. Common alteration types
throughout the deposits are:
silicification, chloritization and
sericitizacion. Local albitization and
K-metasomatism have been also observed.
Silicification began during the time
that exhalative activity took place,
and continued during later metamorphic
phenomena. K-metasomatism is also time-
related to the exhalative activity.
Chloritization and sericitization are
related to the vitrification processes
that place during the first
tectonometamorphic event.

Geochemistry

A comprehensive geochemical study of
Aznalcóllar[11] was carried out in
early 80's. The aim of this study was
focused in establishing whether metal
zoning existed or not at the mine
scale. 576 samples from drill cores
were analyzed for Cu, Pb and Zn. The
results showed no zoning, however, a
reverse behaviour of Cu versus the pair
Pb-Zn was inferred. Log-probability
plots were used to study the
statistical behaviour of Cu, Pb and Zn
(Fig. 10, 11, 12). Copper has a rather
simple distribution and only one
population is observed, with a
background of 0.45 %. Lead and zinc
have distributions displaying two
populations each; their backgrounds
are of 1.7 % and 3.1 %, respectively.

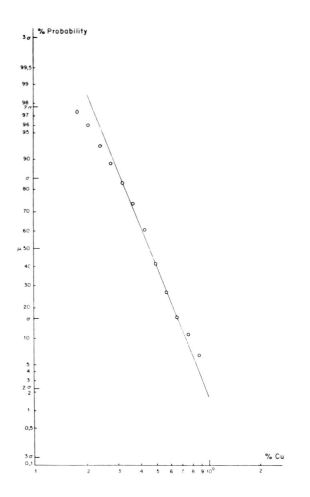

Fig. 10 Log-probability plot of copper.

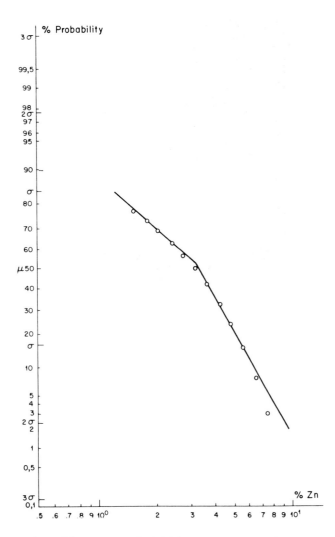

Fig. 12 Log-probability plot of zinc

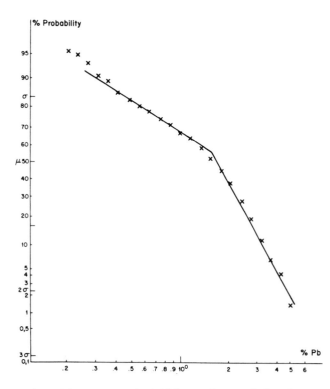

Fig. 11 Log-probability plot of lead

GENETIC ASPECTS

Aznalcóllar is primarily the result of
hydrothermal phenomena related to the
extrusion of the early acid volcanics
(V_1), and can be regarded, to an extent
as an integral part of the CVS
stratigraphic column. The V_1 magmatism
not only gave rise to the extrusion of
the acid lavas and pyroclastics, but
triggered important exhalative activity
that expelled into the seawater the
metals and H_2S. The origin of metals
can be regarded as the consequence of
convective cells of seawater operating
within the volcanic pile, from which
they were leaches, and later
precipitated by exhalative flows. It

44

can be envisaged that a system H$_2$O-NaCl -rock accounted for the initial stage of mineral precipitation at Aznalcóllar. First mineral phases in precipitating[9] would have been of the type greigeite and mackinawite, plus Pb, Cu and Zn sulphides.

At Aznalcóllar, one or more vents fed a continental platform with sulphides (Fig. 13 I). Later, once the pyrite-rich lobes of mineral were formed, gravitational instability induced (Fig. 13 II) the generation of turbidite flows (Fig. 13 III). As shown by the stratigraphic and sedimentalogical record, this activity led to the formation of turbidite lobes of pyrite enclosed by black shales (Fig. 14). Similar to other Kuroko-type deposits, the early metallogenic history of

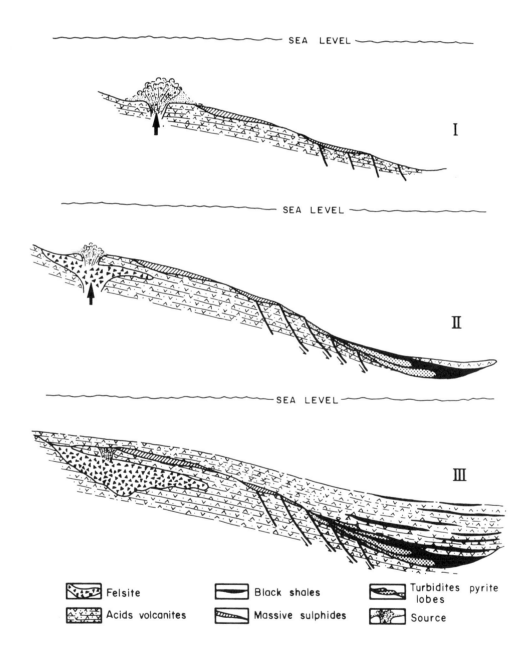

Fig. 13 Genetic model for Aznalcóllar (Sierra et al., 1985).

Fig. 14 Pyrite turbidite lobes.

TABLE 1

MAGMATIC RELATED PHENOMENA		MAIN PHASE OF DEFORMATION (PHASE 1)		PHASE 2 OF DEFORMATION
ORTHOMAGMATIC AND VOLCANOSEDIMENTARY	LATE TO POST MAGMATIC	METAMORPHIC EPISODE	LATE TO POST KINEMATIC	SYN TO POST KINEMATIC
Q1	Q2	Q3	Q4	Q5
Plagioclase	Carbonates	Chlorite (1,2)	Chlorite (3)	Carbonate
Opaques	K-spar	Sericite	Plagioclase	Fe-oxides
Apatite	Plagioclase	Siderite	K-spar.	
Zircon	Rutile	Sphene	Siderite	
Tourmaline	Sphene	Rutile	Muscovite	
			Barite	
			Kaolin	
	Hydrothermal activity	Metamorphic phenomena: recrystalliza-tion, devitrification		

Aznalcóllar seems to be strongly linked to a subvolcanic rhyolitic plug (in this case, the so-called "Felsita"), to which the orebodies are spatially related. Since, as discussed earlier, no clear evidence supports the existence of a "true stockwork" at Aznalcóllar, the orebodies here described can be regarded as "proximal", non-rooted volcanosedimentary massive sulphides. Moreover, this deposit can be genetically considered as a typical Kuroko-type mineralization. Finally, the stratigraphic and mineralogical characteristics of Aznalcóllar allow a classification of this deposit into the type III of Colley[12] (Fig. 15).

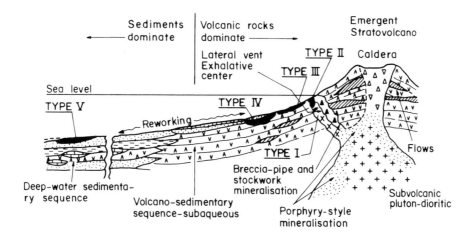

Fig. 15 The five types of Kuroko-like mineralizations (after Colley, 1976).

REFERENCES

1. Fox, K. Aznalcóllar. Mining Magazine, January 1990, p. 20-25.
2. Strauss, G.K., Madel, J., Alonso, F. Exploration practice for strata-bound volcanogenic sulphide deposits in the Spanish-Portuguese Pyrite Belt:Geology, geophysics and geochemistry. In: D.D. Klemm and H.J. Schneider (eds.), Time and Strata-Bound Ore Deposits, Springer Verlag, Berlin, 1977).
3. Sawkins, F.J. and Burke, K. Extensional tectonics and mid-Paleozoic massive sulphide occurrence in Europe. Geologisches Rundschau, vol. 69, 1980, p. 349-360.
4. Sierra, J. Geología, mineralogía y metalogenia del yacimiento de Aznalcóllar_ Litoestratigrafía y tectónica. Bol. Geol. Min., XCV-V, 1984, p. 440-455.
5. Schermerhorn, L.J.G. An outline stratigraphy of the Iberian Pyrite Belt. Bol. Geol. Min., LXXXII, 1971, p. 239-268.
6. Hernández Enrile, J.L. Marco geológico estructural de los yacimientos de sulfuros de Aznalcóllar (Región Oriental de la banda pirítica Ibérica). Reunión de Xeoloxìa e Minería do Noroeste peninsular, 1981.

7. Hofstteter, J.P. L'amas sulfure a Cu-Pb-Zn d'Aznalcóllar (Sevilla), Espagne. Géologie, paleogéographie et metalogenèse de l'extrémité sud-oriental de la ceinture sud Ibérique. Thèse 3eme cycle, Université de Pierre et Marie Curie, France, 1980.
8. García Palomero, F. Caracteres geológicos y relaciones morfológicas y genéticas de las mineralizaciones del anticlinal de Rio Tinto. Acta Salmanticensia, 1977.
9. Sierra, J., Arribas, A. and Gumiel, P. Geología, mineralogía y metalogenia del yacimiento de Aznalcóllar: Metalogenia. Bol. Geol. Min., XCVI, 1985, p. 23-30.
10. Sierra, J. Geología, mineralogía y metalogenia del yacimiento de Aznalcóllar: Mineralogía y sucesión mineral. Bol. Geol. Min., XCV-VI, 1984, p. 553-568.
11. Sierra, J., Astudillo, J. and Lunar, R. Estudio geoquímico del yacimiento de Aznalcóllar (Cinturón Pirítico Hispano Portugués). I Congreso Español de Geología, II, 1984, p. 685-705.
12. Colley, H. A classification and exploration guide for Kuroko-type deposits based on occurrence in Fidji. Trans. Instn. Min. Metall. (Sect. B: appl. earth Sci.), 1977, p. 190-199.

Precious- and base-metal mineralogy of the Hellyer volcanogenic massive sulphide deposit, northwest Tasmania — a study by electron microprobe

A.R. Ramsden
K.M. Kinealy
R.A. Creelman
R.A. Creelman and Associates, Epping, New South Wales, Australia
D.H. French
CSIRO Division of Exploration Geoscience, North Ryde, New South Wales, Australia

SYNOPSIS

The Hellyer volcanogenic massive sulphide deposit, discovered by Aberfoyle Ltd in 1983, is a polymetallic orebody. Although Cambrian, it is relatively undeformed and contains abundant textural evidence to indicate that the sulphides were extremely fine-grained, perhaps even amorphous, at the time of deposition. There are clear affinities with the Tertiary Kuroko deposits of Japan, for which a submarine volcanic-hydrothermal origin is now widely accepted.

As part of a broader metallurgical investigation aimed at developing the optimum treatment strategy for the ore, detailed electron microprobe studies have been carried out to determine the mineralogy of the deposit and provide quantitative data on the mineralogical residences of the metals. Three drill holes have been analyzed to investigate both stratigraphic and lateral variations along the axis of the orebody. The data obtained are relevant not only to the metallurgy, but also lead to a better understanding of the origin of the deposit.

The average Fe-content of sphalerite in the ore (based on 1009 analyses) is 2.7 wt%, but trends within the sphalerite divide the deposit into an upper and lower unit. The boundary, in all three holes, is marked by a thin zone of relatively high-Fe sphalerite (5 - 7 wt%) which is also distinguished texturally by the presence of fragments of concentrically zoned sphalerite crystals carrying abundant sub-micron chalcopyrite. The crystals probably formed in a vent, from which they were expelled to accumulate as clasts on the sea floor. Expulsion occurred after formation of the lower unit, possibly following closure of the original vent system, and resulted in deposition of a 'blanket' of high-Fe sphalerite over part of the pre-existing sulphide mound. The upper unit, which contains most of the precious metals, accumulated above this.

Gold shows three associations: 1) with barite, 2) with arsenopyrite and 3) with pyrite. Gold associated with barite is probably particulate but gold associated with arsenopyrite and pyrite occurs as a trace element in both minerals. The arsenopyrite association is the most important, electron microprobe analyses indicating Au-contents up to 400 ppm for this mineral. The pyrite association is inferred, since the Au-content is generally less than the detection limit for electron microprobe analysis (100ppm). The arsenopyrite is fine-grained (<20μm), well crystallized and shows cryptic oscillatory growth zoning due to small fluctuations in Sb-content. The complexity of the zoning indicates crystallization from a turbulent fluid in which early formed crystals were fragmented and overgrown by more arsenopyrite as crystallization proceeded. The intensity of zoning and fragmentation decreases away from the presumed vent area consistent with crystallization in a quieter part of the plume some distance from the vent.

Silver occurs as a major element in tetrahedrite and as a trace element in galena. The tetrahedrite and galena are closely associated in the ore, and therefore there is an excellent correlation of silver with lead. However, it is tetrahedrite (and freibergite) that are the major hosts rather than galena. Tetrahedrite at some distance from the presumed vent area is less arsenical and more argentian than that from close by. Textural evidence suggests that Ag-enrichment occurred during late stage replacement of the tetrahedrite. Some of the additional silver may have been released from argentian galena at the hanging wall, that is now largely replaced by bournonite and boulangerite. Such argentian tetrahedrite is itself replaced by minor pyrargyrite and carries inclusions of acanthite-argentite.

INTRODUCTION

The Hellyer volcanogenic massive sulphide deposit, N.W. Tasmania, was discovered by Aberfoyle Limited in August 1983, about 5 km northeast of the Que River mine (Fig.1) where a similar polymetallic deposit is being mined by this company. Both deposits are within the Cambrian Mt Read Volcanics and consist of massive sulphides with very little gangue.

The Hellyer orebody comprises a single, faulted, elongate massive sulphide body overlying a well-developed pyrite footwall alteration pipe (McArthur[21]). It shows many features of the Tertiary Kuroko ores of Japan, including the presence of barite overlying massive ore, and an underlying stockwork of stringer mineralization. Unlike the Kuroko deposits, however, copper is only a minor constituent (averaging <1.0 wt%) and there is no gypsum. Precious metal values are concentrated towards the stratigraphic hanging wall, as with other deposits of the gold-zinc association in the Mt Read Volcanics (Large et al.[17]).

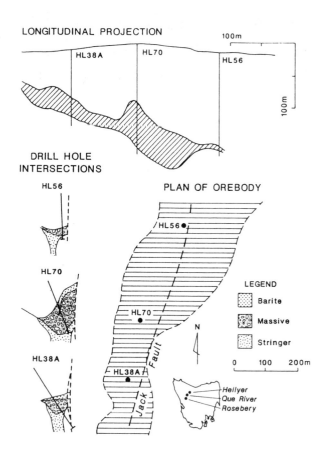

Fig.1 Geographic location of the Hellyer deposit and location of the three drill-holes sampled for the present study. These holes provide stratigraphic cross-sections at interva;s along the axis of the orebody west of the Jack Fault.

The deposit is highly pyritic, comprising on average 54 wt% pyrite, 20 wt% sphalerite, 8 wt% galena, 2 wt% arsenopyrite and 1 wt% chalcopyrite with minor tetrahedrite and other trace sulfosalts. The principal gangue minerals are barite, quartz, ankerite, siderite, chlorite and sericite.

Mineragraphic studies (this work and Fander, pers.comm.) have shown that primary (depositional) banding is a conspicuous feature of the ore, and there is abundant textural evidence (especially with spongy, radiating-fibrous, concentric and colloform pyrite intergrowths) to indicate that the sulphides were extremely fine-grained, perhaps even amorphous, at the time of deposition. The ore is recrystallized in varying degrees, resulting in coarsening and re-mobilization of galena, sphalerite and tetrahedrite.

The work reported in the present paper is based primarily on the results of detailed electron microprobe investigations which were carried out as part of a much broader collaborative project between Aberfoyle Ltd and the Commonwealth Scientific and Industrial Research Organization (CSIRO) aimed at developing the optimum treatment strategy for Hellyer ore. Overall, this project involved metallurgy, minerals engineering, chemistry and geology, with a strong bias towards mineralogy and examination of the ore response to unit processes such as crushing, grinding, flotation, hydro-metallurgy and pyro-metallurgy.

As well as providing data for the process mineralogist, the electron microprobe studies also contribute towards a better understanding of the origins of the mineralization. This aspect of the work, however, must be seen in the context of broader geological studies being carried out by Aberfoyle staff (McArthur, [21]) and others (Large, pers.comm.).

SAMPLES AND METHODS OF INVESTIGATION

Detailed mineralogical studies using electron microprobe techniques were carried out on samples from three drill holes (Fig.1) that provide stratigraphic intersections from hanging wall to foot wall along the axis of the orebody west of the Jack Fault. Hole HL38A intersects the orebody at its southern end where it is overlain by barite ore. Hole HL70 is located some 200m NNE of HL38A and intersects minor barite-enrichment within the massive sulphide (possibly correlating with barite ore overlying the orebody immediately to the east). Hole HL56 is located some 300m further NNE where barite is negligible. The orebody east of the fault has not been sampled in the present study.

Overall, the deposit opens out to the north and plunges in this direction (Fig.1). Thus, the hanging wall is encountered at increasingly greater depth from 152m in HL38A to 223m in HL70 and 300m in HL56.

The drill holes were sampled at 1 to 2 metre intervals, each sample being mounted in epoxy and polished for microscopic and microprobe investigation using standard petrographic procedures. In addition, 9 samples from HL38A were prepared as doubly-polished thin sections, (after Barton[2]), in order to carry out more detailed examination of sphalerite textures in transmitted light. The samples were vacuum coated with approximately 20 nm of carbon to provide a conducting path during analysis in the electron microprobe.

A Cameca Camebax scanning electron microprobe was used for both quantitative and qualitative analysis of the ore minerals. This instrument is fitted with four wavelength dispersive X-ray spectrometers (WDS) analyzing from boron to uranium and a Kevex 7000 energy dispersive X-ray system (EDS) analyzing from sodium to uranium.

Quantitative analysis can be carried out on regions down to about 2 μm in area with semi-quantitative and qualitative analyses possible at sub-micron levels. The WDS system was used for both major and trace element analysis. The EDS system was used primarily for rapid qualitative identification of minerals prior to quantitative analysis, and to monitor the progress of trace element analyses where long counting times are involved (i.e. to ensure that the electron beam remained on the mineral of interest).

Qualitative analysis involved scanning electron microscopy using back-scattered electrons (BSE) to display mineral relationships, and the imaging and profiling of X-ray signals to display element distributions. Images obtained in BSE-mode contain much compositional information, the contrast being determined mainly by differences in average atomic number, so that heavy minerals such as gold or galena appear bright whereas light minerals such as quartz, carbonate and silicate gangue appear much darker. Thus fine-grained minerals (eg micron-sized gold and chalcopyrite) that can be very difficult to distinguish unambiguously with optical microscopy can be readily identified with the electron microprobe, and their compositions determined.

RESULTS

Qualitative mineralogy

Detailed mineragraphic examination has not formed part of the present work, but it is known from other studies (Fander, pers. comm.) that pyrite, sphalerite and galena are the major sulphides with minor arsenopyrite and chalcopyrite and trace amounts of tetrahedrite and other sulphosalts.

The complexity of the ore is illustrated in Figs.2 to 6, using back-scattered electron images to reveal the textures. Several of these are not readily visible with optical microscopy.

Broad-scale features of the ore, arising from the volcaniclastic sedimentary origin of the deposit and remobilization and recrystallization of the more ductile minerals such as galena, tetrahedrite and sphalerite, include possible sedimentary load-structures (Fig.2a) in fine-grained sphalerite ore, bands of coarse (1 mm) concretionary pyrite (Fig.2b), fragmental and brecciated pyritic ore (Figs.2c and 2d) and complex fragmented intergrowths of colloform pyrite and galena (Figs.2e and 2f).

At a much finer scale, primary (depositional) textures are represented mainly by framboidal pyrite and colloform intergrowths of pyrite and sphalerite to 20 μm in diameter (Figs.3a and 3b). The similarity of these textures to those found in hydrothermal sulphide deposits currently forming at the sea floor is striking (Koski et al.,[15]; Paradis et al.,[22]). The close affinities with Kuroko deposits are clearly evident when the microscopic features of sphalerite-rich ore (Fig.3c) are compared with those of black ore from the Shakanai deposit of Japan (Fig.3d).

Sphalerite typically forms coarse (mm-size) patches, commonly full of very

small (<20 μm) inclusions of the other sulphides. Galena also forms coarse patches and veins, but these are comparatively free of other sulphides, occurs as complex intergrowths with pyrite and sphalerite and as minute disseminated grains. Minor chalcopyrite occurs generally as very small (<5 μm) inclusions in sphalerite, and less commonly as coarse (100 μm) patches.

Minor euhedral arsenopyrite occurs as fine (<20 μm) intergranular crystals intimately associated with pyrite (Fig.4a) and as inclusions in sphalerite (Fig.4b). The mineral is auriferous (see below) but its fine grain size makes it difficult to liberate. It commonly displays cryptic oscillatory zoning at the micron and sub-micron scale. Such zoning is clearly related to growth of the crystals (Fig.4c) and many of the larger grains appear to have formed as a result of overgrowth on fragmented cores (Fig.4d). The zoning is most conspicuous in HL38A and least obvious in HL56. Detailed X-ray profiles indicate that it is due to variations in the minor Sb-content (Fig.5).

Argentian tetrahedrite is a trace mineral, concentrated in the upper part of the orebody (the Precious Metals Zone of Aberfoyle) where it occurs as coarse (to 100 μm) grains of complex composition (Fig. 6a) mainly in association with remobilized galena. Micron-sized inclusions of acanthite (Ag_2S) have been observed in such tetrahedrite (Figs.6b, 6c and 6d). The cubic morphology of the inclusions (Fig.6d) indicates they were originally argentite, the high temperature (>177°C) polymorph. Tetrahedrite also occurs as cross-cutting veinlets; less commonly as 5 to 10 μm disseminated grains and, in the barite-ore, as 5 μm inclusions within sphalerite. Very rarely, it occurs intergrown with colloform pyrite.

Other trace minerals found at the hanging wall include boulangerite ($5PbS.2Sb_2S_3$) and bournonite ($2PbS.Cu_2S.Sb_2S_3$) replacing galena, pyrargyrite ($3Ag_2S.Sb_2S_3$) replacing tetrahedrite and gudmundite (FeSbS) — the Sb-analogue of arsenopyrite. These

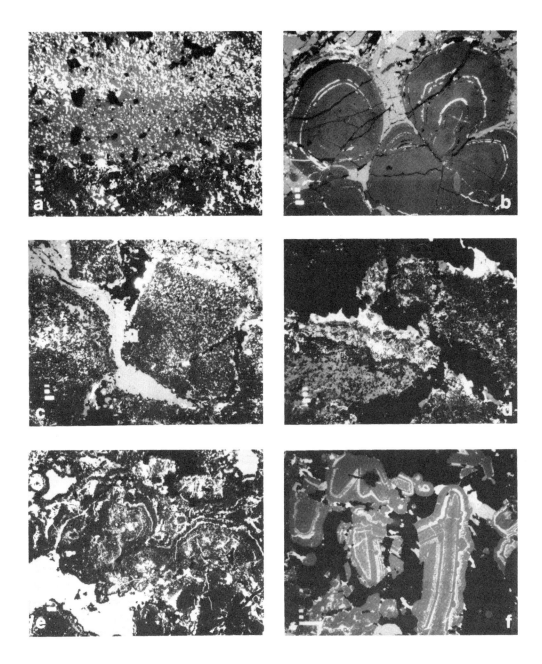

Fig.2 Backscatter electron images showing broad-scale textural
features of the ore. Scale bar = 0.1 mm. a) Sedimentary load
structure in sphalerite ore. Sphalerite (light grey), pyrite
(dark grey), galena (white) and quartz (black). b) Coarse
concretionary pyrite (dark grey) with galena (white) and
sphalerite (light grey). c) Fragments of pyrite ore (dark grey)
containing abundant arsenopyrite (light grey), cemented by
remobilized sphalerite (grey), galena (white) and quartz (black).
d) Fragments of sphalerite ore (grey) and pyrite ore (dark grey)
cemented by quartz (black) and rimmed by remobilized galena
(white) and tetrahedrite (light grey). e) Complex colloform
intergrowths of galena (white), sphalerite (light grey) and
pyrite (dark grey). f) Fragments of colloform galena (white) and
pyrite (grey) in a matrix of calcite (black).

have no economic value but are of genetic interest, their presence being interpreted as the result of late-stage Sb-rich solutions concentrated at the hanging wall.

Quantitative mineral chemistry

Mineralogical residence of gold

On the basis of the current mineralogical results as well as extensive metallurgical testing, there is now widespread agreement among Hellyer geologists and metallurgists that at least 90 % of the gold occurs in auriferous arsenopyrite and/or auriferous pyrite. However, a small proportion of the gold associated with barite-rich ore is amenable to direct cyanidation and therefore probably free (Woodcock, pers.comm.) This conclusion is supported by the observation of a single particle of gold in barite ore (McArthur, pers.comm.). No particulate gold has been observed in the present study.

The Au-content of pyrite is generally below the electron microprobe detection limit of 100 ppm. Electron microprobe trace analyses of the fine-grained arsenopyrite in HL38A and HL56, however, indicate that this mineral is indeed auriferous with Au-contents up to about 400 ppm. In HL56, there is sufficiently good agreement between trends shown by the whole rock Au-assays (Aberfoyle data) and the Au-

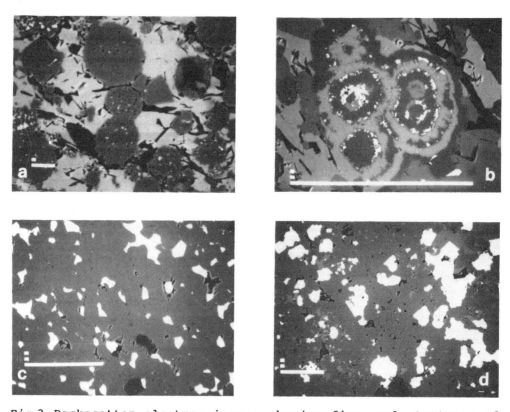

Fig.3 Backscatter electron images showing fine-scale textures of the ore. a) Pyrite framboids (grey) in a matrix of barite (light grey). Minute inclusions of galena (white) are present in the framboids and laths of interstitial chlorite (black) occur with the barite. Scale bar = 10 μm. b) Colloform grains composed of pyrite (dark grey), sphalerite (mid grey), barite (light grey) and galena (white). Flakes of chlorite (black) are visible (upper right). Scale bar = 100 μm. c) Detail of sphalerite-rich ore showing abundant interstitial galena (white) and minor disseminated pyrite (black) in a matrix of sphalerite (grey). Scale bar = 100 μm. d) Detail of Kuroko ore from the Shakanai mine, Hokuroko district, Japan, showing interstitial galena (white), tetrahedrite (light grey) and disseminated pyrite (black) in a matrix of sphalerite (grey). Scale bar = 100 μm.

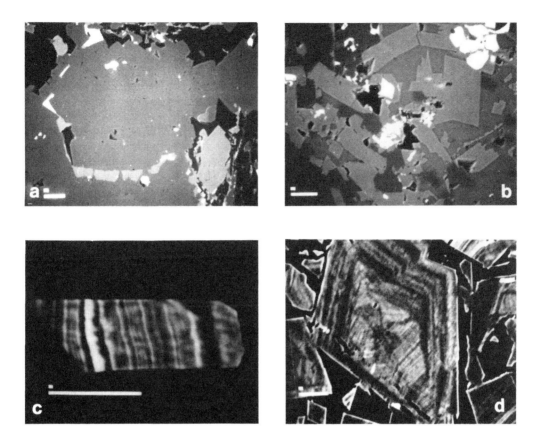

Fig.4 Backscatter electron images of arsenopyrite. a) Fine-grained arsenopyrite (light grey) and minor galena (white) along the grain boundaries of pyrite (grey). b) Coarse-grained arsenopyrite (light grey) and minor galena (white) associated with sphalerite (mid grey) and pyrite (dark grey). c) Detail (high contrast) of cryptic oscillatory growth zones in arsenopyrite. The brighter zones have higher Sb-contents. d) General view (high contrast) of zoned arsenopyrite showing overgrowth on complex fragmented cores. Scale bar = 10 μm.

content of the arsenopyrite (Fig.7) to conclude that this is the dominant host mineral in this section. Overall, however, there is no simple relationship between whole rock As- and Au-contents (see Geochemical Trends), possibly due to the presence of arsenical pyrite.

Mineralogical residence of silver

Tetrahedrite and galena are the principal hosts for silver in the deposit. Tetrahedrite, although only a trace constituent, accounts for about 60 % of the Ag-content of the ore and is the dominant host in the upper part of the orebody (the Precious Metals Zone of Aberfoyle). Galena is dominant in the lower part of the deposit (where tetrahedrite rarely occurs).

Electron microprobe analyses of tetrahedrite show a wide range of compositions, average values for which are summarized in Table I. The number of analyses averaged per sample ranges from 1 to 20 and gives a rough idea of the variation in relative abundance (ease of location) of the mineral. The Ag-content of individual grains varies from 1.9 to 17.9 wt% in HL38A, from 3.0 to 17.9 wt% in HL70 and from 3.0 to 28.3 wt% (freibergite) in HL56. Large variations also occur within individual grains that appear homogeneous under a light microscope. The complex internal structure revealed in backscatter electron images of such grains has been noted earlier (Fig.6a).

Although acanthite-argentite inclusions in tetrahedrite and pyrargyrite replacing tetrahedrite have very high

Ag-contents by definition, these minerals are rare in the Hellyer deposit and have no economic significance. Boulangerite and bournonite from the hanging wall in HL38A are not argentian.

Electron microprobe analyses for trace silver in the galena show a range from less than 100 ppm to about 0.42 wt% (Table II). The highest values are found at the base of the orebody in HL70, and at the base of the the Precious Metals Zone. Similar levels have long been known in galena from the Japanese Kuroko deposits (Yamaoka[38], Shimazaki[27]), and Foord and Shawe[8] have recently reviewed the chemistry of galena from a variety of sources. There can be little doubt that silver in Hellyer galena is in solid solution despite the fact that the solubility of Ag_2S in PbS is insignificant below 500°C (Van Hook[37]). This is confirmed by electron microprobe analyses of galena in Pb-concentrates from the deposit that agree well with the bulk Ag-assays. The presence of antimony accounts for the enhanced solubility (Amcoff[1]).

Reconnaissance electron microprobe analyses for trace silver in the pyrite of HL56 indicate Ag-contents of 250 to 350 ppm in the Precious Metals Zone; values that are well above the detection limit of 100 ppm. Sub-micron inclusions of the associated galena cannot account for such levels because the Ag-content is not high enough (although, of course, the inclusions may be atypical), and inclusions of tetrahedrite are extremely unlikely in view of the scarcity of this mineral. The existence of pyritic silver may well explain the anomalously high Ag-assays reported by Aberfoyle for pyrite concentrates from the hanging wall of the deposit. Analyses of about 200 ppm Ag are typical for such concentrates compared to levels of about 50 ppm Ag from the Footwall.

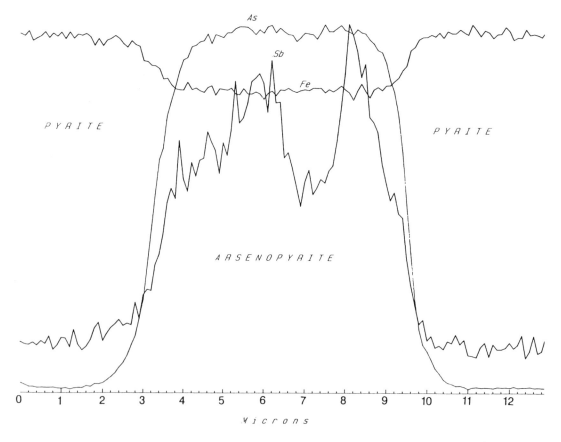

Fig.5 X-ray line-scan showing the distribution of As, Fe and minor Sb across a grain of zoned arsenopyrite in contact with pyrite. Length of scan = 13 μm. Note: maximum for As corresponds to about 42wt% As, maximum for Fe corresponds to about 46wt% Fe and maximum for Sb corresponds to about 2wt% Sb.

56

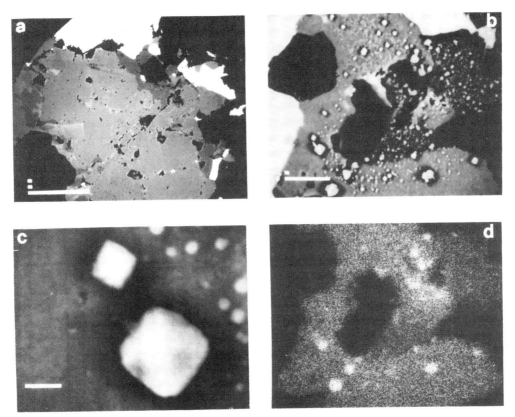

Fig.6 a) Backscatter electron image showing a large grain of complex argentian tetrahedrite (shades of grey) associated with base metal sulphides (black) and galena (white). The different shades of grey reflect variation in the Ag-content which ranges from 11 wt% for the darkest zones to 28 wt% (freibergite) for the lightest. Scale bar = 100 μm. b) Backscatter electron image showing another complex argentian tetrahedrite (dark grey to light grey) associated with remobilized galena (white) and minor pyrite (black). In this case the tetrahedrite also contains abundant inclusions of acanthite-argentite (also white). Scale bar = 10 μm. c) Backscatter electron image at high magnification showing the cubic morphology of the acanthite-argentite inclusions in (b). Scale bar = 1 μm d) X-ray element map showing the distribution of silver in (b).

As-content of pyrite

Reconnaissance data for the pyrite show a wide range in As-content from less than 500 ppm to about 2 wt%, and the distribution of results indicates that several populations (generations) are present. No assessment has been made of relative abundance, but the majority of grains have low to moderate As-contents on the order of 0.5 wt%. Pyrites of very high As-content are represented by the large (1 mm) colloform concretions, illustrated in Fig.2b, that have been observed only in a zone of high As-content in the upper part of HL56 (see

below). Rare zoning has been observed in some crystalline pyrite.

Sb-content and Zn-content of arseno-pyrite

Electron microprobe major element data for arsenopyrite from the arsenical zones of HL38A, HL70 and HL56 are summarized in Table III. The only minor elements of significance are antimony and zinc. The average Sb-contents range from 0.5 to 1.2 wt% but show no systematic trends between the holes. The effects of the cryptic zoning, however, are evident in the variability of the individual analyses which range from about 0.3 to 3.0 wt%

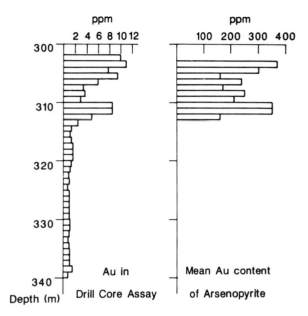

ppm
2 4 6 8 10 12

ppm
100 200 300 400

Au in Drill Core Assay

Mean Au content of Arsenopyrite

Depth (m)

Fig.7 Plot showing the Au-content of arsenopyrite from the Precious Metals Zone in HL56 and correlation with the assay for total gold.

Sb. This variability is greatest in HL38A. The average Zn-contents are generally less than 1 wt% and may, in part, reflect contamination from associated sphalerite.

Fe-content of sphalerite

Detailed electron microprobe analyses of sphalerite in HL38A, HL56 and HL70 are summarized in Tables IV, V and VI, respectively. The composition within individual samples is generally very uniform, and the average values well represent the stratigraphic variation.

The mean Fe-contents for HL38A, HL70 and HL56 are, respectively, 3.0, 2.9 and 2.2 wt%, the average for all sphalerite analyses (1009 in total), including a few samples from other locations within the deposit, being

Table I Average composition (wt%) of Hellyer tetrahedrites

Depth	Cu	Pb	Zn	Fe	Ag	Sb	As	S	Total	N[*]
(m)					**HL38A**					
154.5	21.5	<.1	10.6	4.7	15.7	19.3	3.1	26.9	101.8	5
161.5	39.5	<.1	4.7	3.3	2.1	16.5	9.1	26.8	102.0	3
164.9	37.1	<.1	5.5	3.0	3.1	18.5	7.5	26.3	101.1	10
165.9	38.7	<.1	5.1	3.6	2.6	16.5	9.4	25.9	101.8	3
191.8	33.0	<.1	4.9	5.9	5.3	26.8	0.6	25.3	101.8	4
208.9	36.4	<.1	5.9	2.5	3.3	25.5	2.7	25.3	101.8	10
					HL56					
301.9	27.3	<.1	9.1	3.0	11.0	25.1	0.4	24.5	100.5	3
303.0	22.5	<.1	3.6	3.3	20.3	26.4	0.3	22.7	99.1	10
304.0	24.0	<.1	2.8	3.8	18.2	27.1	0.2	23.3	99.5	20
305.0	32.1	<.1	4.0	3.1	7.7	27.9	0.0	24.6	99.6	5
306.0	32.3	<.1	4.4	3.7	5.9	27.7	0.9	24.5	99.4	10
308.0	33.7	<.1	5.0	2.5	5.2	28.1	0.8	24.9	100.2	1
309.0	30.8	<.1	2.9	4.3	9.0	27.3	1.3	24.3	99.9	2
310.0	30.8	<.1	4.0	3.5	7.5	27.7	0.5	24.2	98.2	10
311.0	26.8	<.1	4.2	3.9	12.6	27.1	1.8	23.8	100.2	5
312.0	32.0	<.1	3.8	6.1	4.2	26.6	1.0	25.5	99.3	5
313.0	33.9	<.1	5.4	4.2	3.0	27.1	0.6	26.2	100.4	1
315.0	35.0	<.1	2.8	6.0	3.4	28.6	0.6	25.5	102.0	1
					HL70					
223.9	32.5	<.1	3.6	3.5	7.4	26.7	1.4	24.8	99.9	5
226.0	31.7	<.1	3.7	3.7	7.7	26.9	1.4	24.7	99.9	10
228.0	32.2	<.1	3.4	4.9	6.0	27.3	1.0	25.0	100.0	10
230.0	24.6	<.1	3.2	4.3	17.5	27.5	0.5	23.2	100.8	1
232.0	32.2	<.1	4.5	3.1	6.6	27.6	1.1	24.7	99.8	11
234.0	31.3	<.1	5.0	2.7	8.2	27.6	1.1	24.9	100.7	11
238.0	27.4	<.1	3.0	5.2	13.6	27.3	0.6	24.0	101.2	2
242.0	25.8	<.1	3.7	3.7	15.2	27.1	0.9	23.8	100.2	10
244.0	29.6	<.1	3.1	5.4	8.1	26.2	1.4	25.3	99.2	11
256.0	34.8	<.1	4.8	2.6	4.1	24.3	3.2	25.2	99.0	4

[*] Note: N = number of analyses.

2.7 wt% Fe. Whilst these average
figures are useful for assessing Zn-
recovery from the ore, they conceal a
marked stratigraphic increase in Fe-
content (to about 7 wt% in HL38A, 5
wt% in HL56 and 6 wt% in HL70) that
is of considerable genetic interest
(see DISCUSSION).

Geochemical trends

Figure 8 summarizes geochemical trends
for Au, Ag, As, Ba, Cu and Pb within
each hole (based on Aberfoyle assays
of the samples used in this study).
Because the orebody is intersected at
progressively greater depth from south
to north, it is convenient to present
the data for each element in such a
way that the profiles for HL38A, HL70
and HL56 are projected onto the one
diagram. It should be emphasized,
however, that there is a spatial
separation along the axis of the
deposit of 500m between HL38A (on the
left) and HL56 (on the right).

As is to be expected in a complex
volcaniclastic-sedimentary environment
where lateral continuity of facies is
limited, much of the detailed down-
hole variability cannot readily be
correlated along the axis of the
deposit. Neverthless, on the broad
scale, trends are present that provide
a useful basis for discussing the
mineralogical results.

Figure 8a shows the distribution of
gold. Assuming an economic cut-off of
4 ppm Au, there are three Au-rich
zones (shaded) in HL38A, two in HL70
and two in HL56. The Au-concentrations
are highest in HL56, where there is a
well defined bimodal distribution in
the upper part of the deposit, and
lowest in HL70, where a bimodal distribution is only broadly discernible.

To assist visual comparison of trends,
these zones have been plotted as
reference intervals on all the
geochemical profiles.

Two Au-associations are well defined,
one with arsenic (arsenopyrite) and
the other with barium (barite). This

Table II Ag-content (ppm) of
Hellyer galena

HL38A		HL56		HL70	
Depth (m)	Ag (ppm)	Depth (m)	Ag (ppm)	Depth (m)	Ag (ppm)
161.5	230	301.9	180	223.9	320
164.9	<100	303.0	170	228.0	290
165.9	150	304.0	290	232.0	150
166.9	590	305.0	620	236.0	230
167.9	190	306.0	170	240.0	1200
171.8	3360	307.0	890	244.0	940
175.9	590	308.0	330	248.0	140
191.8	1100	309.0	120	260.0	970
208.9	260	310.0	270	270.0	540
215.9	380	311.0	370	280.0	4220
		312.0	910	290.0	3510
		313.0	920	300.0	1180
		315.0	110		
		323.0	<100		
		329.0	200		
		338.0	<100		

is clearly seen in HL38A where there
is a good correlation with As-content
(Fig.8b) for the lowermost Au-zone
(where auriferous arsenopyrite is
abundant) and with barite (Fig.8c) for
the uppermost Au-zone (at the hanging
wall). The association with arsenopyrite is also quite clear in HL56
where the Au-content of this mineral
closely parallels the bimodal
distribution of whole rock gold
(Fig.7). In HL70, however, only gold
at the hanging wall can be correlated
with arsenopyrite; the lowermost Au-
zone (which occurs about halfway down
the hole) is associated with barite.

A third association (with pyrite) may
be inferred for the central zone of
Au-enrichment in HL38A. Although there
is a correlation with minor arsenic in
this zone, this cannot explain the
bulk Au-assay, for the arsenopyrite
carries only 300 ppm Au. It is
concluded, therefore, (though not
proved) that auriferous pyrite is
responsible.

Figure 8d shows the Ag-distribution in
the three holes. The Ag-content is
lowest in HL38A and highest in HL70
and HL56. Although there is a good
correlation with bimodal gold (shaded)
at the hanging wall in HL56 this is
not the case further south in HL38A
and HL70. Indeed, there is no
correlation with gold in HL38A, other

Table III Average composition (wt%) of arsenopyrite

Depth	Cu	Pb	Zn	Fe	Ag	Sb	As	S	Total	N[*]
(m)					**HL38A**					
164.9	<.1	<.1	0.5	35.3	<.1	0.5	43.0	20.8	100.1	6
182.9	<.1	<.1	0.3	35.4	<.1	1.2	41.3	20.7	98.9	13
191.0	0.1	<.1	0.7	35.2	<.1	0.9	42.4	20.7	100.1	10
					HL56					
303.0	<.1	<.1	1.4	35.6	<.1	0.3	41.6	22.3	101.3	1
306.0	<.1	<.1	1.2	35.0	<.1	1.2	41.5	21.8	100.8	10
311.0	<.1	<.1	0.6	35.3	<.1	0.5	42.0	21.4	99.9	23
					HL70					
223.9	<.1	<.1	0.3	34.8	<.1	0.7	43.7	21.2	100.7	5
226.0	<.1	<.1	1.9	35.5	<.1	0.4	40.3	23.6	101.8	2
232.0	<.1	<.1	0.5	35.0	<.1	0.8	42.2	20.5	99.1	15
236.0	<.1	<.1	0.0	35.2	<.1	0.9	42.2	20.2	98.5	6

[*] Note: N = number of analyses

Table IV Average compositions (wt%) of sphalerite in HL38A

Depth (m)	Cu	Pb	Zn	Fe	Ag	Sb	As	S	Total	N[*]
154.5	0.1	<.1	65.7	0.6	<.1	<.1	<.1	33.0	99.4	3
160.9	0.2	<.1	65.8	1.0	<.1	<.1	<.1	33.5	100.5	8
161.5	<.1	<.1	65.6	1.2	<.1	<.1	<.1	33.6	100.4	11
162.9	0.1	<.1	64.9	1.3	<.1	<.1	<.1	33.2	99.5	10
164.9	0.1	<.1	66.2	1.4	<.1	<.1	<.1	33.6	101.3	9
165.9	0.1	<.1	65.6	1.5	<.1	<.1	0.1	33.4	100.7	10
166.9	0.1	<.1	62.5	3.1	<.1	<.1	<.1	33.9	99.6	10
167.9	<.1	<.1	66.0	2.2	<.1	<.1	<.1	33.4	101.6	11
170.9	0.3	<.1	61.1	5.0	<.1	<.1	<.1	33.5	99.9	10
171.8	<.1	<.1	60.7	6.1	<.1	<.1	<.1	32.8	99.6	10
172.9	<.1	<.1	60.6	6.7	<.1	<.1	<.1	34.0	101.3	9
174.9	<.1	<.1	65.4	2.6	<.1	<.1	<.1	33.2	101.2	10
175.0	<.1	<.1	63.5	2.8	<.1	<.1	<.1	33.0	99.3	9
176.9	0.3	<.1	63.5	3.5	<.1	<.1	<.1	33.4	100.7	10
178.9	0.1	<.1	62.2	3.6	<.1	<.1	<.1	33.9	99.8	10
182.9	<.1	<.1	63.9	3.4	<.1	<.1	<.1	33.6	100.9	10
184.9	0.2	<.1	62.9	3.7	<.1	<.1	<.1	33.7	100.4	10
186.9	<.1	<.1	62.5	3.9	<.1	<.1	<.1	33.4	99.8	10
188.9	<.1	<.1	62.6	4.2	<.1	<.1	<.1	33.9	100.7	10
191.8	<.1	<.1	63.1	3.5	<.1	<.1	<.1	33.7	100.3	10
192.8	<.1	<.1	64,0	3.1	<.1	<.1	<.1	33.6	100.7	11
194.9	0.1	<.1	63.8	2.3	<.1	<.1	<.1	32.8	99.0	10
196.0	<.1	<.1	63.3	3.3	<.1	<.1	<.1	33.6	100.2	13
200.8	<.1	<.1	64.4	2.0	<.1	<.1	<.1	32.6	98.0	10
202.5	<.1	<.1	64.5	2.1	<.1	<.1	<.1	33.4	100.0	10
204.5	<.1	<.1	63.3	2.2	<.1	<.1	<.1	33.1	98.6	10
207.9	<.1	<.1	63.5	2.9	<.1	<.1	<.1	33.7	100.1	10
208.9	0.1	<.1	62.9	3.7	<.1	<.1	<.1	33.5	100.2	10
209.9	<.1	<.1	63.4	3.7	<.1	<.1	<.1	33.9	101.0	10
211.0	<.1	<.1	62.4	3.7	<.1	<.1	<.1	33.5	99.6	10
213.4	0.2	<.1	62.6	3.3	<.1	<.1	<.1	33.9	100.0	10
217.9	<.1	<.1	64.1	2.4	<.1	<.1	<.1	33.6	100.1	10
All	0.1	<.1	63.6	3.0	<.1	<.1	<.1	33.5	100.2	314

[*] Note: N = number of analyses.

60

Table V Average compositions (wt%) of sphalerite in HL56

Depth (m)	Cu	Pb	Zn	Fe	Ag	Sb	As	S	Total	N*
301.9	0.4	<.1	63.3	2.8	<.1	<.1	<.1	33.6	100.1	10
303.0	1.1	<.1	60.9	3.8	<.1	<.1	<.1	33.8	99.6	9
304.0	0.9	<.1	61.6	3.1	<.1	<.1	0.2	33.6	99.4	8
305.0	0.2	<.1	62.5	3.1	<.1	<.1	<.1	33.6	99.4	10
306.0	0.3	<.1	61.4	3.6	<.1	<.1	0.2	33.4	98.9	16
307.0	0.4	<.1	63.0	4.0	<.1	<.1	<.1	34.0	101.4	8
308.0	0.1	<.1	62.5	3.6	<.1	<.1	<.1	33.1	99.3	10
310.0	0.2	<.1	62.3	3.4	<.1	<.1	<.1	33.2	99.1	10
311.0	0.2	<.1	63.3	3.5	<.1	<.1	<.1	34.7	101.7	7
313.0	0.5	<.1	62.7	3.2	<.1	<.1	<.1	32.5	98.9	18
315.0	0.6	<.1	62.5	3.1	<.1	<.1	<.1	33.7	99.9	10
317.0	0.6	<.1	60.8	3.8	<.1	<.1	<.1	33.4	98.6	7
319.0	0.2	<.1	62.3	3.6	<.1	<.1	<.1	33.6	99.7	8
323.0	<.1	<.1	60.6	5.0	<.1	<.1	<.1	34.1	99.7	10
325.0	<.1	<.1	63.5	3.2	<.1	<.1	<.1	33.9	100.6	10
327.0	<.1	<.1	64.3	2.7	<.1	<.1	<.1	33.1	100.1	10
329.0	0.2	<.1	62.4	3.1	<.1	<.1	<.1	33.7	99.4	10
331.0	0.2	<.1	64.7	2.0	<.1	<.1	<.1	33.1	100.0	9
333.0	<.1	<.1	65.2	1.6	<.1	<.1	<.1	33.0	99.8	9
335.0	<.1	<.1	64.9	1.4	<.1	<.1	<.1	32.7	99.0	10
337.0	0.1	<.1	65.2	1.3	<.1	<.1	<.1	33.1	99.7	10
338.0	0.1	<.1	65.1	1.3	<.1	<.1	<.1	33.4	99.9	10
339.0	<.1	<.1	65.4	1.0	<.1	<.1	<.1	33.3	99.7	10
All	0.3	0.2	62.8	2.9	<.1	<.1	<.1	33.4	99.6	243

*Note: N = number of analyses.

than overall enrichment in the upper part of the deposit (the Precious Metals Zone). In HL70, the zone of highest Ag-concentration also is not associated with gold, although Ag- and Au-enrichment coincide at the hanging wall.

The Pb-distribution (Fig.8e) closely follows that of silver in all three holes, but it is incorrect to conclude that galena is the dominant Ag-bearing mineral. The correlation merely reflects the close physical association between galena and argentian tetrahedrite. Only in the lower part of the deposit in HL38A is the Ag-distribution dominated by the galena.

Copper (Fig.8f) is a very minor constituent of the ore, the dominant host mineral being tetrahedrite rather than chalcopyrite. It is highest in the south (HL38A) where there is a suggestion of layering in the

distribution (due to concentrations of chalcopyrite), and lowest in the north (HL56). The Cu-content increases with depth in HL70, reflecting an increase in chalcopyrite, but is more or less uniform in HL56.

DISCUSSION

On the basis of its general geological setting, ore mineralogy, host rocks and associated alteration (McArthur [21]), there is general agreement amongst Hellyer geologists that the orebody is of volcanic-hydrothermal origin, formed at or near the discharge sites of submarine hydrothermal activity. The characteristic features of such deposits have been reviewed by Franklin et al.,[9]. The results of the present study provide additional information on the processes that have operated to form the Hellyer deposit.

Table VI Average compositions (wt%) of sphalerite in HL70

Depth (m)	Cu	Pb	Zn	Fe	Ag	Sb	As	S	Total	N[*]
223.9	0.1	<.1	62.2	3.1	<.1	<.1	<.1	33.3	98.7	10
226.0	0.1	<.1	63.5	2.6	<.1	<.1	<.1	33.4	99.6	9
228.0	0.1	<.1	62.9	3.0	<.1	<.1	<.1	33.5	99.5	10
230.0	0.1	<.1	62.4	3.4	<.1	<.1	0.1	33.3	99.3	10
232.0	0.1	<.1	62.6	3.3	<.1	<.1	0.2	33.4	99.6	10
234.0	0.2	<.1	63.3	3.2	<.1	<.1	<.1	33.0	99.7	10
236.0	0.1	<.1	63.2	3.4	<.1	<.1	<.1	33.0	99.7	10
238.0	<.1	<.1	63.4	3.1	<.1	<.1	<.1	33.4	99.9	10
240.0	0.1	<.1	62.2	4.2	<.1	<.1	0.1	33.4	100.0	10
242.0	<.1	<.1	62.9	3.7	<.1	<.1	<.1	33.5	100.1	10
244.0	0.5	<.1	60.3	5.6	<.1	<.1	<.1	33.5	99.9	10
246.0	0.1	<.1	64.1	2.3	<.1	<.1	<.1	33.1	99.6	10
248.0	<.1	<.1	64.8	1.3	<.1	<.1	0.1	33.2	99.4	10
250.0	<.1	<.1	64.9	1.6	<.1	<.1	<.1	33.1	99.6	10
252.0	0.1	<.1	64.9	1.7	<.1	<.1	<.1	33.4	100.1	10
253.0	0.1	<.1	64.6	1.6	<.1	<.1	<.1	33.1	99.4	10
254.0	0.1	<.1	64.3	2.1	<.1	<.1	<.1	32.9	99.4	10
256.0	0.1	<.1	64.6	1.3	<.1	<.1	<.1	33.2	99.2	9
258.0	<.1	<.1	64.6	1.8	<.1	<.1	<.1	33.2	99.6	11
260.0	0.1	<.1	64.2	1.9	<.1	<.1	<.1	32.9	99.1	18
264.0	0.1	<.1	63.6	2.5	<.1	<.1	<.1	33.0	99.2	20
268.0	0.1	<.1	64.5	1.7	<.1	<.1	<.1	32.2	98.5	11
270.0	0.3	<.1	64.6	1.2	<.1	<.1	<.1	34.2	100.3	11
272.0	0.1	<.1	64.5	1.8	<.1	<.1	<.1	33.2	99.6	10
264.0	0.1	<.1	64.7	1.9	<.1	<.1	<.1	33.6	100.3	10
276.0	0.1	<.1	64.3	2.2	<.1	<.1	<.1	33.6	100.2	10
278.0	0.1	<.1	62.4	2.9	<.1	<.1	<.1	33.3	98.7	9
280.0	<.1	<.1	64.4	1.9	<.1	<.1	<.1	33.5	99.8	10
282.0	0.1	<.1	64.5	1.6	<.1	<.1	<.1	33.2	99.4	10
284.0	0.1	<.1	64.9	1.5	<.1	<.1	<.1	33.2	99.7	11
286.0	0.1	<.1	64.5	1.4	<.1	<.1	<.1	33.0	99.0	11
288.0	0.1	<.1	64.7	1.5	<.1	<.1	<.1	33.2	99.5	12
290.0	<.1	<.1	64.3	1.6	<.1	<.1	<.1	33.1	99.0	9
292.0	0.3	<.1	64.2	1.6	<.1	<.1	<.1	33.5	99.6	9
294.0	0.1	<.1	64.7	1.4	<.1	<.1	<.1	32.8	99.0	10
296.0	0.3	<.1	64.9	1.5	<.1	<.1	<.1	33.7	100.4	10
298.0	<.1	<.1	64.6	1.7	<.1	<.1	<.1	33.6	99.9	10
300.0	0.1	<.1	64.2	2.0	<.1	<.1	<.1	32.9	99.2	11
302.0	0.1	<.1	63.5	2.5	<.1	<.1	<.1	33.2	99.3	10
302.0	<.1	<.1	63.5	2.3	<.1	<.1	<.1	33.0	98.8	10
304.0	<.1	<.1	64.8	1.4	<.1	<.1	<.1	32.7	98.9	10
All	0.1	<.1	63.9	2.2	<.1	<.1	<.1	33.2	99.4	431

[*]Note: N = number of analyses

Significance of the sphalerite compositions

The fact that a zone of comparatively high-Fe sphalerite (Fig.9a) is encountered at depth in all three holes suggests that this is a widespread horizon. It transgresses the stratigraphy, appearing at progressively greater depth below the hanging wall in HL38A, HL70 and HL56. In the south, it is clearly related to the abundance of sphalerite in HL38A (Fig. 9b) and to a lesser extent also in HL70, but there is no such correlation in HL56 in the north. In HL70 it lies immediately above a zone of minor barite-enrichment (cf.Fig. 8c) that, in this study, has been

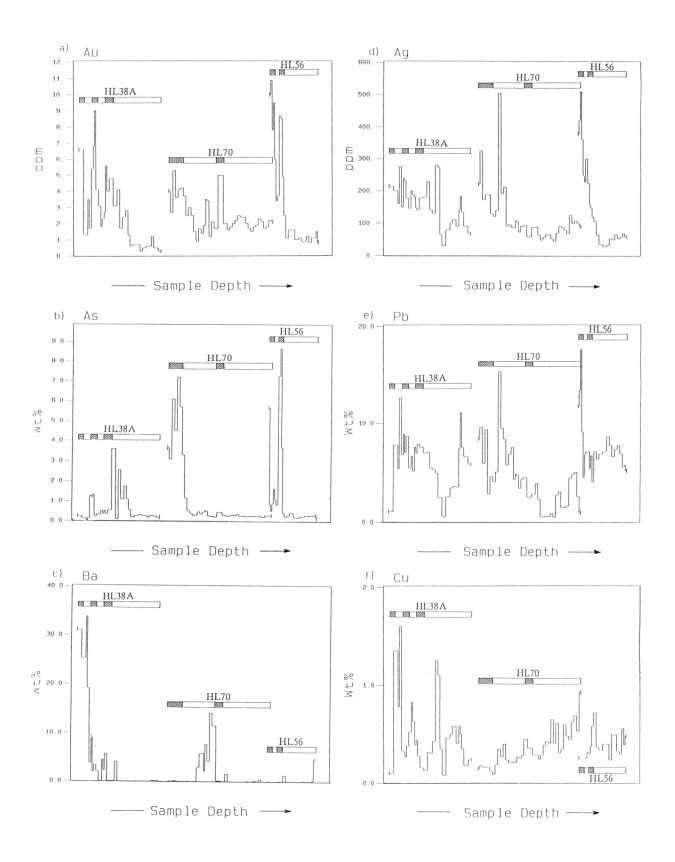

Fig.8 Geochemical profiles for HL38A, HL70 and HL56 illustrating both vertical and lateral trends for a) Au, b) As, c) Ba, d) Ag, e) Pb and f) Cu. Note that there is a separation of 500m along the axis of the deposit between HL38A in the south and HL56 in the north. Shaded sections indicate intervals containing more than 4 ppm Au.

interpreted as an extension of a barite capping immediately to the east (Fig. 1).

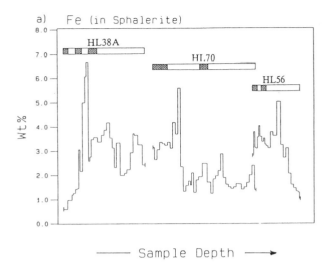

a) Fe (in Sphalerite)

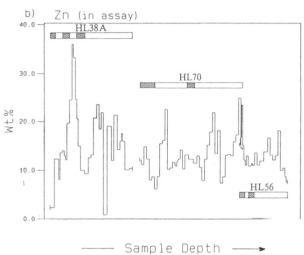

b) Zn (in assay)

Fig.9 Down hole profiles illustrating both the vertical and lateral trends in a) the average Fe-content of sphalerite and b) the Zn-content of the ore. Shaded sections indicate intervals containing more than 4 ppm Au.

Examination in transmitted light of polished thin-sections from HL38A shows that sphalerite in the ore is commonly homogeneous and varies in colour from straw yellow to almost colourless as the Fe-content decreases from about 3.0 wt% (typical of the deposit) to about 0.5 wt% Fe (in the barite zone at the hanging wall). Sphalerite in the zone of high Fe-content, however, is markedly different and consists of concentrically zoned crystals and crystal

fragments embedded in a matrix of yellow-brown sphalerite. The zoned crystals contain abundant sub-micron inclusions of chalcopyrite, described as 'chalcopyrite disease' by Barton[2]. Similar crystals have been reported in the Kuroko ores of Japan, although without evidence of stratigraphic control, and Eldridge et al.,[6] have described them as facies 2 sphalerite (recrystallized by 250°C fluids) in terms of their maturation model for Kuroko deposits.

Figure 10a shows an unbroken zoned crystal viewed in transmitted light; opaque 'growth' zones are conspicuous. Detailed electron microprobe analyses across this crystal, however, show no difference in composition between translucent and opaque zones, and it is concluded that the opacity arises from 'chalcopyrite disease' concentrated along growth planes below the

Fig.10 a) Photomicrograph in transmitted plane polarized light showing a crystal of concentrically zoned high-Fe sphalerite in a matrix of high-Fe sphalerite from the base of the upper unit. b) Backscatter electron image of the same crystal showing minor chalcopyrite (dark grey) concentrated on the growth planes. Other minerals visible are pyrite (black) and galena (white). Scale bar = 100 μm.

surface of the section. These intersect the surface as very thin concentric zones as can be seen from the backscatter electron image (Fig.10b). The width of the opaque zones in transmitted light is thus determined by the inclination of these planes to the surface of the thin section.

This interpretation is consistent with the conclusions of Barton[2] and Scott[26] that 'chalcopyrite disease' is brought about by the reaction of Cu-bearing solutions with the FeS component of sphalerite. Perhaps copper in excess of the equilbrium solubility was concentrated in regions of high dislocation densities as suggested by Tauson et al.,[33]. Such regions may be expected to exist at the crystal surface. The zonal pattern suggests episodic growth which could indicate that the crystals formed in a vent.

Although data from only three holes is not sufficient to prove lateral continuity of the high-Fe sphalerite zone, nevertheless it provides a strong indication that a marker horizon is present which divides the orebody into lower and upper units. A plausible interpretation can be developed as follows:-

1. The lower unit was derived from a submarine hydrothermal vent (or vents) somewhere between HL70 and HL56, where extensive stringer mineralization underlies the deposit (Fig.1). The mound of ore that formed on the sea floor above this had a barite capping that thinned over the area now intersected by HL70 (Fig.1). That such mounds can form in this environment is by now well established from study of modern day hydrothermal activity (Goldfarb et.al.,[10],Graham et al.,[11]; Hannington and Scott,[13]; Marchig et al.,[20]; Thompson et al.,[34]). They grow and mature by way of a complex succession of precipitation and replacement processes in response to the waxing and waning of a thermally intensifying system; more or less as envisaged in the Kuroko model of Eldridge et al.,[7].

2. At some stage the activity ceased and a new vent (or vents) opened up further south. A new mound with a barite capping formed above this. The zone of high-Fe sphalerite contains crystals expelled from this vent and deposited on the sea floor as clasts. These spread over the slopes of the original mound, burying attenuated sections of the barite capping in the vicinity of HL70 and thinning out further north beyond HL56.

A model that well fits this hypothesis is deposition from a plume of buoyant sulphide forming solutions, as proposed by Solomon and Walshe[29]. Maturation of the deposit, as envisaged in the Kuroko model of Eldridge et al.,[6] and Pisutha-Arnond and Ohmoto[24]), appears to have been limited to the formation of facies 2 assemblages (i.e.to recrystallization and coarsening by fluids at about 250°C). Other models for the flow of hot saline solutions from vents on the sea floor have been reviewed by Turner and Gustaffson[35].

At low temperatures and pressures, phase relations in the Fe-Zn-S system (Scott[26]) indicate that the Fe-content of sphalerite should be highest at low sulphur activity (the onset of solfataric activity) and decrease as the sulphur activity increases during construction of the mound. The relatively high Fe-content of sphalerite at the base of the upper unit, therefore, is to be expected from resurgence of hydrothermal activity at a new vent, and is consistent with the generally held belief (Moles[22]) that Fe-rich sphalerites of submarine hydrothermal origin have precipitated from the metalliferous solutions during their expulsion from the hydrothermal vents and carried in suspension in the laterally spreading brines.

Decrease in Fe-content of sphalerite with stratigraphic height in Kuroko deposits has been reported by Urabe[36], who ascribed the trend largely to a decline in temperature, and by Bryndzia et al.,[3]. However, as noted above, this trend may also be related to an increase in total sulphur concentration (and oxygen fugacity) which also favours the formation of barite (Finlow-Bates[7]). Such a systematic decrease in Fe-content through to the barite ore is clearly seen within the upper unit of the Hellyer deposit.

The precious metals

With the exception of minor gold at the top of the lower unit (associated with arsenopyrite in HL38A and with barite in HL70) most of the precious metals mineralization is found within the upper unit.

Gold

Cathelineau et al.,[5] have shown, by means of ion microprobe analysis, that the Au-content of hydrothermal arsenopyrites commonly ranges up to 1000 ppm, and that in most cases the gold is heterogeneously distributed within growth zones or overgrowths in the crystals due to an antipathetic relationship between Au and the Sb-content. Concentrations as high as 0.44 wt% Au in arsenopyrite have been reported by Cabri et al.,[4], but even at these levels the gold is invisible to high resolution electron microscopy indicating that it is probably in solid solution. The average Au-content in arsenopyrite is thus highly variable within a deposit, as a function of the distribution of Au-rich grains. Cathelineau et al.,[5] have also shown that Au-rich arsenopyrites crystallize for the most part at quite low temperatures (170 - 250°C).

The results of the present study find ready interpretation in the light of these data. The Au-contents of the arsenopyrite determined with the electron microprobe fall within the range cited by Cathelineau et al.,[5] and the patchy distribution of gold noted by these authors can be well understood in terms of the complex oscillatory variation in Sb-content described in the present paper.

Thermodynamic modelling of the hydrothermal fluid (Large[16], Eldridge et al.,[6], Pisutha-Arnond and Ohmoto[24]) indicates that gold in the upper part of zinc-rich massive sulphide lenses

was likely transported as Au(HS)$_2^-$ and that high gold grades are promoted by near neutral oxidised and low temperature (150 - 275°C) conditions. Huston and Large[14] have proposed further that partial oxidation of reduced sulphur caused by mixing with seawater may remobilize this gold along the seawater interface to distal parts of the deposit in the form of AuS$_n^-$ (n>2). This model can well explain the higher gold grades in HL56 and would indicate that the auriferous arsenopyrite crystallized within the upper parts of the mound close to the sea floor rather than within the buoyant plume itself.

Although the average Sb-content of the arsenopyrite appears not to vary significantly from hole to hole (Table III) there is a very clear diminution in both the frequency of the cryptic zoning and its compositional range from south to north. This is consistent with a change from turbulent conditions of crystallization near the hydrothermal vent (HL38A) to quiet conditions in remote parts of the plume (HL56). Overall, the comparatively uniform conditions of the distal environment would be expected to favour incorporation of gold into the arsenopyrite compared with the turbulent proximal environment where co-precipitation of gold would be inhibited, by the antipathetic behaviour of gold and antimony, during the frequent excursions to higher Sb-content, .

Silver

Natural tetrahedrite shows complex compositional variation in which As and Sb display extensive mutual substitution, and elements such as Fe, Zn, Hg and Ag commonly substitute for Cu. Because of this complexity, it is not possible to use the composition as a guide to the physico-chemical conditions of formation, despite extensive experimental studies (Skinner et al.,[26], Luce et al.,[18], Tatsuka and Morimoto[31,32], Makovicky and Skinner[19]), since few of the equilibrium reactions can be directly applied to natural mineral assemblages. Nevertheless, systematic compositional variations within the Hellyer tetrahedrites imply that Ag-enrichment at the hanging wall is the result of late stage Sb-metasomatism.

On average, the most argentian tetrahedrites are found at the top and bottom of the upper unit (Fig.11a). They are also quite arsenical in HL38A, whereas further north, in HL70 and HL56, the As-content is negligible (Fig.11b). When the individual analyses are taken into account, three populations can be distinguished in HL38A: 1) tetrahedrite, 2) argentian tetrahedrite and 3) arsenical tetrahedrite (Fig.12); the argentian and arsenical varieties occur only in the upper unit. These distinctions cannot be made in HL70 and HL56 where the tetrahedrite shows a continuous range of compositions through to freibergite (Fig.12). The division between freibergite and argentian tetrahedrite is here taken at 20 wt%Ag, following the classification of Riley[25]. As in HL38A, trace tetrahedrite in the lower unit contains only minor silver in HL70. No tetrahedrite was observed in the lower unit in HL56.

Figure 13 shows the tetrahedrite analyses plotted in terms of the principal 'end member' components. Such plots, adapted here from the "Modified Johnson Spinel Prism" described by Stevens[30] and Haggerty[12], provide a useful summary of the complex compositions. In HL38A, for example, the composition of the arsenical tetrahedrite trends along the back plane of the projection and can be understood primarily in terms of substitution of As for Sb. This population is substantialy diminished in HL70 and is absent in HL56. The argentian tetrahedrites, which trend along the front triangular surface, have a more complex composition involving significant Fe as well as Ag subsitution in HL38A but comparatively simple compositions with negligible Fe subsitution in HL70 and HL56. Similar trends are apparent with regard to Zn substitution (not plotted).

In HL38A, the most argentian tetrahedrite occurs as 5 μm inclusions in low-Fe sphalerite in the barite-ore, implying that a primary trend towards Ag-enrichment existed during accumulation of the deposit.

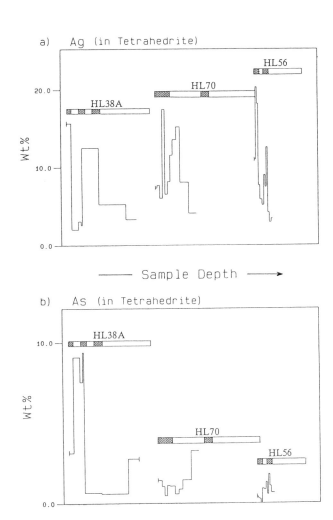

Fig.11 Down hole profiles showing both vertical and lateral variations in a) the average Ag-content and b) the average As-content of tetrahedrite. Shaded sections indicate intervals containing more than 4 ppm Au.

However, in HL70 and HL56, grains of highly argentian tetrahedrite and freibergite in the Ag-rich parts of the deposit are coarse (100μm) and associated with recrystallized galena. Such grains are inhomogeneous and show large variations in Ag-content due to complex zoning (Fig.6a). Similar variability has been reported for tetrahedrites in the Kuroko deposits of Japan (Shimazaki[27]). The source of additional silver in the Hellyer deposit is believed to have been argentian galena that has been replaced by non-argentian boulangerite and bournonite at the hanging wall. The present bimodal distribution of silver and its

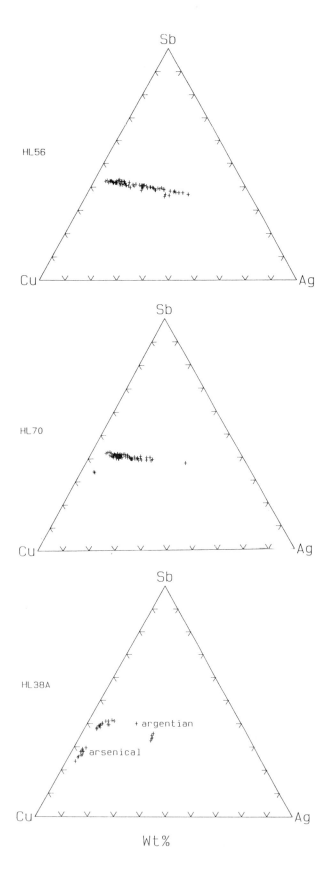

Fig.12 Tetrahedrite compositions plotted in the ternary diagram Cu-Ag-Sb showing variation in trends from south (HL38A) to north (HL56) in the orebody.

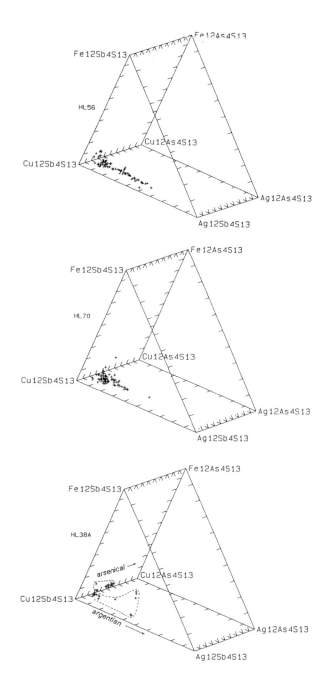

Fig.13 Tetrahedrite compositions plotted in the 6-component prism $Cu_{12}Sb_4S_{13}$ - $Fe_{12}Sb_4S_{13}$ - $Ag_{12}Sb_4S_{13}$ - $Cu_{12}As_4S_{13}$ - $Fe_{12}As_4S_{13}$ - $Ag_{12}As_4S_{13}$ showing variation in trends from south (HL38A) to north (HL56) in the orebody.

68

northwards enrichment at the hanging wall may be a consequence of metasomatic solutions moving along the top and bottom contacts of the upper unit.

The Ag-content of galena in HL56 parallels the bimodal distribution shown by the Ag-content of the tetrahedrite. The maximum Ag-content, however, is much lower (<0.1 wt%) than in the presumed vent area to the south despite the higher Ag-content of the associated tetrahedrite. This is consistent with the hypothesis, outlined above, that silver has been released from the galena through replacement by Sb-rich minerals.

CONCLUSIONS

It is concluded that:

1) The Hellyer volcanogenic massive sulphide deposit can be divided into upper and lower units, representing accumulation of sulphides from different hydrothermal vents. A zone of high-Fe, in part clastic, sphalerite at the base of the upper unit over-rides the lower unit and marks the onset of hydrothermal activity from a new vent further south.

2) Most of the precious metals mineralization is associated with the upper unit.

3) Arsenopyrite is the dominant host for the gold but an association with pyrite is inferred and independent evidence suggests that minor free gold may occur in the barite ore. The arsenopyrite is believed to have crystallized at the unconsolidated surface of the mound close to the sea water interface where gold may have been present as $Au(HS)_2^-$, in the proximal facies, or AuS_n ($n>2$), in the distal facies. Cryptic oscillatory zoning of the arsenopyrite reflects fluctuations in the Sb-content of the fluids. These fluctuations were greatest in the proximal facies and least in the distal facies. This may, in part, explain the lower gold grades of ore in the proximal facies, co-

precipitation of gold in arsenopyrite of higher Sb-content being inhibited by the antipathetic behaviour of gold and antimony.

4) Tetrahedrite is the dominant host for silver, but galena is also argentian. Recrystallization associated with late stage Sb-metasomatism resulted in formation of highly argentian tetrahedrite and freibergite at the hanging wall. Replacement of argentian galena by non-argentian bournonite and boulangerite provided the additional silver.

ACKNOWLEDGEMENTS

The authors wish to thank the Directors of Aberfoyle Limited for providing access to the Hellyer deposit at an early stage of exploratory drilling and development, and for financial assistance towards conduct of the research. Drs W.L. Griffin, J.R. Wilmshurst and Mr G.J. McArthur are thanked for reviewing the manuscript.

REFERENCES

1. Amcoff O. The solubility of silver and antimony in galena. _Neues Jahrbuch fur Mineralogische Monatshefte_, Heft 6, 1976, p. 247-261.

2. Barton P.B.Jr. Some ore textures involving sphalerite from the Furotobe mine, Akita Prefecture, Japan. _Mining Geology_, vol. 28, 1978, p. 64-72.

3. Bryndzia L.T., Scott S.D. and Farr J.E. Mineralogy, geochemistry and mineral chemistry of siliceous ore and altered footwall rock in the Uwamuki 2 and 4 deposits, Kosaka mine, Hokuroko district, Japan. In: _The Kuroko and related volcanogenic massive sulphide deposits_. Economic Geology Monograph No. 5, 1983, p. 507-522.

4. Cabri L.J., Chryssoulis S.L., De Villiers J.P.R., Laflamme J.H.G. and Buseck P.R. The nature of "invisible" gold in arsenopyrite. _Canadian_

Mineralogist, vol. 27, 1989, p. 353-362.

5. Cathelineau M., Boiron M.C., Holliger Ph. and Marion Ph. Gold-rich arsenopyrite: Crystal chemistry, gold location and state, physical and chemical conditions of crystallization. In: Bicentennial Gold 88, Symposium sponsored by The Australian Institute of Mining and Metallurgy, Melbourne, 1988, p. 235-240.

6. Eldridge C.S., Barton P.B.Jr. and Ohmoto H. Mineral textures and their bearing on the formation of the Kuroko orebodies. In: The Kuroko and related volcanogenic massive sulphide deposits. Economic Geology Monograph No. 5, 1983, p. 241-281.

7. Finlow-Bates T. The chemical and physical controls on the genesis of submarine exhalative orebodies and their implications for formulating exploration concepts. A review. Geologische Jahrbuch, vol. D40, 1980, p. 131-168

8. Foord E.E. and D.R. Shawe. The Pb-Bi-Ag-(Hg) chemistry of galena and some associated sulfosalts: a review and some new data from Colorado, California and Pennsylvania. Canadian Mineralogist, vol. 27, 1989, p. 363-381.

9. Franklin J.M., Lydon J.W. and Sangster D.F. Volcanic-associated massive sulphide deposits. Economic Geology, 75th Anniversary Volume, 1981, p. 484-627.

10. Goldfarb M.S., Converse D.R., Holland H.D. and Edmond J.M. The genesis of hot spring deposits on the East Pacific Rise 21°N. In: The Kuroko and related volcanogenic massive sulphide deposits. Economic Geology Monograph No. 5, 1983, p. 184-197.

11. Graham U.M., Bluth G.L. and Ohmoto H. Sulfide-sulfate chimneys on the East Pacific Rise, 11° and 13° latitudes. Part 1: Mineralogy and paragenesis. Canadian Mineralogist, vol. 26, 1988, p. 487-504.

12. Haggerty S.E. Opaque mineral oxides in terrestrial igneous rocks. In: Oxide Minerals, Mineralogical Society of America, Short Course Notes, 1976, p. 101-300.

13. Hannington M.D. and Scott S.D. Mineralogy and geochemistry of a hydrothermal silica-sulfide-sulfate spire in the caldera of axial seamount, Juan de Fuca ridge. Canadian Mineralogist, vol. 26, 1988, p. 603-625.

14. Huston D.L. and Large R.R. A chemical model for the concentration of gold in volcanogenic massive sulphide deposits. Ore Geology Reviews, vol. 4, 1989, p. 171-200.

15. Koski R.A., Shanks W.C., Bohrson W.A. and Oscarson R.L. The composition of massive sulfide deposits from the sediment-covered floor of Escanaba Trough, Gorda Ridge: Implications for depositional processes. Canadian Mineralogist, vol. 26, 1988, p. 655-673.

16. Large R.R. Chemical evolution and zonation of massive sulfide deposits in volcanic terrains. Economoic Geology, vol. 72, 1977, p. 549-572.

17. Large R.R., Huston D.L., McGoldrick P.J. and Buxton P.A. Gold distribution and genesis in Australian volcanogenic massive sulfide deposits and their significance for gold transport models. Economic Geology, Monograph 6, 1990, p. 512-518.

18. Luce F.D., Tuttle C.L. and Skinner B.J. Studies of sulfosalts of copper: V. Phases and phase relations in the system Cu-Sb-As-S between 350° and 500°C. Economic Geology, vol. 72, 1977, p. 271-289.

19. Makovicky E. and Skinner B.J. Studies of the sulfosalts of copper. VI. Low-temperature exsolution in synthetic tetrahedrite solid solution $Cu_{12+x}Sb_{4+y}S_{13}$. Canadian Mineralogist, vol. 16, 1978, p. 611-623.

20. Marchig V., Rosch H., Lalou C., Brichet E. and Oudin E. Mineralogical zonation and radiochronological relations in a large sulfide chimney from the East Pacific Rise at 18°25'S. Canadian Mineralogist, vol. 26, 1988, p. 541-554.

21. McArthur G.J. Hellyer. In: Geology and Mineral Resources of Tasmania Chapter 4, 1989, p. 144-148.

22. Moles N.B. Sphalerite composition in relation to deposition and

metamorphism of the Foss stratiform Ba-Zn-Pb deposit, Aberfeldy, Scotland. Mineralogical Magazine, vol. 47, 1983, p. 487-500.

23. Paradis S., Jonasson I.R., Le Cheminant G.M. and Watkinson D.H. Two zinc-rich chimneys from the plume site, souther Juan de Fuga ridge. Canadian Mineralogist, vol. 26, 1988, p. 637-654.

24. Pisutha-Arnond V. and Ohmoto H. Thermal history, and chemical and isotopic composition of ore-forming fluids responsible for the Kuroko massive sulphide deposits in the Hokuroko district in Japan. In: The Kuroko and related volcanogenic massive sulphide deposits. Economic Geology Monograph, No. 5, 1983, p. 523-558.

25. Riley J.F. The tetrahedrite-freibergite series with reference to the Mt Isa Pb-Zn-Ag orebody. Mineralium Deposita, vol. 9, 1974, p. 117-124.

26. Scott S.D. Chemical behaviour of sphalerite and arsenopyrite in hydrothermal and metamorphic environments. Mineralogical Magazine, vol. 47, 1983, p. 427-435.

27. Shimazaki Y. Ore minerals of the Kuroko-type deposits. In: Geology of Kuroko Deposits. Mining Geology, Tokyo Special Issue No.6, 1974, p. 311-322.

28. Skinner B.J., Luce F.D. and Mackovicky E. Studies of the sulfosalts of copper. III. Phase and phase relations in the system Cu-Sb-S. Economic Geology, vol. 67, 1972, p. 924-938.

29. Solomon M. and Walshe J.L. The formation of massive sulphide deposits on the sea floor. Economic Geology, vol. 74, 1979, p. 797-813.

30. Stevens R.E. Composition of some chromites of the Western Hemisphere. American Mineralogist, vol. 29, 1944, p. 1.

31. Tatsuka K. and Morimoto N. Tetrahedrite stability relations in the Cu-Sb-S system. Economic Geology, vol. 72, 1977, p. 258-270.

32. Tatsuka K. and Morimoto N. Tetrahedrite stability relations in the Cu-Fe-Sb-S system. American Mineralogist, vol. 62, 1977, p. 1101-1109.

33. Tauson V.L., Makeyev A.B., Akimov V.V. and Paradis L.F. Copper distribution in zinc sulfide minerals. Geokhimya, No. 4, 1988, p. 492-505.

34. Thompson G., Humphris S., Schroeder B., Sulanowska M. and Rona P.A. Active vents and massive sulfides at $26^{O}N$ (TAG) and $23^{O}N$ (SNAKEPIT) on the mid-Atlantic ridge. Canadian Mineralogist, vol. 26, 1988, p. 697-711.

35. Turner J.S. and Gustaffson L.B. The flow of hot saline solutions from vents on the sea floor - some implications for exhalative massive sulfide and other ore deposits. Economic Geology, vol. 73, 1978, p. 1082-100.

36. Urabe T. Iron content of sphalerite co-existing with pyrite from some Kuroko deposits. In: Geology of Kuroko Deposits. Mining Geology, Tokyo Special Issue No.6, 1974, p. 377-84.

37. Van Hook H.J. The ternary system $Ag_2S-Bi_2S_3-PbS$. Economic Geology, vol. 55, 1960, p. 759-788.

38. Yamaoka K. Metallic minerals of the Kuroko deposits in Northeast Japan. In: Proceedings Symposium on "Mineral Constituents of Kuroko Deposits and Geochemistry of Sulfide Minerals from Hydrothermal Deposits", Morioka, 1969, p. 1-38.

Mineralogy and petrology of the lead-zinc-copper sulphide ores of the Viburnum Trend, southeast Missouri, U.S.A., with special emphasis on the mineralogy and extraction problems connected with cobalt and nickel

Richard D. Hagni
Department of Geology and Geophysics, University of Missouri-Rolla, Rolla, Missouri, U.S.A.

SYNOPSIS

The ore deposits of the Viburnum Trend in the Southeast Missouri Lead District form the world's largest lead producing district. The district is comprised of eleven mines, ten of which are currently in production. The principal minerals are few in number, simple in composition, and form simple superposition and replacement textures. The economically valuable minerals are the sulphides, galena, sphalerite, and chalcopyrite; silver is recovered from galena and cadmium from sphalerite at the lead smelter and zinc refinery. The most abundant gangue minerals are dolomite, pyrite, marcasite, calcite, and quartz. The Viburnum ores, however, contain a diverse assemblage of minerals uncommon in most Mississippi Valley-type ore deposits. These include: siegenite, bravoite, vaesite, fletcherite, nickelean carrollite, bornite, gersdorffite, tennantite, enargite, luzonite, millerite, polydymite, chalcocite, digenite, anilite, djurleite, covellite, blaubleibender covellite, pyrrhotite, magnetite, anhydrite, and dickite.

The valuable ore minerals exhibit a variety of open space and replacement textures. Most grains of galena, sphalerite, and chalcopyrite are sufficiently coarse to provide an adequate separation by flotation at a grind of about 60% minus 200 mesh. Most mills in the district produce three flotation concentrates: lead, zinc, and copper. Portions of the ores are fine-grained or intricately intergrown and present a variety of recovery, concentrate grade, and deleterious constituent problems. Beneficiation problems that result from the character of the ore textures include zinc in the lead concentrates; magnesia, cobalt, and nickel in the zinc concentrates; lead in the copper concentrates; lead, zinc, and copper losses to the tailings. Ore microscopic study of the ores and beneficiation products contribute to a better understanding of the mineralogical and textural characteristics that cause these metallurgical problems.

Cobalt and nickel minerals are present in small amounts in the Viburnum Trend ores, and they comprise the second largest reserve of cobalt in the United States. Cobalt occurs primarily in siegenite; nickel is present mainly in siegenite and bravoite. Although small amounts of cobalt and nickel are present in the ores at all of the mines, their concentrations vary throughout the district. The greatest concentration occurs in the Fredericktown subdistrict where the ores average 0.23% cobalt. Within the Viburnum Trend, cobalt and nickel are most abundant in its central portion at the Magmont and Buick mines.

In recently discovered ores at the Magmont-West mine, nickel is locally concentrated in nickel-arsenic sulphide ores along portions of a linear breccia ore deposit. This deposit is unique in that it occurs two miles west of the main Viburnum Trend and in its location close to sulfide deposits in the underlying Precambrian basement. The Magmont-West ores are predominantly collomorphic in character and composed of millerite, gersdorffite, polydymite, and vaesite, partly replaced and intergrown with chalcopyrite, sphalerite, galena, and quartz.

Although cobalt and nickel formerly were recovered in the Fredericktown subdistrict, the metals have not yet been recovered at mines in the Viburnum Trend nor elsewhere in the Southeast Missouri Lead District. The recovery problems are related to the fine-grained character, coating and intergrowth textures, and flotation properties of siegenite. Current research efforts to recover cobalt and nickel involve fine grinding, flotation, and separation by pyrometallurgical procedures.

The Southeast Missouri Lead District has been the largest producer of lead in the United States since 1907, except for 1962 when a strike disrupted production. It has been the world's largest producer since 1970. More than 90% of United States and about 15% of world lead production currently is derived from the district. A total of about 380,000 short tons of recoverable lead metal is recovered annually from the district.[1]

Important amounts of additional metals are recovered as by-products from the lead ores at the mill and smelter. About 40,000 short tons of zinc metal recovered from the district places Missouri third in U.S. zinc production. Nearly 15,000 tons of copper metal places Missouri sixth among copper-producing states. The recovery at the smelters of about 1.2 million troy ounces of silver annually places Missouri seventh among silver-producing states. The annual value of these four metals currently amounts to about 400 million U.S. dollars.

The ore deposits are distributed in five subdistricts as shown in Figure 1 from Kisvarsanyi.[2] The most productive subdistricts

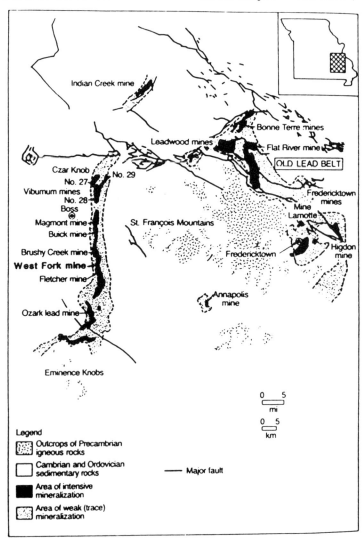

Fig. 1. Map of Southeast Missouri Lead District showing subdistricts, mines, mineralized areas, and Precambrian outcrops (modified from Kisvarsanyi[2]).

74

are the old Lead Belt that was mined from 1865 to 1972, and the Viburnum Trend or new Lead Belt that was first mined in 1960 and from which all of the current production is derived. Smaller subdistricts include Indian Creek, Fredericktown, and Annapolis. Current exploration is centered on the Winona area, southwest of the Viburnum Trend.

Total production from the older subdistricts in the Southeast Missouri Lead District was 220 million metric tons of ore.[3] Production to date from the Viburnum Trend totals about 140 million metric tons of ore. In contrast to ores from the old Lead Belt, the principal district prior to the discovery of the Viburnum Trend, that had an average grade of about 2.8% lead,[3] the ores from the Viburnum Trend average 5.8% lead, 0.8% zinc, 0.14% copper, and about 1/4 ounce silver per ton of ore.[4]

The Viburnum Trend extends for a distance of about 72.5 km in a north-south direction. The town of Viburnum is about 8 km south of the northern end. Ten mines currently are operating in the Viburnum Trend. Their initial mining dates are given below beginning at the north end of the Trend. Mining in the Trend began in 1960 at the Viburnum No. 27 mine, a deposit that subsequently has been exhausted. The Viburnum 28 and 29 mines were opened in 1962 and 1964. The Casteel or Viburnum 35 mine, where mining began in 1983, has mined unusually copper-rich ores. Although the Magmont mine began in 1968, mining to the west from the Magmont-West mine began in 1979, and the latter has produced from a zinc-rich deposit. The Buick mine opened in 1969. Brushy Creek was opened in 1973 and was temporarily shut down from 1986-1988. West Fork, the newest mine in the Trend, was opened in 1985. Mining began at the Fletcher mine in 1967. The Sweetwater mine, formerly called the Ozark Lead mine and the Milliken mine and located at the south end of the Trend, was opened in 1968, closed in 1983, and reopened in 1987. The West Fork and Sweetwater mines are operated by ASARCO, Magmont and Magmont-West mines by Cominco American, and all other mines by Doe Run Company. Doe Run Company, created in 1986, has been owned 57.5% by Fluor Corporation and 42.5% by Homestake Mining, but Fluor

currently is purchasing all of Homestake's holdings.

The stratigraphy, structure, forms of ore bodies, and ore controls recently have been reviewed by Hagni[4] who included numerous references on specific aspects. The exploration history in the district has been reviewed by Ohle and Gerdemann.[5] Recent descriptions of the geology of individual mines include: Viburnum 27 (Grundmann[6]), Viburnum 28 (Pettus and Dunn[7]), Casteel (Dunn and Grundmann[8]), Magmont (Sweeney et al.[9]), Magmont-West (Bradley and Krolak[10]), Buick (Rogers and Davis[11]; Mugel and Hagni[12]), Brushy Creek (Evans[13]) West Fork (Dingess[14]), Fletcher (Paarlberg and Evans[15]), Sweetwater (Mouat and Clendenin[16]; Larsen et al.[17]; Walker[18]). The sedimentary facies and their influence have been discussed by Larsen[19] and Gregg and Gerdemann.[20]

MINERALOGY

The principal sulphide minerals in the ores of the Viburnum Trend are galena, sphalerite, chalcopyrite, marcasite, and pyrite. All of the sulphide minerals were repetitively deposited and occur in a variety of forms. Galena occurs as massive replacements of the hostrock dolomite, as replacements of earlier deposited sulphide minerals, and as euhedral crystals deposited in vugs. Sphalerite occurs dominantly as relatively fine-grained replacements of hostrock dolomite, but small, subsequently deposited, vug-lining crystals also are present. Chalcopyrite occurs commonly as massive to colloform replacement masses, but also as subhedral crystals disseminated in host rock dolomite and as euhedral crystals in vugs. Marcasite and pyrite form small disseminated crystals, large colloform masses, and euhedral crystals in vugs.

The main gangue minerals, in addition to marcasite and pyrite, are dolomite, quartz, and calcite. The host rock Cambrian Bonneterre limestone has been dolomitized nearly everywhere that it contains ore deposits. Dolomite also was deposited as euhedral crystals in vugs in association with the sulfides. Cathodoluminescence microscopic study of those dolomite crystals by Voss and Hagni[21] and Voss et al.[22]

shows that they have been deposited during four periods that are closely associated with the periods of deposition of the ore sulphides. Quartz is most abundant at the Magmont and Buick Mines where it occurs as jasperoid replacements of dolomite and as euhedral crystals formed in vugs.[23] Quartz also was deposited during the span of sulfide mineral deposition, and it contains fluid inclusions, studied by Roedder[24] and Hagni,[25] that indicate it was deposited from ore fluids with elevated temperatures. Calcite forms euhedral crystals that were deposited in vugs after the sulphide mineral deposition had largely ceased.

Although the principal minerals are few in number and simple in composition, many minor minerals of more complex compositions also are locally present in the Viburnum ores. Many of these minerals are ones that are not commonly associated with Mississippi Valley-type ore deposits. They include: siegenite, bravoite, vaesite, fletcherite, nickelean carrollite, bornite, gersdorffite, tennantite, enargite, luzonite, millerite, polydymite, chalcocite, digenite, anilite, djurleite, covellite, blaubleibender covellite, pyrrhotite, magnetite, anhydrite, and dickite.

Zinc sulfide with wurtzite morphology has recently been found to occur in bands interlayered with fine-grained normal sphalerite.[26] Such schalenblende zinc sulfides are moderately abundant at the new West Fork mine in the Viburnum Trend where they occur as one to three mm thick yellow bands that are deposited on the surface of colloform masses of sphalerite. They are composed of radiating acicular crystals with much porosity between the crystals.

Although barite is absent from the Southeast Missouri Lead District, a major barite district, the Washington County Barite District, is located immediately to the north of the lead district. Recent descriptions of the barite deposits include Wharton[27] and Kaiser[28].

All of the sulfide minerals and calcite are locally leached, but the most prominent leaching is that of galena.[25,29] Galena that belongs to the octahedral generation has been leached after the deposition of crusts of chalcopyrite,

marcasite, and dolomite crystals. The octahedral galena may be partially or completely leached and leave behind the coatings with empty casts.

ORE TEXTURES AND RELATED MINERAL BENEFICIATION PROBLEMS
The Viburnum ores exhibit a wide variety textures that result from combinations of open space filling and replacement processes. Open space filling textures are characterized by the simple superposition of one mineral upon another in vugs, fractures, and spaces between breccia fragments. Replacement textures include those in which host rock dolomite crystals are replaced by sulphides and those in which earlier sulphide minerals are replaced by later sulfides. Typical open space and replacement textures have been illustrated previously by Hagni and Trancynger.[30] Partial replacements, fine sulphide grain sizes, and other intricate intergrowths tend to develop locked particles when these ores are ground prior to beneficiation. The partitioning of such particles is difficult to control during the subsequent flotation steps. Beneficiation practice at the mills in the Viburnum Trend has recently been reviewed by Watson,[31] and the role of mineralogy in beneficiation problems in the Viburnum Trend has been discussed by Hagni.[32] Six examples of current beneficiation problems in the Viburnum Trend, which are a function of ore mineral textures, are given in the following paragraphs.

Galena, the most abundant ore mineral in the Viburnum Trend, is deposited during two main periods, an earlier octahedral generation, and a later cubic generation.[22,32] Both generations of galena were deposited after most sphalerite grains, and locally galena exhibits a preference to replace the earlier sphalerite grains. Partially replaced sphalerite grains form intricately locked sphalerite-galena grains that result in increased zinc contents of the lead concentrates and losses of lead to the zinc concentrates.

Most of the sphalerite grains were deposited early in a disseminated texture that formed principally by replacement of the host rock dolomite.[30] This feature, together with the fact that sphalerite forms smaller grains than those of galena, causes abundant locked sphalerite-dolomite particles in the zinc concentrate.[33] Because the magnesium from the dolomite is a deleterious constituent at the electrolytic zinc refineries, the refineries limit the magnesium content to 0.6% in the zinc concentrates. Until recently, all zinc concentrates in the Viburnum Trend had to be treated by sulfuric acid digestion to diminish their magnesium contents. Some of the final zinc concentrates, such as those from the Magmont mine and mill, have been reground and subjected to additional flotation in order to avoid costly acid treatment.

Chalcopyrite was the earliest of the economically valuable minerals to be deposited. Chalcopyrite commonly was followed directly by siegenite, the principal Co-Ni mineral. Siegenite formed crusts upon the surfaces of the earlier chalcopyrite and usually partly replaced the underlying chalcopyrite grains or crystals.[30] Those partial replacements lead to the formation of chalcopyrite-siegenite locked particles that cause cobalt and nickel to be concentrated in the copper concentrate. In addition, very fine-grained inclusions of siegenite may occur within chalcopyrite in some copper concentrates.[34,35] The fact, however, that the flotation properties of siegenite also are similar to those of chalcopyrite causes part of the free siegenite particles to also partition into the copper concentrates.

In some ore bodies in the Viburnum Trend, such as Magmont East, siegenite tends to replace sphalerite.[33] Where this is the case, partial replacements lead to the formation of sphalerite-siegenite locked particles during grinding and flotation. Those locked particles constitute an important beneficiation problem because their presence raises the cobalt and nickel contents in the zinc concentrates above those allowable at the zinc refinery. Cobalt and nickel are deleterious constituents in zinc concentrates at the refineries, and it is necessary to hold cobalt below 0.13%, and to restrict cobalt plus nickel below 0.25%. At times, some of the zinc ores have not been mined or the zinc

concentrates from portions of some ore bodies in the Viburnum Trend have had to be stockpiled because of cobalt and nickel contents.

Fine-grained sulphide replacements of host rock dolomite are locally present. Galena, sphalerite, and chalcopyrite may form grains as small as 10 μm that are disseminated through dolomite crystals. Although these intergrowths are not common, where they are present they form locked particles upon grinding. During flotation these locked particles carry lead, zinc, and copper losses to the tailings. Other metallic losses to the tailings, however, are in the form of free galena, sphalerite, and especially chalcopyrite.[33] Research by the U.S. Bureau of Mines to recover the metallic losses to the Viburnum tailings by reflotation using a variety of flotation reagents has produced concentrates with acceptable lead contents but with significant contents of impurity elements.[35]

Zinc-rich ores at some of the Viburnum mines have recently been found to float poorly with conventional treatment. It is suspected that the presence of abundant organic matter or clays with the zinc ores result in excessive consumption of flotation reagents and result in very significant zinc losses to the tailings. Mineralogical studies in progress are designed to investigate the mineral phases responsible for that mill problem.

COBALT-NICKEL MINERALOGY
The mineralogy of the cobalt and nickel phases in the Viburnum Trend is of special interest because the Trend is the second largest reserve of cobalt in the United States, behind that of the Blackbird district in Idaho. Also, cobalt is a strategic metal important to the welfare of the country. Although the cobalt and nickel occurs in the Viburnum Trend ores that are currently being mined, the metals are not being recovered. The mineralogy of cobalt and nickel in the district has been investigated by several individuals: Hagni[23,25], Pignolet-Brandom[37], Horrall[38], and Jessey.[39]

The most important cobalt-nickel mineral is siegenite. Siegenite forms octahedral crystals that typically form coatings on chalcopyrite grains and crystals. The thickest coating of siegenite occur at the Madison mine in the Fredericktown subdistrict, south of the Old Leadbelt, where the ores averaged 0.23% cobalt.[40] The metals were commercially recovered from that subdistrict during World Wars I and II by producing an iron sulfide flotation concentrate that contained much of the cobalt and nickel. The concentrate was then subjected to a roast and the cobalt and nickel were recovered by hydrometallurgy. The Madison mine was dewatered in the late 1970's, shortly after a disruption of cobalt supplies from Zaire and Zambia had caused the price of cobalt to escalate to about $74 per pound, but operations subsequently ceased when the price returned to about $8 per pound. Although some cobalt and nickel production was made from concentrates formerly developed and stored on the property, no new mine production was achieved before the price declined.

Siegenite also occurs in smaller amounts as coatings of chalcopyrite in all of the Viburnum mines. Larger amounts of cobalt and nickel and higher cobalt-nickel ratios occur at the Magmont and Buick mines in the central portion of the Viburnum Trend and appear to indicate that a significant portion of the ore fluids entered the Trend in that vicinity.[23,39] Because larger amounts of cobalt and nickel and higher cobalt:nickel ratios also occur at the Sweetwater mine at the southern end of the district, another important source of fluid probably entered the Trend from the south.

Recent research on the recovery of cobalt and nickel from the ores in the Viburnum Trend has centered on efforts to recover those metals from the copper concentrate at the Magmont mine.[34] Due to the fine intergrowths between chalcopyrite and siegenite, it has been found that an extremely fine grind to a particle size of 10 μm is necessary to achieve separation of siegenite from chalcopyrite.

Bravoite is the second most important cobalt-nickel mineral in the ores from the Fredericktown and Viburnum Trend subdistricts. Bravoite occurs as disseminated pyritohedral crystals that are closely associated with siegenite and other ore minerals. The bravoite crystals typically are chemically zoned with the brown, less reflective zones carrying the largest nickel content, up to 12.0%. Although bravoite is an iron-nickel sulphide, it may contain significant amounts of cobalt replacing the nickel. Bravoite crystals from the lead district have been found to contain as much as 9.2% cobalt.[36,41] Bravoite was an abundant constituent in the cobalt-nickel concentrates from the Fredericktown ores.

A nickel-rich mineral association is locally present in some of the mines; it is known especially at the Buick mine, but it is present locally at most of the mines. The association consists of prismatic to acicular crystals of millerite that were deposited after octahedral galena but before cubic galena.[25,32] Where the crystals were enclosed by later cubic galena deposition, they are unaltered. Most millerite (NiS) crystals were not coated by galena and they exhibit all degrees of alteration to polydymite (Ni_3S_4). The volume change involved in this pseudomorphic alteration results in the development of prominent shrinkage cracks aligned largely along the basal cleavage. The subsequent deposition of vaesite (NiS_2) occurred on the surface of the polydymite crystals and within the shrinkage cracks within the polydymite. Nickel for the deposition of vaesite was supplied by the excess nickel from the pseudomorphic alteration of millerite to polydymite and by the introduction of additional nickel from the ore fluids.[23,25,32]

Chalcopyrite-bornite pods and lenses that locally are present in most of the mines in the Viburnum Trend also contain cobalt-nickel minerals.[32] The pods are horizontal lenses that are from about four inches to about four feet thick and may be 15 feet or more in lateral extent. They are composed of massive sulfides that have totally replaced the host rock dolomite at an early time in the paragenetic sequence and commonly are located low in the Viburnum Trend orebodies. The mineralogy and paragenetic sequence for these ores has been discussed by Hagni.[32] The chalcopyrite and bornite exhibit a wide variety of textural intergrowths that include exsolution, replacement, and colloform relationships. Gersdorffite (NiAsS) crystals were deposited early and were successively altered to the copper-arsenic minerals, tennantite and enargite, with the introduction of the copper-rich ore fluids that deposited massive bornite and chalcopyrite. Compositionally zoned, cobaltian pyrite cubes were subsequently deposited in these ores.[35] Fletcherite (Cu(Ni, Co)$_2$S$_4$) is locally present in the chalcopyrite-bornite pods and it was also replaced by tennantite and enargite. Vaesite crystals formed euhedral crystal replacements of chalcopyrite and bornite, commonly in the vicinity of enargite crystals. Nickelean carrollite occurs in some chalcopyrite-bornite pods.[36,38] Late veinlets composed of chalcocite, digenite, anilite, djurleite, covellite, and blaubleibender covellite traverse the massive chalcopyrite-bornite pods.

MAGMONT-WEST ORES

A drift, more than 2 km long and initiated in 1979, provided access to the new Magmont-West orebody. Since 1985, that deposit has provided more than half of the production from the Magmont mine.[10] The Magmont-West ore body differs in many respects from typical Viburnum Trend deposits. It lies about two miles west of the main ore bodies in the Trend. The position of the ore is controlled by a narrow linear breccia that is associated with fracturing. In contrast to the main ore bodies in the Viburnum Trend that trend northly, the Magmont-West ore body trends north-northwest and locally to the east-northeast. Its stratigraphic position is rather high, and its greatest concentration of mineralization is centered upon the False Davis horizon within the Bonneterre Formation. The ore body is unusually zinc-rich and averages about 3% zinc, approximately three times that of the average deposit in the Viburnum Trend. A portion of those ores, especially along the easterly side of the ore trends, contain nickel-arsenic-rich ores that are unlike nickel-rich ores elsewhere in the district. Because the mineralogy of the Magmont-West nickel-arsenic-rich ores has not previously been described, some emphasis is given here to the mineralogy and paragenetic sequence of those ores.

The nickel-arsenic-sulphide-rich ores were deposited dominantly as millerite that formed during four generations of crystalline and colloform nickel sulfide as shown in Figure 2. The first generation of millerite (generation I) was deposited as coarse prismatic crystals. Millerite I was followed shortly by two generations of colloform millerite (generations II and III) deposition. Millerite II was deposited in three pulses (Millerite IIA, IIB, and IIC). Millerite IIB constitutes the main portion of millerite II deposition, and IIA and IIC were composed of thin layers of acicular millerite that were deposited before and after millerite IIB and were especially susceptible to replacement by subsequently deposited ore minerals. Millerite II consisted of two pulses of deposition, an earlier main pulse, and a later thin layer of acicular millerite crystals that commonly has been partially to completely replaced by later ore minerals. The fourth generation of millerite (millerite IV) crystals consist of thin prismatic crystals.

All four generations of millerite have been intensely altered to and replaced by especially the three minerals, polydymite, gersdorffite, and vaesite, but also by the subsequently deposited sphalerite, chalcopyrite, and galena. The millerite crystals of generations I and IV have been partially to completely altered to polydymite. Shrinkage cracks developed by this alteration are partially filled with vaesite. The mineralogy and textures of this mineral assemblage are similar to that noted earlier for ores elsewhere in the district, such as at the Buick mine. The colloform millerite of subsequent generations, however, has not been found elsewhere in the district. This fine-grained millerite is especially susceptible to replacement and only small remnants of the millerite usually remain to attest to its original mineralogical character. Gersdorffite is the most common mineral that replaces the millerite, but millerite IIA, IIC, and IIIB may be almost completely replaced by vaesite that forms prismatic pseudomorphs after the original millerite.

After the deposition of millerite III, a coating of euhedral pyritohedral vaesite

Figure 2.

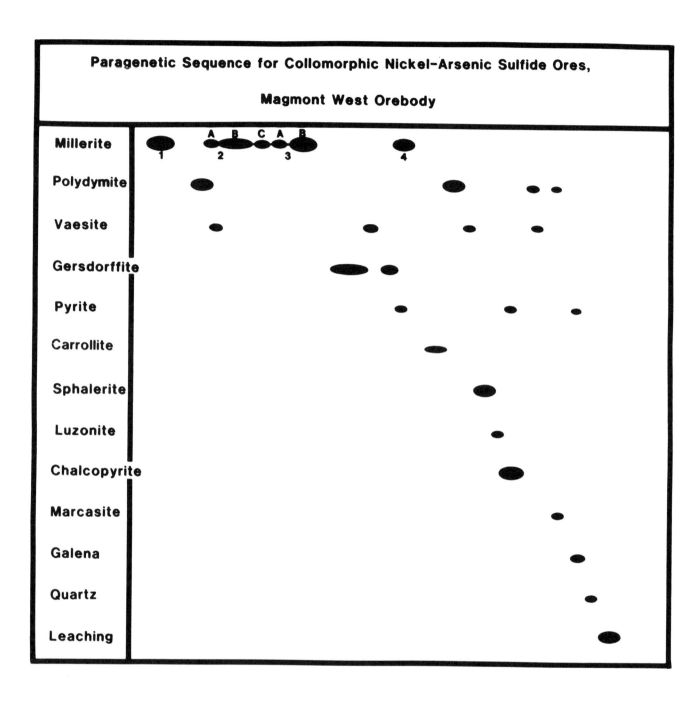

Paragenetic Sequence for Collomorphic Nickel-Arsenic Sulfide Ores,

Magmont West Orebody

crystals were deposited upon the surface of colloform millerite III, and these crystals were followed by a coating of cubic and pyritohedral crystals of gersdorffite. Locally, those gersdorffite crystals form thin veinlets that traverse the earlier generations of colloform millerite II and III. The small crystals of vaesite and gersdorffite were both deposited before the deposition of millerite IV.

Millerite crystals of generation IV are not as coarse as those belonging to generation I.

They are more intensely altered to polydymite and vaesite. Small octahedral crystals of carrollite commonly were deposited on the surfaces of the millerite IV prisms. Another generation of vaesite crystals, in turn, were deposited on the carrollite.

The copper, lead, and zinc sulfides were deposited after all of the above minerals in the Magmont-West nickel-arsenic sulphide ores. Sphalerite was the first to be deposited and it filled small vugs between the millerite IV crystals. It also locally completely replaced layers of millerite IIA, IIB, and IIIB. Chalcopyrite also was deposited in the spaces between millerite IV crystals, selectively replaced layers of millerite IIA, IIB, and IIIB, and locally veined and replaced sphalerite. The initial deposition of copper was in the form of luzonite that occurs at the margins of chalcopyrite areas. Pyrite and marcasite commonly are associated with the deposition of chalcopyrite. Galena formed in the spaces between millerite IV crystals and locally replaced both earlier deposited sphalerite and chalcopyrite. Locally, galena exhibits prismatic forms where it pseudomorphically replaced vaesite that is itself pseudomorphic after prismatic crystals of millerite IIA. Sphalerite, chalcopyrite, and galena preferentially replaced carrollite crystals that were deposited before vaesite. That replacement resulted in octahedral pseudomorphs of sphalerite, chalcopyrite, and galena that are contained within vaesite.

The paragenetic sequence for those ores was completed with the deposition of quartz and significant leaching of the earlier deposited ore minerals. Because the major portion of these ores are unlike those found elsewhere, it is difficult to fit their deposition into the paragenetic sequence previously constructed for Viburnum ores found throughout the Trend.[22,32] Although the millerite crystals of generation IV are similar in many respects to those found at the Buick mine and elsewhere in the district where they were deposited late in the mineral sequence, other features suggest that the nickel-arsenic sulfide ores in the Magmont-West orebody have been deposited at an earlier time. The fact that all of the nickel-arsenic sulphide minerals were deposited before sphalerite, chalcopyrite, and galena supports this interpretation. The fact that the most abundant nickel-arsenic sulphide mineral is gersdorffite, a mineral found to be deposited early in the paragenetic sequence elsewhere in the Trend also supports this idea.[32]

The Magmont-West ore deposit occurs within about 0.3 km of the Boss-Bixby deposit, a deposit in the underlying Precambrian rocks,

that has a mineralogy that is comparable to that of the Olympic Dam deposit in Australia.[42-44] The very close proximity of these two deposits raises the question of their genetic relationship. The Boss-Bixby is a major iron deposit that has been thoroughly drilled but has not been opened to mining. If this deposit is mined, copper and cobalt will be recovered as by-products. One interpretation is that the copper and cobalt sulphides in the Boss-Bixby deposit were deposited after the deposition of the Precambrian iron ore and that they were deposited at the same time and from the same ore fluids that deposited the Viburnum ores. This interpretation appears not to be the case because Precambrian granite dikelets appear to traverse the copper-cobalt sulphide ores.[45] Another interpretation for the unusually nickel-rich character of the Magmont-West deposit is that the copper, cobalt, and nickel may have been leached out from the underlying Precambrian Boss-Bixby ore deposit, by the saline connate brines that deposited the Viburnum Trend ores, and redeposited in the overlying Bonneterre Formation. The fact that the Boss-Bixby ores are cobalt-rich but are nickel- and arsenic-poor in contrast to the Magmont-West ores that are nickel- and arsenic-rich and cobalt poor, however, does not appear to support this interpretation. It appears from the information known to this point, that the genetic connection between the two deposits is a more remote one, if any at all.

References

1. Esparza L.E. United States Bureau of Mines, Mineral Industry Surveys, Washington, D.C., 1989, 3 pp.
2. Kisvarsanyi K. The Role of the Precambrian Igneous Basement in the Formation of the Stratabound Lead-Zinc-Copper Deposits in Southeast Missouri. Economic Geology, vol. 72, 1977, p. 435-442.
3. Wharton H.M. Letter to Paul A. Gerdemann on lead-zinc production totals for Southeast Missouri Lead District, Missouri Department of Natural Resources, Division of Geology and Land Survey, 1981, 2 pp.
4. Hagni R.D. The Southeast Missouri Lead

District: A review. In: _Mississippi Valley-Type Mineralization of the Viburnum Trend, Missouri_, Society of Economic Geologists, Guidebook Series, vol. 5, R.D. Hagni, and R.M. Coveney Jr., eds., Fort Collins, Colorado, 1989, p. 12-57.

5. Ohle, E.L. and Gerdemann, P.E. Recent exploration history in Southeast Missouri, In: _Mississippi Valley-Type Mineralization of the Viburnum Trend, Missouri_, Society of Economic Geologists, Guidebook Series, vol. 5, R.D. Hagni, and R.M. Coveney, Jr, eds, Fort Collins, Colorado, 1989, p. 1-11.

6. Grundmann, W.H. Jr. Geology of the Viburnum No. 27 mine, Viburnum Trend, Southeast Missouri. _Economic Geology_, vol. 72, 1977, p. 349-364.

7. Pettus, J.R. Jr. and Dunn R.G. Jr. Geology of the St. Joe Minerals Corporation Number 28 mine, In: _Sediment-Hosted Pb-Zn-Ba Deposits of the Midcontinent_, Hagni, R. D. et al., eds. Society of Economic Geologists Field Guidebook, 1988, p. 63-76.

8. Dunn, R.G. Jr. and Grundmann, W.H. Jr. Geology of the Casteel mine: Copper-rich ore in a MVT setting, in Mississippi Valley-Type Mineralization of the Viburnum Trend, Missouri, In: _Mississippi Valley-Type Mineralization of the Viburnum Trend, Missouri_, Society of Economic Geologists, Guidebook Series, vol. 5, R.D. Hagni, and R.M. Coveney, Jr, eds, Fort Collins, Colorado, 1989, p. 58-83.

9. Sweeney, P.H., Harrison, E.D. and Bradley, M.F. Geology of the Magmont Mine, Viburnum Trend, Southeast Missouri. _Economic Geology_, vol. 72, 1977, p. 365-371.

10. Bradley, M. F. and Krolak, T.E., 1989, The Magmont-West lead-zinc-copper mine (Cominco American Incorporated - Dresser Minerals), Viburnum Trend, Southeast Missouri, In: _Mississippi Valley-Type Mineralization of the Viburnum Trend, Missouri_, Society of Economic Geologists, Guidebook Series, vol. 5, R.D. Hagni, and R.M. Coveney, Jr, eds, Fort Collins, Colorado, 1989, p. 84-95.

11. Rogers R.K. and Davis J.H. Geology of the Buick mine, Viburnum Trend, Southeast Missouri. _Economic Geology_, vol. 72, 1977, p. 372-380.

12. Mugel D.N. and Hagni, R.D. Mineralogy and geology of the blanket lead-zinc deposit, Buick mine, Viburnum Trend, Southeast Missouri, In: _Applied Mineralogy: Proceedings of the Second International Congress on Applied Mineralogy in the Minerals Industry_, Park, W. C., Hausen, D. H. and Hagni R. D., eds., AIME, New York, N.Y., 1984, p. 965-986.

13. Evans L.L. Geology of the Brushy Creek mine, Viburnum Trend, Southeast Missouri. _Economic Geology_, vol. 72, 1977, p. 381-390.

14. Dingess P.R. Geology of the ASARCO West Fork mine, in Mississippi Valley-type mineralization of the Viburnum Trend, Missouri, In: _Mississippi Valley-Type Mineralization of the Viburnum Trend, Missouri_, Society of Economic Geologists, Guidebook Series, vol. 5, R.D. Hagni, and R.M. Coveney, Jr, eds, Fort Collins, Colorado, 1989, p. 96-110.

15. Paarlberg N. L. and Evans L. L. Geology of the Fletcher mine, Viburnum Trend, Southeast Missouri. _Economic Geology_, vol. 72, 1977, p. 391-397.

16. Mouat M. M. and Clendenin, C.W. Geology of the Ozark Lead Company mine, Viburnum Trend, Southeast Missouri. _Economic Geology_, vol. 72, 1977, p. 398-407.

17. Larsen, K.G. Stratigraphic and facies nomenclature of the Viburnum Trend, Southeast Missouri. In: N. Paarlberg, ed., _The 26th Annual Field Trip Guidebook_, Association of Missouri Geologists, 1979, p. 15-19.

18. Walker W.B. Geology of the Sweetwater mine, Viburnum Trend, Southeast Missouri, In: _Mississippi Valley-Type Mineralization of the Viburnum Trend, Missouri_, Society of Economic Geologists, Guidebook Series, vol. 5, R.D. Hagni, and R.M. Coveney, Jr, eds, Fort Collins, Colorado, 1989, p. 111-131.

19. Larsen K.G. (1977) Sedimentology of the Bonneterre Formation, Southeast Missouri. _Economic Geology_, vol. 72, 1977, p. 408-419.

20. Gregg J.M., and Gerdemann P.E., Sedimentary facies, diagenesis, and ore distribution in the Bonneterre Formation (Cambrian), Southeast Missouri, In: _Mississippi Valley-Type Mineralization of the Viburnum Trend, Missouri_, Society of Economic Geologists, Guidebook Series, vol. 5, R.D. Hagni, and R.M. Coveney, Jr, eds, Fort Collins, Colorado, 1989, p. 141-154.

21. Voss R.L. and Hagni R.D. The application of cathodoluminescence microscopy to the study of sparry dolomite from the Viburnum Trend, Southeast Missouri, Chapter 5 in D.M. Hausen and O.C. Kopp, eds., Mineralogy--Applications to the Minerals Industry, Proceedings of the Paul F. Kerr Memorial Symposium, AIME, New York, NY, 1985, p. 51-68.

22. Voss R.L. Hagni, R.D. and Gregg J.M., Sequential deposition of zoned dolomite and its relationship to sulfide mineral paragenetic sequence in the Viburnum Trend, Southeast Missouri, Special issue on the "Relationships between dolomitization, basin development, and Mississippi Valley-type ore deposits, Carbonates and Evaporites, vol 4, no. 2, 1989, p. 195-209.

23. Hagni, R.D. Minor Elements in Mississippi Valley-type ore deposits: Chapter 7, In: Cameron Volume on Unconventional Mineral Deposits, Shanks W.C. III, ed, Society of Economic Geologists - Society of Mining Engineers, AIME, New York, 1983, p. 71-88.

24. Roedder, E. Fluid inclusion studies of ore deposits in the Viburnum Trend, Southeast Missouri. Economic Geology, vol. 72, 1977, p. 474-479.

25. Hagni, R.D. Ore microscopy, paragenetic sequence, trace element content, and fluid inclusion studies of the copper-lead-zinc deposits of the Southeast Missouri Lead District, In: International Conference on Mississippi Valley Type Lead-Zinc Deposits, Proceedings Volume, Kisvarsanyi G., Grant S.K., Pratt W.P., Koenig J. W., eds., University of Missouri-Rolla, Rolla, Missouri, 1983, p. 243-256.

26. Mavrogenes, J.A. Hagni, R.D., Dingess, P.R., Metal sulfide mineral zoning at ASARCO's West Fort mine, Viburnum Trend, Southeast Missouri (abs.), Geological Society of America, North-Central Section annual meeting, vol. 22, no. 5, Notre Dame University, South Bend, Indiana, Apr. 1989, p. 14.

27. Wharton, H.M., Total mine production and value from the Viburnum Trend mines from the State of Missouri in 1964, Missouri Department of Natural Resources, Division of Geology and Land Survey, 1986, 1 p.

28. Kaiser C.J., Kelly W.C., Wagner R.J., and Shanks W.C. III Geologic and geochemical controls of mineralization in the Southeast Missouri Barite District. Economic Geology vol. 82, 1987, p. 719-734.

29. Clendenin C.W., 1977, Suggestions for interpreting Viburnum Trend mineralization based on field studies at Ozark Lead Company, Southeast Missouri. Economic Geology, vol. 77, 1977, p. 465-473.

30. Hagni, R.D. and Trancynger T.C. Sequence of deposition of the ore minerals at the Magmont mine. Economic Geology, vol. 72, 1977, p. 451-464.

31. Watson J.L. South East Missouri Lead Belt - A review. Minerals Engineering, vol. 1, no. 2, 1988, p. 151-156.

32. Hagni, R.D. Ore microscopy and paragenetic sequence of the ores of the Southeast Missouri Lead District, International Mineralogical Association, 14th General Meeting, Abstracts with Program, Stanford University, Stanford, California, 1986, p. 118.

33. Hagni, R. D., 1983, Applications of the ore microscope to ore dressing problems in the Viburnum Trend, Southeast Missouri, U.S.A., ICAM 81: Proceedings of the First International Congress on Applied Mineralogy, de Villiers J.P.R. and Cawthorn, P.A., eds., Special Publication No. 7, The Geological Society of South Africa, Marshalltown, South Africa, 1983, p. 209-215.

34. Cornell W.L. and O'Connor, W.K., 1987, Ore microscopy in support of applied beneficiation research to recover cobalt from Missouri lead ores. In: Process Mineralogy VII: Applications to Mineral Beneficiation Technology and Mineral Exploration, with Special Emphasis on Disseminated Carbonaceous Gold Ores, Vassiliou A.H., Hausen D.M., and Carson D.J.T., eds., American Institute of Mining, Metallurgical and Petroleum Engineers, Warrendale, PA, p. 603-615.

35. Cornell W.L., Wethington A.M., Holtgrefe D.C. and Sharp F.H., Continuous flotation testing to recover cobalt from Missouri lead ores, U.S. Bureau of Mines Report of Investigations 9072, 1987, 11 p.

36. Frank Sharp, 1980, oral communication.

37. Pignolet-Brandom, S., Mineralogy, paragenesis and electron microprobe analysis of

the cobalt-nickel minerals in the Mississippi Valley-type deposits of Southeast Missouri, Unpub. Ph.D. dissertation, University of Missouri-Rolla, Rolla, Missouri, 1988, 221 pp.

38. Horrall, K.B. Mineralogical, textural, and paragenetic studies of selected ore deposits of the Southeast Missouri Lead-Zinc-Copper District and their genetic implications, Unpub. Ph.D. dissertation, University of Missouri-Rolla, 1982, 650 pp.

39. Jessey, D.R. Mineralogical and Compositional Variations of the Nickel-Cobalt Mineralization in the Southeast Missouri Mining District. In: Process Mineralogy: Extractive Metallurgy, Mineral Exploration, Energy Resources, Hausen D.M. and Park W.C., eds, AIME, New York, 1981, p. 159-177.

40. Brooke G.L., Madison cobalt mine, Madison County, Missouri, American Institute of Mining, Metallurgical and Petroleum Engineers, Preprint 82-324, Warrendale, PA, 1981, 3 p.

41. Moh, G.H. and Ottemann J. X-ray fluorescence and electron-probe analysis of some pyrite-type minerals, Year Book, 1963-64, Carnegie Inst. of Washington, 1964, p. 214-216.

42. Brandom R.T. and Hagni R.D., A Comparison of the mineralogy of the Boss-Bixby, Missouri Copper-Iron deposit, and the Olympic Dam (Roxby Downs) copper-uranium-gold deposit, South Australia, Geological Society of America Annual Meeting, Orlando, Florida, 1985, p. 530-531.

43. Hagni R.D. and Brandom R.T., Mineralogical similarity of the Boss-Bixby, Missouri copper-iron-cobalt deposit to the Olympic Dam, South Australia ore deposit: potential for additional deposits in the Midcontinental United States (abs.), 28th International Geological Congress Program, Washington, D.C., vol. 2, 1989, p. 7-8.

44. Hagni, R.D. and Brandom R.T., The Mineralogy of the Boss-Bixby, Missouri Copper-Iron-Cobalt Deposit and a Comparison to the Olympic Dam Deposit at Roxby Downs, South Australia. In: Olympic Dam-Type Deposits and Geology of Middle Proterozoic Rocks in the St. Francois Mountains Terrane, Missouri, Society of Economic Geologists, Guidebook Series, vol. 4, V.M. Brown, E.B. Kisvarsanyi, R.D. Hagni, eds, Fort Collins, Colorado, 1989, p. 82-92.

45. Brandom R.T. and Hagni R.D., Ore Microscopy of the Boss-Bixby, Missouri, Cobalt-Copper-Iron Ores. Society of Mining Engineers of AIME Transactions, vol. 280, 1987, p. 2139-2142.

Ore processing and mineralogy

Principles and practice of sulphide mineral flotation

R. Herrera-Urbina
J. S. Hanson
G. H. Harris
D. W. Fuerstenau
Department of Materials Science and Mineral Engineering, University of California at Berkeley, Berkeley, California, U.S.A.

SYNOPSIS

This paper presents both the fundamentals of sulphide mineral flotation and the practical aspects of sulphide ore flotation. The fundamental aspects of the flotation of sulphide minerals both in the absence and presence of thiol collectors are discussed in relation to their crystal structure, surface chemistry and electrochemical characteristics. Since the electrochemistry of the sulphide mineral/aqueous solution interface and of various sulphide mineral/aqueous thio-compound systems has been amply investigated over the past two decades, the results of these modern electrochemical studies are briefly reviewed. Mechanisms of the action of various flotation reagents, including collectors, activators and depressants, commonly used in sulphide mineral flotation are discussed in terms of the chemical and electrochemical principles presented within this paper. Simplified flow sheets of representative flotation plant operations are given to illustrate how these principles are applied for the commercial separation of sulphide minerals from their ores. Various novel techniques that take advantage of controlling the electrochemical and physical behaviour of the flotation system are also presented.

The practice of sulphide minerals flotation is as old as the flotation process itself. In the United States, for example, the first froth flotation plant began operation in 1911 at the Butte and Superior Copper mill in Basin, Montana and recovered sphalerite from slimes (material 92% finer than 150 mesh).[1] Although in 1911 sphalerite was the only sulphide mineral concentrated commercially by flotation, over the years the industrial flotation of sulphides has grown impressively in size, complexity and importance. According to statistics compiled recently by the United States Bureau of Mines, the largest tonnage of ores treated by flotation corresponds to sulphide ores, which also consume the largest amount of reagents.[2] Today, froth flotation of sulphide ores constitutes the single most important operation for the concentration of sulphide minerals. Current flotation practice often involves the separation of three or four sulphides from the same ore. Industrially important sulphide minerals include galena, chalcopyrite, sphalerite, pyrite, chalcocite and molybdenite. Traditionally, these minerals have been separated from each other using relatively simple flotation schemes to process high-grade, coarse-grained ores. Because of the depletion of these easy-to-process ores, however, flotation scientists and engineers are now confronted with the need to exploit existing low-grade sulphide deposits with complex mineralogical composition and finely disseminated values. As a result of this need, in recent years the flotation of complex sulphide ores has been the topic of a number of symposia and workshops.[3-5]

In contrast to the extensive amount of progress that has been gained industrially, a clear understanding of the basic principles of the flotation of sulphide minerals has not yet been achieved though most of the early fundamental flotation research involved investigation of the flotation behaviour of this group of minerals. The way by which those sulphides that do not exhibit natural flotability are made to respond to flotation has been perhaps the aspect investigated more intensively and in the greatest detail. This aspect involves the interaction of the mineral with amphiphatic surfactants, which are called collectors in flotation terminology, that adsorb at the solid/solution interface and make the attachment of an air bubble to the mineral surface thermodynamically favourable.

Since the discovery in 1925 by Keller[6] that the alkali xanthates act as collectors of sulphide minerals, these thio-compounds have become the universal collectors for sulphide minerals other than molybdenite. In spite of over half a century of continued research on the interaction of this thiol collector and sulphide minerals, this system is nowyielding both new results and problems as new experimental techniques are introduced in modern flotation research and ultimately into flotation plant practice. In addition to xanthates, such thio-compounds as dithiophosphates (patented by Whitworth in 1926) and thiocarbamates (patented by Harris and Fischback in 1954) are also common sulphide minerals collectors. Although the interaction of thiol collectors and sulphide minerals is not completely understood, both its chemical and electrochemical nature has been amply documented. Galena, bornite and chalcocite, for example, interact chemically with thiol collectors producing a surface metal thiolate.[7] On the basis of modern electrochemical studies, on the other hand, it has been possible to propose a mixed-potential model to rigorously explain the electro-chemical nature of this interaction. This model states that the hydrophilic/hydrophobic transformation of the sulphide mineral surface by the interaction with a thio-compound involves simultaneous electrochemical reactions. The cathodic step is usually the reduction of oxygen, while the anodic step involves the oxidation of the collector.[8]

Apart from the issue of how thiol collectors interact with sulphide minerals making them amenable to flotation, various aspects of these flotation systems remain unclear. For example, the role of oxygen has become the most controversial issue mainly because of lack of adequate oxygen control in many investigations. Based on the most recent results on electrochemical research, however, it is now widely recognized that an electron transfer is associated with the adsorption of a thiol collector to make most sulphide surfaces hydrophobic. Industrially, oxygen acts as the electron acceptor but can be substituted by other oxidizing agents such as sodium hypochlorite,[9] or by anodic oxidation. After many years of being denied as a real phenomenon, the flotation response of sulphide minerals in the absence of collectors and under certain conditions is now well documented.[10] Mildly oxidizing environments render some sulphide minerals amenable to flotation without collectors.

This paper presents both the fundamentals of sulphide minerals flotation and the practical aspects of sulphide ores flotation. Since the crystal structure and surface chemistry of sulphides determine their response to flotation, the nature of the sulphide mineral surface will be examined first to delineate the effect of crystal structure on the inherent hydrophobicity of some of these minerals. For sulphide minerals that do not exhibit natural flotability, approaches to render their surfaces hydrophobic and the phenomena involved are discussed. How the surfaces of sulphides interact with thiol collectors, depressants and activators used commercially for the concentration of sulphide minerals by flotation are analysed. Over the past two decades, the electrochemical aspects of these interactions have received considerable attention. Results of these studies will be reviewed briefly and discussed to the extent that various working hypotheses can be applied to plant practice. To illustrate how these fundamental aspects of flotation chemistry are applied for the separation of sulphide minerals from their ores, simplified flowsheets of representative commercial operations will be given. In addition, various novel techniques that take advantage of controlling the electrochemical and physical aspects of the system are also presented.

CRYSTAL STRUCTURE AND SURFACE CHEMISTRY OF SULPHIDES

Surface chemistry is the very basis of flotation. Therefore, a complete understanding of the physical-chemical characteristics of solid surfaces is essential to better understand the complex phenomena taking place during flotation. These surface characteristics are determined by such crystal properties as the types of forces that hold the solid together, that is the chemical bonding, and its atomic or molecular structure. Surfaces created by the rupture of bonds other that van der Waals bonds contain very reactive, ionic sites, while those surfaces created by the rupture of van der Waals bonds are capable of reacting with the environment only through van der Waals type interactions. Furthermore, the interaction between the surfaces of solids and their environment depends on the solid's cleavage characteristics, which are determined by the type of bonds that are most susceptible to breakage during size reduction. Using such techniques as X-ray diffraction and Θ-Θ solution diffractometry, Lindqvist and Stridh[11] have demonstrated that the flotation behaviour of galena, pyrite and sphalerite is directly related to a

Table I. Classification of sulphide minerals based on the relative abundance of surface sulphide sites.

Sulphur-rich surfaces	Sulphur-deficient surfaces
Pyrrhotite	Wurtzite
Marcasite	Sphalerite (Blende)
Arsenopyrite	Niccolite
Pyrite	Bismuthinite
Cinnabar	Stibnite
Molybdenite	Tetrahedrite
Covellite	
Chalcocite	
Bornite	
Galena	
Chalcopyrite	
Realgar	

specific crystallographic cleavage plane. For galena and pyrite, this plane corresponds to the (100) plane, while for sphalerite it corresponds to the (110) plane.

The atomic or molecular arrangement of the solid, on the other hand, determines the position of the different chemical species and their distance from the surface. In the case of sulphides, their surfaces can be chemically classified as sulphur-deficient and sulphur-rich depending on the relative abundance of sulphur atoms. Table I lists as class A those sulphide minerals whose surfaces contain exposed sulphur atoms, and as class B those sulphides with surfaces characterized by the absence of sulphur atoms, which are located inside the elementary cell.[12] Those sulphide minerals listed in class A have been found to respond well to flotation with thiol collectors, while those in class B respond poorly to flotation or do not float at all, such as sphalerite. According to this classification, it seems reasonable to assume that in order for this zinc sulphide to respond to flotation with short-chain xanthates, the superficial layer of zinc atoms must be removed to expose the hidden internal sulphur atoms.[12]

Another aspect related to both the crystal structure and surface chemistry of sulphides that has been found to influence their flotation behaviour is the solubility of the mineral. Theoretically, the bonding between metal atoms and sulphur is mostly covalent, and is responsible for the low solubility of sulphide minerals in water. Under common working conditions, however, it is well known that most sulphides are not as chemically inert as predicted by classical thermodynamics, which does not take into consideration surface phenomena related to the instability of sulphide surfaces in the presence of oxygen-containing fluids such as water or air. In these environments, the surfaces of sulphide minerals undergo an oxidation process and, since the solubility of these oxidation products is higher than that of their sulphide counterparts, the solubility of the mineral increases. As a result, oxidation phenomena alter significantly the surface and aqueous chemistries of sulphide minerals and therefore directly affect their flotation behaviour.

Surface oxidation

Oxidizing environments, chemically or electrochemically induced, are known to greatly influence the flotation response of sulphide minerals, both in the absence and the presence of collectors, through oxidation of their surfaces. Surface oxidation phenomena are closely related to the interfacial electrochemical potential, which determines both the type and the rate at which a reaction may take place. During the last decade, modern sulphide flotation research has been increasingly involved with delineating the effect of potential on the extent of surface oxidation under different solution conditions. Such techniques as linear potential sweep voltammetry, intermittent galvanostatic polarization, and A-C impedance have been invaluable in achieving our present understanding of surface oxidation phenomena and their relation to flotability in sulphide mineral flotation systems. Woods and Richardson[8] have obtained linear potential sweep voltammograms for fresh surfaces of various sulphide minerals, and these are reproduced in Figure 1, which shows the potential regions in which surface oxidation occurs.

Characterization of oxidized sulphide surfaces has been undertaken to identify the chemical nature of the oxidation products. Electrochemical techniques, however, do not allow identification of the chemical composition and structure of sulphide mineral surfaces subjected to oxidizing conditions. For this purpose, flotation scientists have used such spectroscopy techniques as X-ray photoelectron spectroscopy (XPS). Among these researchers, Buckley and Woods[13] applied XPS to identify the chemical species formed when the surfaces of galena, pyrrhotite, pyrite and chalcopyrite are oxidized. Their results indicate that the initial surface product in these systems is a metal-deficient (or sulphur-excess) sulphide. Further oxidation can result in the formation of sulphide species of progressively lower metal content and eventually elemental sulphur.[14]

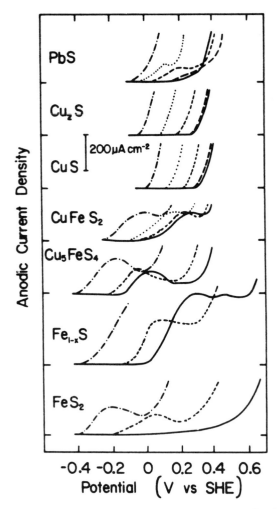

Fig. 1. Linear potential sweep voltammograms for fresh surfaces of sulfide minerals: positive-going scans at 20 mV/s in solutions of pH —— 4.6; — — — 6.8; – – – 9.2; •••• 11; -•-•-• 13. (After Ref. 8)

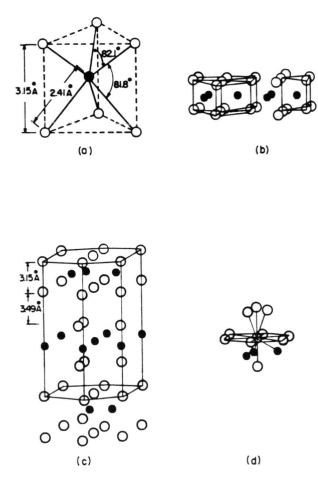

Fig. 2: The crystal structure of Molybdenite.
   (a) Environment of the molybdenum atom - the trigonal prismatic coordination polyhedra.
   (b) The S-Mo-S sandwich layer.
   (c) The hexagonal (2H₁) crystal molybdenite.
   (d) The environment of the sulfur atom.

Guy and Trahar[15] have summarized the different oxidation mechanisms that may take place when a sulphide mineral is exposed to aqueous solutions under acidic or alkaline conditions. In both cases, elemental sulphur is suggested to form on the surface; under acidic conditions metal ions pass into solution, while under alkaline conditions a metal hydroxide is formed. It is also suggested that further oxidation may lead to the formation of oxy-sulphur species according to the following reaction:

$$x\ M_zS + (zxn + y)H_2O = zx\ M(OH)_n + S_xO_y^{2-}$$
$$(2y + zxn)H^+ + (2y + zxn-2)e^-$$

The subsequent flotation behaviour of an oxidized sulphide mineral will depend upon the nature of the surface products formed. If sulphur or a sulphur-like species is formed at the surface, the mineral will respond to flotation in the absence of collectors. In the case of strong oxidizing conditions, the hydrophilic oxy-sulphur species present at the interface will suppress flotability.

NATURAL FLOTABILITY OF SULPHIDES

Many sulphides were once thought to exhibit natural or inherent flotability. It is now well known, however, that this surface property arises from the inherent hydrophobicity displayed only by those sulphide minerals with special crystallographic structures, of which molybdenite is a typical example. Gaudin et al.,[16] had long recognized that the natural flotability of a solid arises when a new surface is created without breakage of interatomic bonds other than residual van der Waals bonds. Two sulphide minerals exhibit natural flotability: molybdenite, MoS_2, and stibnite, Sb_2S_3. As can be seen in Figure 2, molybdenite exhibits a layered crystal structure consisting of S-Mo-S sheets held together by van der Waals bonds. Rupture of these bonds results in a hydrophobic surface, while rupture of the mostly covalent Mo-S bond produces hydrophilic sites. This breakage pattern produces predominantly hydrophobic faces and to a lesser extent edges which are hydrophilic. As a consequence, it has been reported that

90

the flotation response of molybdenite is determined by the surface made up of faces.[17] In the case of stibnite, its crystalline structure consists of chains where the Sb and S atoms are held together by weak bonds of unknown character.[17] Another characteristic of naturally flotable sulphides is that, on oxidation, they generally form anionic species that remain as individual sites. In contrast, non-naturally flotable sulphides form cations that may hydrolyze to form insoluble oxides (or hydroxides) on the particle surface. These oxides can grow as patches even on the hydrophobic part of the surface rendering it hydrophilic. In addition, those sulphides that exhibit natural flotability float under reducing conditions, while other sulphide minerals do not. By exploiting this phenomenon, it is possible to separate chalcopyrite from molybdenite.

COLLECTORLESS FLOTATION OF SULPHIDE MINERALS

The response of sulphide minerals to flotation without conventional collectors was reported by Gaudin as early as 1932.[6] At that time, however, lack of a satisfactory explanation led to the suggestion that contamination of the mineral surface was responsible for this unusual flotation behaviour of sulphide minerals, and the subject was not investigated further. It was not until the mid 1970s that the flotation of sulphide minerals without collectors was again reported.[9,18,19] Flotation of sulphide minerals other than molybdenite without the addition of a conventional thiol collector is now termed "collectorless flotation," and is used to describe flotation systems where a mineral acquires some degree of hydrophobicity in the absence of a collector due to chemical changes at the surface.[20] These chemical changes, which are closely related to the interfacial electrochemical potential of the system, strongly affect the nature of the sulphide mineral surface and therefore control its collectorless flotation. For example, the more cathodic or anodic the rest potential and the less easily oxidized the sulphide surface, the more readily it floats without collector. This appears to be due to the formation of sulphur and sulphur-rich mineral patches at mildly oxidizing potentials.

Chalcopyrite[9] and galena[15] have been shown to float in the absence of collectors under mildly oxidizing conditions. Chalcocite, on the other hand, does not float significantly under similar conditions. More recently Shannon and Trahar[10] have also shown that chalcopyrite exhibits flotability in the absence of

collectors, a behaviour closely resembled by galena and pyrrhotite, while pyrite and sphalerite float poorly without collectors. The trend on the collectorless flotation of these common sulphide minerals and the greatest response of the sulphide of copper and iron has been confirmed by Leroux et al.[20] The flotation of oxidized chalcopyrite without collectors has been attributed to the formation of a hydrophobic surface species, which has not been identified with certainty,[10] but is thought to be a sulphur-enriched surface species,[14] or a metal polysulphide.[21]

Collectorless flotation has been observed to be a very real property of several common sulphide minerals and to occur in typical sulphide flotation environments, that is, in air saturated solutions ranging from moderately acidic (pH 3) to moderately alkaline (pH 11) conditions.[10] When a frother is added to the flotation pulp, collectorless flotation is more pronounced. Recently, Hayes et al.,[22] have reviewed several aspects of classical surface oxidation studies through to modern electrochemical, surface chemical and spectroscopic studies relevant to the collectorless flotation of sulphide minerals.

In addition to electrochemical techniques, X-ray photoelectron spectroscopy (XPS) has been used to determine the nature of the sulphur enrichment at sulphide mineral surfaces and its role in flotation, a subject recently reviewed by Buckley and Walker,[23] who concluded that the collectorless flotation of sulphide minerals is controlled by the concentration of surface oxides as well as the degree of sulphur enrichment at the surface, regardless of the nature of the surface sulphur. Metallic compounds precipitated on the surface of the mineral have been found to be especially effective for the supression of collectorless flotation. As can be seen in Figure 3, which shows the collectorless flotation of chalcopyrite in the absence and presence of copper salts reported by Guy and Trahar,[15] the recovery of chalcopyrite is significantly reduced as the concentration of added copper(II) species increases. The depressant action of this compound is closely related to the precipitation of copper hydroxide at the mineral surface.

The industrial impetus towards collectorless flotation is the major reduction in reagent requirements, particularly reagents for pH control.

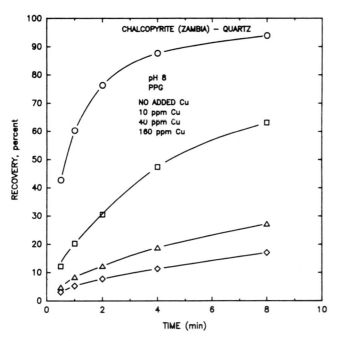

Fig. 3. Influence of added copper on the collectorless flotation of chalcopyrite.

CHEMICAL AND ELECTROCHEMICAL ASPECTS OF THE SULPHIDE MINERAL/THIOL COLLECTOR INTERACTION

Sulphide minerals other than molybdenite are inherently hydrophilic under common flotation conditions. To effect the hydrophilic/hydrophobic transformation of the surfaces of these minerals, and hence make them amenable to flotation, a collector must be added to the system. Since 1925, when alkali xanthates were first used as collectors, these compounds have become the standard collectors in the flotation of sulphide minerals.[6] Dithiophosphates, whose use as collectors was patented a year later,[6] have remained the next most commonly used collectors in the flotation of sulphide minerals. Over the years numerous chemical reagents have also been reported to have collecting properties. Of these chemical compounds, those that possess a polar group containing at least one sulphur atom not bonded to oxygen are the most common sulphide mineral collectors, and they belong to the type of thio-compounds. These thiol collectors are listed in Table II, which also gives their structural formula and trade name.

The role played by oxygen in the adsorption of thiol collectors onto sulphide minerals has been perhaps the most controversial issue in sulphide mineral flotation investigations. Dewey[24] recognized in 1933 that oxygen must be present in order for xanthate to adsorb onto a sulfide mineral surface. The theories proposed to

Table II. Thio-compounds commonly used as collectors in sulphide mineral flotation.

Collector (Trade Name)	Structural Formula
Mercaptan	$R-S^{\ominus}$
Alkyl disulfide	$R-S$ | $R-S$
O-alkyl dithiocarbonates or xanthates (Cyanamid AET 300, DOW Z-3 to Z-12 Series)	$R-O-C{\nwarrow}^{S}_{S}{}^{\ominus}$
Dialkyl disulfide or dixanthogens	$R-O-C{<}^{S}_{S}-S{>}^{S}_{C-O-R}$
Xanthic esters	$R-O-C{<}^{S}_{S-R'}$
Xanthogen formate (Minerec)	$R-O-C{<}^{S}_{S}-C{<}^{O}_{O-R'}$
Aryl dithiophosphoric acids or dialkyl dithiophosphates (Cyanamid AEROFLOAT 200 Series)	$R-O{\atop R-O}P{<}^{S}_{S}{}^{\ominus}$
Dialkyl dithiocarbamate	$R{\atop R}N-C{<}^{S}_{S}{}^{\ominus}$
Isopropyl thionocarbamate (Dow Z-200)	$(CH_3)_2CH-O-\overset{S}{\overset{\|}{C}}-N{<}^{H}_{C_2H_5}$
Mercaptobenzothiazole (Cyanamid AERO 400 Series)	
1,3-diphenyl 2-thiourea or Thiocarbanilide (Cyanamid AERO 130 Series)	
Diphenyl Guanidine	

account for oxygen's role in the adsorption process fall basically into two categories. The first and oldest mechanism proposed is the metathetical, ion-exchange type reaction. Although this hypothesis of oxygen's role in flotation has been used for many years, it is often found to be inconsistent with experimental observation. The past thirty years have seen the development of another approach to explain the adsorption of flotation collectors on sulfide minerals, namely the use of electrochemical principles to account for the role of oxygen. The electrochemical mechanism is still the focus of numerous investigations and undergoes continual refinement.

<u>Early physical-chemical theories of sulphide flotation</u>
The interaction of sulphide minerals and thiol collectors was explained at the beginning of flotation research in terms of an ordinary chemical reaction between these chemical species,[25] and by means of an adsorption mechanism that led to the formation of non-Daltonian compounds.[26,27] Taggart's chemical theory[25] proposes that because of surface oxidation the sulfide mineral surface is actually composed of metal oxides, carbonates and sulfates, which upon addition of a thiol collector such as xanthate react in a metathetical, ion-exchange reaction. Metathesis between these surface oxidized compounds, $MA_{(s)}$, and xanthate ions in solution, $X^-_{(aq)}$, results in the formation of a metal xanthate, $MX_{(s)}$, on the sulphide mineral and the transfer of the oxidized surface anions, A^-, into solution:

$$MA_{(s)} + X^-_{(aq)} = MX_{(s)} + A^-_{(aq)}$$

The chemistry of thiol collector/sulphide mineral interaction has been investigated more recently using such techniques as infrared spectroscopy[28] and calorimetry.[29] In the case of the galena/ethyl xanthate system, the infrared spectra obtained by Leja et al.[28] and reproduced as Figure 4 reveal that indeed a chemical compound, namely lead ethyl xanthate, forms at the surface of lead sulphide immersed in solutions of ethyl xanthate. Spectrum (a) in this figure is for bulk lead ethyl xanthate, while spectrum (b) is for a lead sulphide film oxidized under atmospheric conditions. After exposing this film to a xanthate solution, spectrum (c) was obtained. As can be seen, this spectrum is virtually identical to spectrum (a) thus confirming that a new surface chemical compound is formed, most probably lead ethyl xanthate. Finally, after this sample was washed with a strong solvent, namely pyridine, spectrum (e) was obtained. This spectrum shows that this washing procedure completely removes the lead ethyl xanthate leaving behind only an oxidized lead sulphide surface. The heats of reaction measured by Mellgren,[29] on the other hand, clearly indicate that the uptake of ethyl xanthate by oxidized galena, which had been treated with 0.1 M H_2SO_4 and washed, is

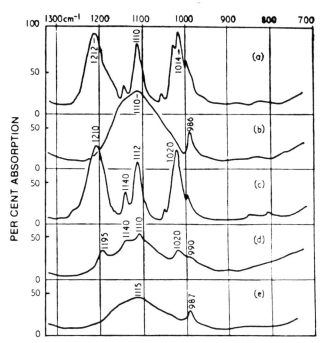

Fig. 4. Infrared spectra of: a) bulk lead ethyl xanthate solid as mujol mull, b) freshly evaporated PbS film after atmospheric oxidation, c) freshly evaporated PbS film treated with aqueous ethyl xanthate solution, d) freshly evaporated PbS film after prolonged washing in ether, and e) freshly evaporated PbS film after washing in pyridine. (After Ref. 28).

energetically equivalent to the metathetical, ion-exchange type reaction proposed by Taggart for the formation of lead ethyl xanthate on galena immersed in xanthate solutions.

<u>Electrochemical aspects of sulphide flotation</u>
The difficulty in accurately evaluating the amount of dissolved oxygen in solution and assessing the degree of oxidation of the sulphide mineral surface make electrochemical techniques ideal for investigating the adsorption of thiol collectors on mineral surfaces. Further complicating the adsorption process is the ease in which the sulfhydryl collectors oxidize. This phenomenon often obscures the effect of oxygen in the flotation system because the adsorption mechanism may be altered if the mineral surface is not well characterized or the collector itself has undergone some degradation. In most electrochemical studies the role of oxygen as electron acceptor is replaced by an external electric circuit.

Electrochemical techniques were used in sulphide flotation research as early as 1931.[30] It was not until 1954, however, that the sulfide mineral/thiol collector interaction was explained in electrochemical terms, and a sulfide flotation model, known today as the mixed-potential model, was proposed.[31] This model, in its most general form, considers that sulphide minerals, being good electronic conductors, interact with thiol collectors through a mechanism that involves simultaneous electrochemical reactions at the mineral surface. The important role played by electrochemical reactions during the interaction of thiol collectors with sulphide minerals has been demonstrated by the results of numerous electrochemical investigations reported in the last three decades. These recent results, which have been thoroughly reviewed and summarized by Woods[32,33] and Woods and Richardson,[8] have helped in identifying the simultaneous electrochemical reactions that occur in these systems. The cathodic step usually involves the reduction of oxygen,

$$O_2 + 2H_2O + 4e^- = 4OH^-$$

and it is coupled with an anodic step involving oxidation of either the collector or the mineral. The products of the anodic reaction depend on the mineral and collector used, and the pretreatment of the mineral surface. They have been identified as chemisorbed collector, dithiolates and metal/collector compounds. The reactions for the anodic oxidation of the collector can be written as follows:[8]

(a) Charge transfer chemisorption of a thiol ion (X^-):

$$X^- = X_{(ads)} + e^-$$

(b) Oxidation of a thiol ion to its disulfide:

$$2X^- = X_2 + 2e^-$$

(c) Formation of a thiol compound with a metal component of the mineral:

$$MS + 2X^- = MX_2 + S^o + 2e^-$$

The mixed-potential model does not actually contradict earlier theories of sulphide flotation. Wark's and Gaudin's adsorption model would be equivalent to reaction (a) while Taggart's chemical reaction theory would be represented by the anodic process given by reaction (c), which can be a two-step reaction involving oxidation of the mineral and metathesis.

The product resulting from the interaction of sulphide minerals and thiol collectors has been identified and correlated with the rest potential of the mineral. Allison et al.[7] identified the surface products in sulphide/ethyl xanthate systems, and a selection of their results is reproduced in Table III. These results clearly indicate that, with the exception of covellite, sulphide minerals that display a rest potential below the corresponding reversible potential for the oxidation of the collector to its disulphide will react with thiols forming metal thiolates. Those sulphides whose rest

Table III. Rest potential and reaction product of minerals in xanthate solutions (After Ref. 7).

Mineral	Rest Potential, V	Reaction Product
Sphalerite	-0.15	NPI
Stibnite	-0.13	NPI
Galena	+0.06	MX
Bornite	+0.06	MX
Chalcocite	+0.06	NPI (MX)
Chalcopyrite	+0.14	X_2
Molybdenite	+0.16	X_2
Pyrrhotite	+0.21	X_2
Pyrite	+0.22	X_2

Potential for $X_2 + 2e^- \rightarrow 2X^-$ at 0.625 mM KEX = 0.13 volt

potentials are above this value oxidize the thiol to its disulphide, and this product is now accepted as the collecting species in these systems. Similar results have been reported for sulphide/diethyl dithiocarbamate systems.[34]

A necessary condition for sulphide mineral flotability is the thermodynamic stability of the hydrophobic entity formed at the surface. If this compound forms at a rate too slow compared to the rate at which the sulphide mineral oxidizes, however, the mineral will not float because surface oxidation may result in the formation of oxide-type materials that may not only impede electron transfer from the collector to the mineral but is also extremely hydrophilic. In addition, if the hydrophobic species is decomposed by oxidation, strong oxidizing conditions would be detrimental to flotation. Knowledge of the oxidation potential (Eh) of the system is therefore vital in sulphide mineral flotation. It is possible to predict the oxidizing and reducing conditions necessary for a substance to be thermodynamically stable by constructing the appropriate Eh-pH diagram. These diagrams, first constructed for sulphides/thiol collector systems by Toperi and Tolun[35], have been used for comparing experimental results with thermodynamic equilibria.[36,37] Figure 5 presents the Eh-pH diagram for the chalcocite/diethyl dithiophosphate/water system. This diagram shows the stability domains of the different mineral and collector species. It can be noted that copper diethyl dithiophosphate, CuDTP, the species responsible for chalcocite flotation, is unstable in alkaline solutions regardless of the oxidation potential. In acidic, neutral and mildly alkaline solutions, as the potential becomes increasingly positive, chalcocite first

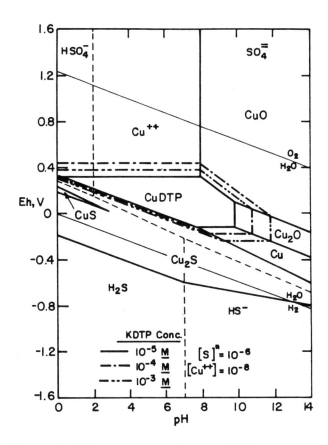

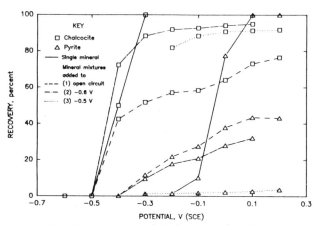

Fig. 6. Flotation recovery of chalcocite and pyrite at pH 9.2 as a function of potential for single and mixed mineral beds. (After Ref. 42).

Fig. 5. Eh-pH diagram for the chalcocite/diethyl dithiophosphate system.

oxidizes to CuDTP, which then oxidizes to Cu^{2+} (or CuO). Strongly oxidizing conditions leading to the formation of Cu^{2+} or CuO have a deleterious effect on the flotation of chalcocite with diethyl dithiophosphate as the collector.

It is now widely accepted that the flotation of sulphide minerals with thiol collectors is controlled by the interfacial electrochemical potential. Using in-situ electrodes in microflotation cells, Gardner and Woods[38] and Chander and Fuerstenau[39] were the first to independently and simultaneously demonstrate that the electrochemical potential can control the flotation of sulphide minerals. More recently, Walker et al.,[40] developed an electrochemical microflotation cell that uses a packed bed of conducting sulphide particles as the working electrode to correlate interfacial electrochemical reactions with flotation response. Richardson and Walker[41] used this cell to investigate the xanthate flotation of chalcocite, bornite, chalcopyrite and pyrite as a function of the electrochemical potential. The flotation of these sulphide minerals was found to be strongly dependent on the conditioning potential and to occur only under moderately oxidizing conditions. At a certain reducing potential, whose value is mineral

dependent, flotation ceases. Identification of the products of the sulphide mineral/ethyl xanthate interaction was achieved using such techniques as linear sweep voltammetry and ultraviolet spectrophotometry.[41] The species responsible for imparting surface hydrophobicity appear to be metal xanthates or surface analogues of metal xanthates formed by chemisorption on chalcocite and bornite, dixanthogen on pyrite, and both metal xanthate and dixanthogen on chalcopyrite.

Gebhardt, et al.,[42] working with mixed beds of equal weights of chalcocite and pyrite, have achieved flotation separations of these sulphide minerals using xanthate as the collector by controlling the electrochemical potential of the flotation cell. Their results are reproduced in Figure 6, which presents the flotation recovery of both single-component mineral beds and the components of mineral mixtures for three different methods of addition of the mineral mixture to the cell. This figure clearly shows that in addition to the general xanthate flotation behaviour of sulphides, which are depressed at reducing potentials and floated at oxidizing potentials, the flotation of chalcocite and pyrite from the mixture does not follow that of the single mineral. The presence of chalcocite enhances pyrite flotation in the potential range of -0.4 to -0.1 V and promotes pyrite depression above 0.1 V. Pyrite flotation below the reversible xanthate/dixanthogen potential, namely -0.13 V,[7] was attributed to activation by Cu(II) ions produced by the anodic dissolution of chalcocite.

Through control of the electrochemical potential at the solid/liquid interface, Chander and Fuerstenau[43] have also been able to separate sulphides, namely chalcocite from molybdenite, by flotation using diethyl dithiophosphate as the collector. Choosing the appropriate flotation potential, chalcocite can be depressed and molybdentie floated, or vice versa. In this investigation, molybenite was found to be relatively electrochemically inactive, and therefore its flotation response was not suppressed by applying low potentials. Chalcocite flotation is affected by the rate of oxidation, which was controlled by changing the sweep rate to alter the value of the conditioning potential to that of flotation.

ACTIVATION AND DEPRESSION IN SULPHIDE MINERAL FLOTATION

With the exception of sphalerite and pyrrhotite, sulphide minerals of the base metals respond well to flotation with short-chained xanthate collectors. Upon the addition of copper salts, however, sphalerite responds readily to flotation with xanthates. This means of enhancing the flotability of sphalerite has become the classic example of activation in flotation technology, and since 1913, when Leslie Bradford discovered in Australia that copper sulphate enhances the flotability of zinc sulphide,[6] this inorganic salt has proved to be indispensable for effectively concentrating zinc sulphide ores by flotation. The general consensus on the activation of sphalerite by Cu(II) ions has been that this process involves an ion exchange mechanism in which cupric ions are exchanged for zinc ions, according to the following reaction:

$$ZnS_{(s)} + Cu^{2+} = CuS_{(s)} + Zn^{2+}$$

The stoichiometric 1:1 Cu/Zn exchange and the coating of sphalerite by covellite were observed by Cooke[44] in a long-term (50 days), high-temperature (100° C) study of the activation of sphalerite with copper chloride. The results reported by Gaudin et al.,[45] on the uptake of Cu(II) ions by sphalerite also indicate that an ion exchange reaction occurs in this process, and that there is a 1:1 Cu/Zn exchange. Using ^{64}Cu as a radioactive tracer, and starting with 3 micromoles Cu(II), these researchers reported that 2.8 micromoles Cu(II) were abstracted and 2.9 micromoles Zn(II) were displaced in this reaction. This chemical exchange reaction was also reported by Mellgren et al.,[46] to be responsible for the activation of sphalerite by copper ions, and through calorimetry they measured the heat of adsorption of Cu(II) on sphalerite. The enthalpy change for the exchange reaction was found to be 62.8 kJ/mole Cu at 25° C.

Salts of metal ions other than copper have also been shown to activate sphalerite for xanthate flotation. Through contact angle measurements and flotation experiments, Sutherland and Wark[47] examined the activation behavior of sphalerite with various metal salts. Their results indicate that successful activators are salts of metals which themselves or their sulphides respond readily to flotation with thiol collectors. Salts of lead, cadmium, silver and mercury fall within this category. Since the flotation behaviour of sphalerite activated with metal salts is similar to that of the activator metal sulphide mineral, there appears to be a direct correlation between flotation response and the solubility product of the metal xanthate. The chemistry of sphalerite activation has been thoroughly reviewed by Finkelstein and Allison.[48]

Cu(II) ions have also been found to activate other sulphide minerals for xanthate flotation. Using electrochemical techniques and a rotating disc electrodes, Nicol[49] has recently investigated the activation mechanism of galena and pyrrhotite with copper salts; the results show that CuS forms on the surface of these activated minerals, which in turn exhibit flotation behaviour typical of copper sulphides. The activation of sulphide minerals with heavy metals has been recently reviewed by Wang et al.,[50] who point out that the ion-exchange mechanism postulated to explain this process is oversimplified because it takes in to consideration neither surface oxidation phenomena nor metal ion hydrolysis. An electrochemical approach was taken by these researchers to propose that the overall activation process can be explained in terms of a mixed-potential mechanism.

A number of different approaches have been found to be effective for sulfide depression[33] which can be accomplished by preventing the interaction of thiol collectors with the sulfide mineral resulting in the formation of a hydrophobic surface:

1. enhancement of sulphide surface oxidation,
2. inhibition of oxygen reduction,
3. decrease in oxygen concentration,
4. inhibition of collector oxidation,
5. complexation of surface sites.

In practice, flotation engineers achieve selectivity by changing the pH of the pulp and by adding to the system such depressant-acting reagents as sodium salts of cyanide, sulfide, sulfite, and potassium dichromate. The addition of these chemicals alters the electrochemical potential of the system thus affecting mineral flotation. The mechanisms by which these depressants inhibit

Table IV. Mechanism of Depressant Action on Sulfide Mineral Flotation:

DEPRESSANT	GALENA	PYRITE	CHALCOCITE	CHALCOPYRITE
Hydroxyl	Due to the formation of lead hydroxy species at about pH 11	Xanthate oxidation to dixanthogen does not occur above pH 11. Also $Fe(OH)_3$ is the stable species above pH-11.	Cuprous ethyl xanthate is so stable that hydroxide ions can function as a depressant only at very high pH values for both chalcocite and chalcopyrite.	
Cyanide	No effect because there is no lead cyanide complex.	Formation and chemisorption of $Fe(CN)_6^{-4}$ at the surface	In absence of Fe only high concentrations of CN will depress, due to the formation of $Cu(CN)_2^-$	Similar to pyrite
Sulfide	PbS is more stable than PbX_2 so collection reaction is reversed.	Reduces the oxidation potential so that the formation of X_2 is inhibited. Also there can be chemisorption of sulfide ion.	Due to the formation of Cu_2S at the surface.	Similar to pyrite
Chromate	Due to the formation of lead chromate at the surface.			
Sulfite		Reduces oxidation potential below that is required to form dixanthogen.		

flotability are summarized in Table IV. In terms of the oxidation potential of the system, the collector species responsible for flotation, and the species leading to mineral depression, Chander[51] has classified sulphide minerals as reversible and passivated. This classification is presented as Figure 7, which shows the Eh-pH conditions for the stability domain of sulphides, metal/collector compounds for reversible sulphides and disulphides for passivated sulphides. This figure also gives the depressant species for both groups of minerals. The group of reversible sulphides include galena and chalcocite, while the group of passivated sulphides may include pyrite, chalcopyrite and bornite. The reactions of passivated sulphides are irreversible. Using this type of information, the appropriate conditions for selectively separating sulphide minerals by flotation with thiol collectors can be predicted.

INDUSTRIAL PRACTICE IN THE FLOTATION OF SULPHIDE ORES

Flotation plant practice at the Magmont Mill, Southeast Missouri, will be described to illustrate a typical separation of lead, zinc and copper sulphides from a sulphide deposit containing galena, sphalerite, chalcopyrite, pyrite and marcasite.[52] The gangue in this ore is dolomite. In this operation, the first flotation step aims at recovering all of the lead and copper minerals while depressing iron and zinc sulphides, and dolomite. This is achieved by flotation at pH 8.5 to 9.0 using amyl xanthate as the collector, lime for pH control, and very small quantities of cyanide (0.02-0.03 kg/tonne) for pyrite, sphalerite and marcasite depression. In addition, zinc sulphate (1.6 to 20 kg/tonne) is used to help enhance sphalerite depression. Separation of galena from chalcopyrite is possible using liquid sulphur dioxide and starch as galena depressants while floating chalcopyrite. At the Magmont Mill, this method was chosen because the copper content of the bulk lead-copper concentrate is usually low (2 to 7% Cu).[52] The tailings from the lead-copper flotation circuit are treated with copper sulphate to activate sphalerite for its subsequent recovery. Lime is used to maintain a pH of 11.0 to 11.5 in the cleaners and 9.0 to 9.5 in the roughers during sphalerite flotation with xanthate. Cyanide is fed to the cleaners to depress the iron and unrecovered copper minerals. The flowsheet of this operation is shown in Figure 8.

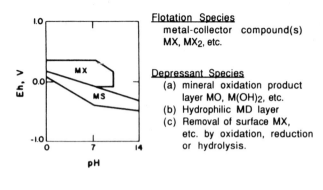

REVERSIBLE SULFIDES

Flotation Species
  metal-collector compound(s)
  MX, MX$_2$, etc.

Depressant Species
  (a) mineral oxidation product
      layer MO, M(OH)$_2$, etc.
  (b) Hydrophilic MD layer
  (c) Removal of surface MX,
      etc. by oxidation, reduction
      or hydrolysis.

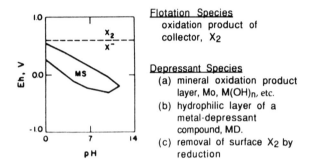

PASSIVATED SULFIDES

Flotation Species
  oxidation product of
  collector, X$_2$

Depressant Species
  (a) mineral oxidation product
      layer, Mo, M(OH)$_n$, etc.
  (b) hydrophilic layer of a
      metal-depressant
      compound, MD.
  (c) removal of surface X$_2$ by
      reduction

Fig. 7. Conditions for flotation and depression of reversible and passivated sulphide. (After Ref. 51).

An extensive and detailed study of the Climax molybdenite ore (0.3-0.4% MoS$_2$) flotation has been reported by Hoover and Malhotra.[53] The Climax Mill flowsheet is presented in Figure 9. It consists of a single-stage grind to 35-40% plus 100 mesh for rougher flotation followed by several stages of fine grinding restricted to 3-6% of the total feed. Satisfactory recovery at an extremely coarse grind is reportedly possible by the effective flotation of a large amount of low grade quartz-molybdenite middlings due to the high flotability of molybdenite. Overall recovery at the Climax Mill historically has been 89-90%. Molybdenite ore flotation at Climax uses a battery of reagents carefully designed to ensure maximum recovery and grade. Molybdenite, similar in flotation behaviour to unoxidized, higher-rank coal, is hydrophobic and is floated with oily collectors like vapor oil (light kerosene). An emulsifying agent, called Syntex, is used to enhance molybdenite flotation by more effectively distributing the collector. In addition to vapor oil and Syntex, molybdenite ore flotation requires other reagents to depress impurities: NaCN to depress pyrite, Nokes reagent to control galena and copper flotation, and sodium silicate to depress the silicate gangue.

Research on the electrochemistry of sulphide mineral flotation with thiol collectors has been conducted at a rapid pace during the last thirty years on a variety of minerals by numerous investigators. However, the only industrial use of electrochemical control was recently described by Heimala et al.,[54] who report that Outokumpu Oy has developed an industrial cell incorporating electrochemical control to effect a better separation and concentrate grade in the flotation of a copper/nickel ore. Their results show that by simply monitoring and controlling the electrochemical potential of the flotation cells, both the amount of collector and the amount of reagents used for pH adjustment can be reduced significantly, while improving the grade and recovery of the concentrate.

Controlling the electrochemical potential of the flotation system may also be accomplished on an industrial scale by controlling the oxygen concentration in the slurry. One technique to help keep the electrochemical potential down is to use nitrogen as the flotation gas. This approach is being used industrially in copper-molybdenum separations, using hydrosulphide in closed flotation cells. Closed flotation cells are extremely effective in reducing reagent cost by excluding oxygen. Dissolved oxygen not only raises the potential but when hydrosulphide is present it increases the consumption of this reagent. At Anamax, the use of a closed flotation system reduced the molybdenite depressant by 68%, with no nitrogen make up being necessary.[55]

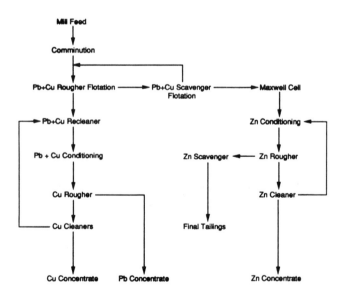

Fig. 8. Flowsheet of the copper-lead-zinc sulphide separation by flotation at the Magmont Mill.

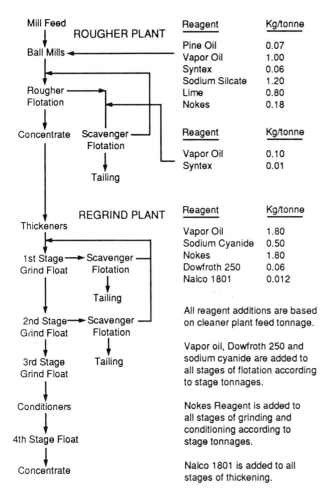

Mill Feed		
ROUGHER PLANT	Reagent	Kg/tonne
	Pine Oil	0.07
	Vapor Oil	1.00
	Syntex	0.06
	Sodium Silcate	1.20
	Lime	0.80
	Nokes	0.18

	Reagent	Kg/tonne
	Vapor Oil	0.10
	Syntex	0.01

REGRIND PLANT	Reagent	Kg/tonne
	Vapor Oil	1.80
	Sodium Cyanide	0.50
	Nokes	1.80
	Dowfroth 250	0.06
	Nalco 1801	0.012

All reagent additions are based on cleaner plant feed tonnage.

Vapor oil, Dowfroth 250 and sodium cyanide are added to all stages of flotation according to stage tonnages.

Nokes Reagent is added to all stages of grinding and conditioning according to stage tonnages.

Nalco 1801 is added to all stages of thickening.

Fig. 9. Flowsheet of the copper-molybdenum separation by flotation at the Climax Mill.

SUMMARY

The results of numerous investigations on both pure and impure systems have helped to further our understanding of the complex flotation behaviour of sulphide minerals with thiol collectors. It is now well accepted that the flotation response of sulphides is determined by three characteristics: 1) their low solubility, 2) their surface oxidation behaviour and 3) their semiconducting properties. Modern electrochemical studies have made possible the establishment of a mixed-potential model to rigorously explain the hydrophilic/hydrophobic transformation of a sulphide mineral surface upon its interaction with a thiol collector. This model states that this surface transformation involves simultaneous electrochemical reactions; the cathodic step usually being the reduction of oxygen, while the anodic step involves the oxidation of the collector or the formation of a metal thiolate. This information helps mineral processing engineers develop effective schemes to selectively recover sulphide minerals from complex ores.

References

1. Fuerstenau D.W. In: Froth Flotation. D.W. Fuerstenau, Ed. New York: American Institute of Mining, Metallurgical and Petroleum Engineers, Inc., 1962, 677pp.
2. Martin T.W. and others. In: Minerals Yearbook 1985. Washington: Bureau of Mines, vol. 1, 1987. p. 7-65.
3. Flotation of Sulphide Minerals. K.S.E. Forssberg, Ed. New York: Elsevier, 1985, 480pp.
4. Complex Sulphide Ores. M.J. Jones, Ed. London: The Institution of Mining and Metallurgy, 1980, 278pp.
5. Complex Sulphides. A.D. Zunkel and others, Eds. Warrendale: The Metallurgical Society, Inc., 1985, 938pp.
6. Gaudin A.M. Flotation. 2nd ed. New York: McGraw-Hill, 1957, 573pp.
7. Allison S. A. and others. A determination of the products of reaction between various sulphide minerals and aqueous xanthate solutions, and a correlation of the products with electrode rest potentials. Metallurgical Transactions, vol. 3, 1972, p. 2613-2618.
8. Woods R. and Richardson P.E. The flotation of sulphide minerals-electrochemical aspects. In: Advances in Mineral Processing. P. Somasundaran, Ed. Littleton: Society of Mining Engineers, 1986. p. 154-170.
9. Heyes G.W. and Trahar W.J. The natural flotability of chalcopyrite. International Journal of Mineral Processing, vol. 4, 1977, p. 317-344.
10. Shannon L.K. and Trahar W.J. The role of collector in sulphide ore flotation. In: Advances in Mineral Processing. P. Somasundaran, Ed. Littleton: Society of Mining Engineers, 1986. p. 408-425.
11. Lindqvist O. and Stridh K. Structural studies on adsorption of collectors on mineral surfaces. In: XVI International Mineral Processing Congress. E. Forssberg, Ed. New York: Elsevier, 1988, p. 717-725.
12. Scordamaglia R. and others. Correlation of collecting power of functionalized organic molecules and their molecular structure by quantitative chemical and statistical methods of calculation. In: Reagents in the Minerals Industry. M.J. Jones and R. Oblatt, Eds. London: Institution of Mining and Metallurgy, 1984. p.189-192.
13. Buckley A.N. and Woods R. In: International Symposium on Electrochemistry in Mineral and Metal Processing. P.E. Richardson, S. Srinivasan, and R. Woods, Eds. Electrochemical Society, 1984. p. 286-302.

14. Buckley A.N. and others. Investigation of the surface oxidation of sulphide minerals by Linear Potential Sweep Voltammetry and X-Ray Photoelectron Spectroscopy. In: Flotation of Sulphide Minerals. K.S.E. Forssberg, Ed. New York: Elsevier, 1985. p. 41-59.

15. Guy P.J. and Trahar W.J. The effects of oxidation and mineral interaction on sulphide flotation. In: Flotation of Sulphide Minerals. K.S.E. Forssberg, Ed. New York: Elsevier, 1985. p. 91-109.

16. Gaudin A.M. and others. In: 2nd International Congress of Surface Activity. London: Butterworths, 1957. p. 202-210.

17. Chander S. and others. On native floatability and the surface properties of naturally hydrophobic solids. In: Advances in Interfacial Phenomena of Particulate/Solution/Gas Systems. Applications to Flotation Research. P. Somasundaran and R.G. Grieves, Eds. New York: American Institute of Chemical Engineers, vol. 71, no. 150, 1975. p. 183-188.

18. Lepetic V.M. Flotation of chalcopyrite without collector after dry autogenous grinding. Canadian Institute of Metallurgy Bulletin, 1974, p. 71-77.

19. Finkelstein N.P. and others. Natural and induced hydrophobicity in sulphide minerals systems. In: Advances in Interfacial Phenomena of Particulate/Solution/Gas Systems. Applications to Flotation Research. P. Somasundaran and R.G. Grieves, Eds. New York: American Institute of Chemical Engineers, vol. 71, no 150, 1975. p. 165-175.

20. Leroux M. and others. Collectorless flotation in the processing of complex sulphide ores. In: Advances in Coal and Mineral Processing Using Flotation. S. Chander and R.R. Klimpel, Eds. Littleton: Society for Mining, Metallurgy and Exploration, Inc., 1989. p. 65-70.

21. Luttrell G.H. and Yoon R.H. The collectorless flotation of chalcopyrite ores using sodium sulphide. International Journal of Mineral Processing, vol. 13, 1984, p. 271-283.

22. Hayes R.A. and others. Collectorless flotation of sulphide minerals. Mineral Processing and Extractive Metallurgy Review, vol. 2, 1987, p. 203-234.

23. Buckley A.N. and Walker G.W. Sulphur enrichment at sulphide mineral surfaces. In: XVI International Mineral Processing Congress. K.S.E. Forssberg, Ed. New York: Elsevier, 1988. p. 589-599.

24. Dewey F.D. Reactions of some sulphur-bearing collecting agents with certain copper minerals. Ph. D. Dissertation, Montana School of Mines, 1933.

25. Taggart A.F. and others. Chemical reactions in flotation. Technical Publication of the American Institute of Mining and Metallurgical Engineers, no. 312, 1930, p. 3-33.

26. Gaudin A.M. Flotation mechanism, a discussion of the functions of flotation reagents. Technical Publication of the American Institute of Mining and Metallurgical Engineers, no. 4, 1927, 27pp.

27. Wark I.W. The chemical basis of flotation. Proceedings. Australasian Institute of Mining and Metallurgy, no. 90, 1933, p. 83-123.

28. Leja J. and others. Transactions. Institution of Mining and Metallurgy, vol. 72, 1962-63, p. 414-423.

29. Mellgren O. Heat of adsorption and surface reactions of potassium ethyl xanthate on galena. Transactions. American Institute of Mining and Metallallrgical Engineers, vol. 235, 1966, p.46-60.

30. Kamienski B. So-called flotation. Przem. Chem., vol. 15, 1931, p.201-2. (Chemical Abstracts, vol. 26, 1932, p. 53).

31. Salamy S.G. and Nixon J.C. Relation between a mercury surface and some flotation reagents: an electrochemical study. Australian Journal of Chemistry, vol. 7, 1954, p. 146-156.

32. Woods R. Electrochemistry of sulphide flotation. In: Flotation. A.M. Gaudin Memorial Volume. M.C. Fuerstenau, Ed. Vol. 1, 1976. p. 298-333.

33. Woods R. Electrochemistry of sulphide flotation. In: Principles of Mineral Flotation. The Wark Symposium. M.H. Jones and J.T. Woodcock, Eds. Parkville: The Australasian Institute of Mining and Metallurgy, symposia series no. 40, 1984. p. 91-115.

34. Finkelstein N.P. and Goold L.A. The reaction of sulphide minerals with thiol compounds. Nat. Institute of Metallurgy. South Africa, report no. 1439, 1972.

35. Toperi D. and Tolun R. Electrochemical study and thermodynamic equilibria of the galena-oxygen-xanthate flotation system. Transactions. The Institution of Mining and Metallurgy, vol. 78, 1969, p. C191-C197.

36. Chander S. and Fuerstenau D.W. Electrochemical reaction control of contact angles on copper and synthetic chalcocite in aqueous potassium diethyl dithiophosphate solutions. International Journal of Mineral Processing, vol. 2, 1975, p. 333-352.

37. Pritzker M.D. and Yoon R.H. Thermodynamic calculations on sulphide flotation systems: I. Galena-ethyl xanthate system in the absence of metastable species. International Journal of Mineral Processing, vol. 12, 1984, p. 95-125.

38. Gardner J.R. and Woods R. An electrochemical investigation of contact angle and of flotation in the presence of alkylxanthates. I. Platinum and gold surfaces. Australian Journal of Chemistry, vol. 27, 1974, p. 2139-2148.

39. Chander S. and Fuerstenau D.W. Effect of potential on the flotation and wetting behaviour of chalcocite and copper. Transactions. Society of Mining Engineers, vol. 258, 1975, p. 284-285.

40. Walker G.W. and others. Electrochemical flotation of sulphides: reactions of chalcocite in aqueous solution. International Journal of Mineral Processing, vol. 12, 1984, p. 55-72.

41. Richardson P.E. and Walker G.W. The flotation of chalcocite, bornite, chalcopyrite, and pyrite in an electrochemical-flotation cell. In: XVth International Mineral Processing Congress, tome II, 1985. p. 198-210.

42. Gebhardt J.E. and others. Electrochemical conditioning of a mineral particle bed electrode for flotation. United States Bureau of Mines. Report of Investigation 8951, 1985, 10pp.

43. Chander S. and Fuerstenau D.W. Electrochemical flotation separation of chalcocite from molybdenite. International Journal of Mineral Processing, vol. 10, 1983, p. 89-94.

44. Cooke S.R.B. Flotation. Advances in Colloid Science, vol. 3, 1950, p. 357- 374.

45. Gaudin A.M. and others. Activation and deactivation studies with copper on sphalerite. Transactions. American Institute of Mining, Metallurgical and Petroleum Engineers, vol. 214, 1959, p. 430-436.

46. Mellgren O. and others. Thermochemical measurements in flotation research. In: 10th International Mineral Processing Congress. London: The Institution of Mining and Metallurgy, 1974. p. 451-472.

47. Sutherland K.L. and Wark I.W. Principles of flotation. Melbourne: Australasian Institute of Mining and Metallurgy, 1955. 489pp.

48. Finkelstein N.P. and Allison S.A. The chemistry of activation, deactivation and depression in the flotation of zinc sulphide: a review. In: Flotation. A.M. Gaudin Memorial Volume. M.C. Fuerstenau, Ed. New York: American Institute of Mining, Metallurgical and Petroleum Engineers, Inc., vol. 1, 1976. p. 414-457.

49. Nicol M.J. In: International Symposium on Electrochemistry in Mineral and Metal Processing, The Electrochemical Society, 1984. p. 152-168.

50. Wang X. and others. The aqueous and surface chemistry of activation in the flotation of sulphide minerals-a review. Part 1: an electrochemical model. Mineral Processing and Extractive Metallurgy Review, vol. 4, nos. 3-4, 1989, p. 135-165.

51. Chander S. Oxidation/reduction effects in depression of sulphide minerals-a review. Minerals and Metallurgical Processing, vol. 2, no. 1, 1985, p. 26-35.

52. Sharp, F.H. Lead-zinc-copper separation and current practice at the Magmont Mill. In: Flotation. A.M. Gaudin Memorial Volume. M.C. Fuerstenau, Ed. New York: American Institute of Mining, Metallurgical, and Petroleum Engineers, Inc., vol. 2, 1976. p. 1215-1231.

53. Hoover R.M. and Malhotra D. Emulsion flotation of molybdenite. In: Flotation. A.M. Gaudin Memorial Volume. M.C. Fuerstenau, Ed. New York: American Institute of Mining, Metallurgical, and Petroleum Engineers, Inc., vol. 1, 1976. p. 485-505.

54. Heimala S. and others. New potential controlled flotation methods developed by Outokumpu Oy. In: XVth International Mineral Processing Congress, tome III, 1985. p. 88-98.

55. Onstott K.Y. and Person P.L. By-product molybdenum flotation from copper sulphide concentrate with nitrogen gas in enclosed Wemco nitrogen flotation machines. Preprint No. 84-65. Littleton: Society of Mining Engineers of AIME, 1984, 8pp.

Chelating reagents for flotation of sulphide minerals

A. Marabini
M. Barbaro
Istituto per il Trattamento dei Minerali, C.N.R., Rome, Italy

Chelating reagents, because of their metal specificity, can function as effective mineral processing reagents.

A brief review of the work carried out in this area is presented in particular, with regard to:

- Calculation method for the selection of chelating functional groups
- Criteria for designing and synthetizing chelate-type collectors with optimal structure for a given metallic mineral
- Comparison between conventional collectors and new synthetic aliphatic-aromatic structures

The application of chelating reagents in sulphide mineral flotation is reviewed in particular, with respect to influence of molecular structure of chelating reagents on their collecting power (effect of alkyl substituent and chain branching).

INTRODUCTION

The beneficiation of sulphide minerals is an extremely difficult problem [1] first, because a sulphide ore contains many valuable minerals with similar surface properties and secondly, because of the high number of parameters that play an important role in the flotation process. It is interesting to note that in 1955 Sutherland and Wark [2] listed 32 variables in sulphide flotation, of which only 22 can usually be controlled by the metallurgist.

A complete evaluation of the variables in sulphide mineral flotation is still a challenge.

As to the first problem, this is particularly evident for ores containing several commercially useful mineral components finely intergrown with the gangue and with one another, or for ores characterized by a very fine size analysis. This is the case of the so-called complex sulphide ores which have been defined as those ores from which it is difficult to recover one or more selective products of acceptable quality and economic value with minimal losses and at reasonable cost. Complex sulphide ores are fine-grained, intimate associations of chalcopyrite ($CuFeS_2$,) sphalerite (ZnS) and galena (PbS), disseminated in dominant pyrite, and which contain valuable amounts of silver and, in some cases, gold.

As to the second problem it is known that the surface that sulphide minerals present to collectors is generally not rather well defined; nor does it correspond closely to the bulk mineralogical species. In fact, it is known that sulphides are often strongly oxidized and this is considered to influence adsorption of collectors. Despite numerous studies, [1-68] the adsorption mechanism of known collectors on sulphide surfaces is not yet completely clarified, [1-14] and it has not yet been established to what extent the adsorption is [15-30] affected by surface oxidation.

The difficulties in clarifying the interaction mechanism of collectors are demonstrated by the number of studies carried out on their adsorption. Spectroscopic techniques [31-34] and, in particular, infrared absorption spectrometry [35-36] and attenuated total reflection infrared spectrometry (ATR-IR) [37-42] or X-ray photoelectron spectroscopy (XPS) [43-44] have been used, together with solution equilibria studies [56-57] and calorimetric studies. [58]

Generally, the collectors employed are thiols, reported in (Fig. 1), and the most common are xanthates. [5] In practice, however, xanthates are active towards the whole class of sulphide minerals rather than towards an individual mineral.

Thus, in order to float a given mineral from a mixture of minerals belonging to the same sulphide class, modifiers always have to be used in order to render the action of the collector more specific.

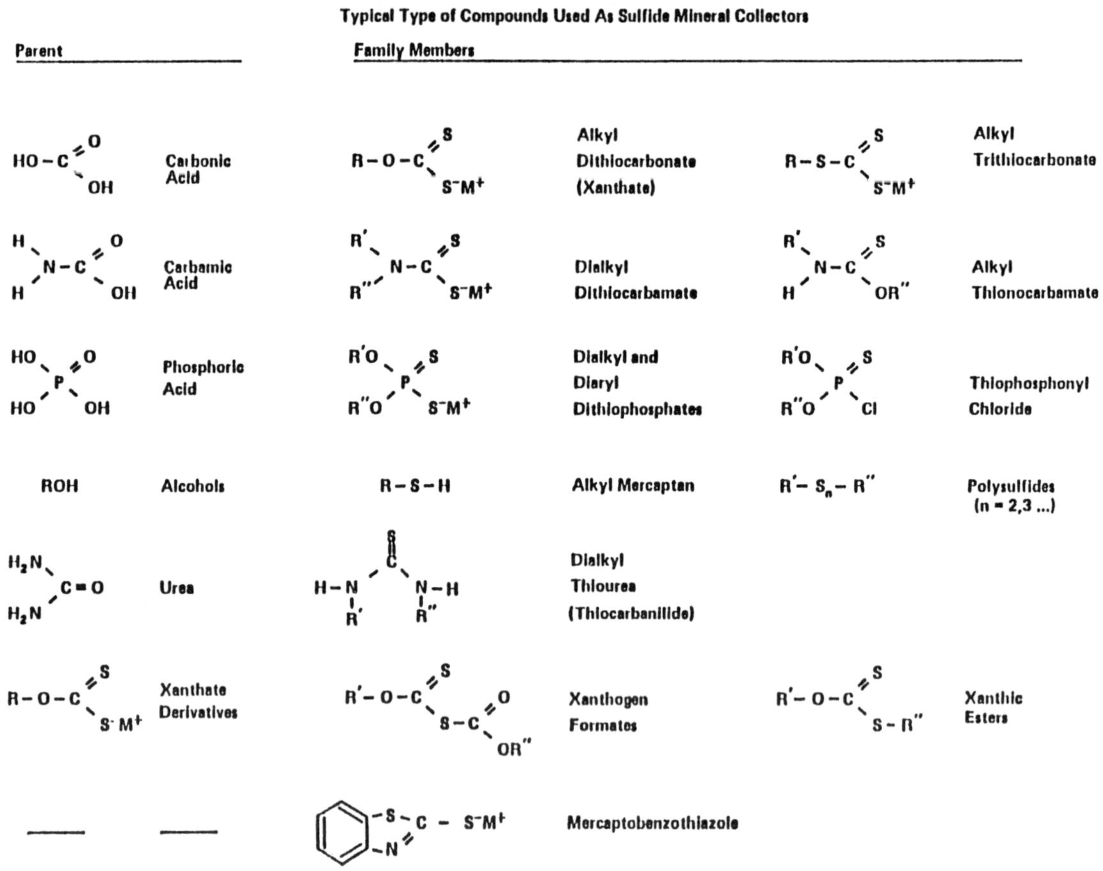

Typical Type of Compounds Used As Sulfide Mineral Collectors

Parent		Family Members			
$HO-C(=O)OH$	Carbonic Acid	$R-O-C(=S)S^-M^+$	Alkyl Dithiocarbonate (Xanthate)	$R-S-C(=S)S^-M^+$	Alkyl Trithiocarbonate
$H,H-N-C(=O)OH$	Carbamic Acid	$R',R''-N-C(=S)S^-M^+$	Dialkyl Dithiocarbamate	$R',H-N-C(=S)OR''$	Alkyl Thionocarbamate
$HO,HO-P(=O)OH$	Phosphoric Acid	$R'O,R''O-P(=S)S^-M^+$	Dialkyl and Diaryl Dithiophosphates	$R'O,R''O-P(=S)Cl$	Thiophosphonyl Chloride
ROH	Alcohols	$R-S-H$	Alkyl Mercaptan	$R'-S_n-R''$	Polysulfides (n = 2,3 ...)
$H_2N,H_2N-C=O$	Urea	$H-N(R')-C(=S)-N(R'')-H$	Dialkyl Thiourea (Thiocarbanilide)		
$R-O-C(=S)S^-M^+$	Xanthate Derivatives	$R'-O-C(=S)S-C(=O)OR''$	Xanthogen Formates	$R'-O-C(=S)S-R''$	Xanthic Esters
——	——	benzothiazole $-S-C(-S^-M^+)=N-$	Mercaptobenzothiazole		

Fig. 1 Typical type of Compounds used as sulfide mineral collectors [91]

Many studies have been performed on modulation of flotation with xanthates [59-68] in order to improve separation efficiency.

However, there are often considerable difficulties involved in this procedure and the desired results are not always obtained, especially in the case of minerals of a complex composition–hence the importance of seeking collectors capable of linking selectively with given minerals. Selective linkage is possible if the collector structure incorporates active groups with a specific affinity for certain cations characteristic of the mineral surface.

The decision to investigate the possibility of utilizing chelate-forming reagents for flotation was first taken by Gutzeit, [69] who used chelate-forming compounds in the anionic flotation of oxidized iron ores, their role being to "complex" the heavy metal ions so as to prevent quartz from floating.

CHELATING AGENTS AND CHELATE COMPOUNDS

Chelating agents are those particular complexing reagents which are formed of large organic molecules capable of bonding to the metal ion via two or more functional groups with the formation of one or more rings. The compounds formed in this way are known as "chelates" or "chelate compounds".

The bonds that exist between the chelating agent and metal vary according to the nature of the metal and that of the functional groups of the chelating agent, and include coordination bonds, covalent bonds, ionic bonds and bonds that are partially ionic and partially covalent.

In any case, every chelating agent is characterized by the presence of at least one group capable of donating a pair of electrons to the metal to which it bonds, with the formation of a coordination covalent bond. In practice, this virtually restricts the choice to groups containing N, O and S.

Regarding metal ions capable of forming chelate compounds, the tendency for chelate formation with donor molecules may be interpreted as the tendency for metal ions to fill up unoccupied orbitals and thereby approach the electronic configuration of the nearest inert gas (Table I).

Tab. I Periodic classification of chelate-forming metals [70]

Common chelate forming metals _____
Metals which form few chelates - - - - - - - -

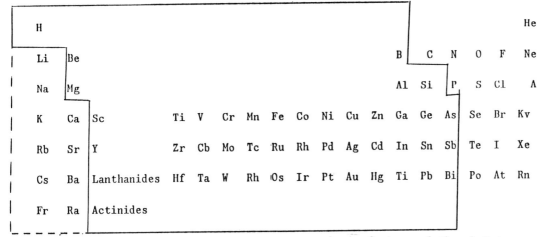

It is generally true that the farther removed the configuration of the metal ions is from that of the rare gases, the greater will be their tendency to accept pair of electons to approach the structure of the nearest rare gas to form chelates.

The atoms of the alkali metals and the alkaline-earth metals instead, which on the loss of one and two electrons, respectively, have the exact configuration of the nearest rare gas, have little or no tendency to accept a pair of electrons or to form chelate compounds.

The heavy metals and the transition metals in particular, have come to be considered as having the strongest tendency to form chelates.

On the basis of the points made so far it is apparent that chelating compounds form a class of reagents that can be used for the flotation of metallic ores. Their inability to form bonds with silicon, aluminium, sodium and potassium guarantees their complete selectivity as regards all siliceous and silicate gangues, whereas their very slight tendency to link calcium also ensures good differential action for carbonate gangues.

PROPERTIES OF CHELATES

The main properties of chelate compounds are their high stability and the selectivity or specificity of their formation reactions. A basic feature of the chelate non-electrolytes is also their very low electrolytic dissociation, their insolubility or low solubility in water and their solubility in non-polar solvents.

The factors governing the stability of a chelate are, essentially, the structure of the chelating agent and the nature of the metal ion involved.

For a given metal ion the stability of a chelate depends on the number and size of the chelate rings, stability increasing with the number of rings up to a maximum of five or six. The basic strength and the degree of dissociation of a ligand are also decisive factors for the stability of chelates.

Other factors that have a marked influence on the degree of stability of a chelate are steric factors and resonance.

Martell [70] has summarized the effect of the variation of structure of the chelating agent on the chelating force (Table II).

Table II Ligand structure influence on chelate stability

Basicity of the ligand
Resonance effects
Number of metal chelate per ligand
Size of chelate ring
Steric effects
Specific effects

The stability of the chelate naturally depends not only on the structure of the chelating agent but also on the nature of the metal ion to be held. The influence of the metal ion is, in turn, governed principally by two factors - the influence of the donor atoms and that of the properties of the metals themselves.

As regards the influence of the donor atoms, the main donor atoms in chelate structures are, in practice, limited to O, N and S, as was noted earlier.

Metals can be divided into three groups according to their tendencies to combine with oxygen (usually through a normal covalent bond) or with nitrogen (usually through a coordinate covalent bond).

These three groups are:
- Group 1 - Bonding to oxygen more strongly than to nitrogen
- Group 2 - Bonding to oxygen and nitrogen comparable in strenght
- Group 3 - Bonding to oxygen less strongly than to nitrogen (Table III)

Table III Metal tendencies for combination with oxygen or nitrogen [70]

$\overline{O} \rangle N$ Mg, Ca, Sr, Ba, Ga, In, Tl, Ti, Zr, Th, Si, Ge, Sn, V^V, V^{IV}, Cb^V, Ta^V, Mo^V, U^{VI}, U^{IV}, Fe^{III}, Co^{II}

$\overline{O} \simeq N$ Be, Cr^{III}, Fe^{II}, platinum metals

$N \rangle \overline{O}$ Cu^I, Ag^I, Au^I, Cu^{II}, Cd, Hg, V^{III}, Co^{III}, Ni^{II}

The second influence consists in the properties of the metals themselves and particularly, the ionic forces that are related to the charge and to the radius of the metal ion, as already noted. For instance, in the case of the transition metals and for most metals in general, stability of the chelate increases as the ionic radius decreases and as the metal charge increases.

Contrary to what happens in the case of common chemical compounds or non-chelated complex salts, numerous and diverse factors influence chelate stability. It is precisely the multiplicity and complexity of these factors, and especially their interrelationships, that determine the extent to which chelating agents are selective and, in certain cases, specific in their action.

It would be as well to replace the expression "specific or selective reagent" with the more appropriate term "specific or selective reaction", thus clearly demonstrating the importance of the reaction medium and reaction conditions in eliminating interference caused by other metal ions.

CHELATING REAGENTS AND FUEL OIL AS COLLECTORS FOR METALLIC MINERALS.

At this point, from a purely theoretical aspect, it may be thought that chelating agents are compounds with characteristics ideal for the flotation of given metal ores. Not only should such reagents react in a highly selective if not actually specific manner on a given mineral but they should also form a highly insoluble coating on the surface thereof by nullifying the ionic charge of the surface ion thereby preventing all form of hydration of the mineral. However, although the resulting chelated particle is fairly hydrophobic, it is, nevertheless, not sufficiently aerophilic to ensure flotation because the chelating agents commercially available are almost all without long hydrocarbon chains.

The line followed by Marabini et al. in their early investigations was that of rendering particles hydrophobic by making contemporaneously available for this purpose long-chain organic groups (as fuel oil or oily frother) capable of adhering to the surface chelate formed by chemical reaction.

From 1970 systematic research was carried out by Marabini et al. to evaluate effectivenes of chelating agents and fuel oil for metallic mineral flotation. [71] Most of these studies however were performed on oxidized minerals. [72-75]

The first application on sulphide minerals was in 1973. A chelating agent, 8-hydroxy quinoline (Fig. 2), with fuel oil was employed to float mixed oxide-sulphide minerals of Zn and Pb. [73] Table IV reports the results of this work. Tests in laboratory cell on an ore containing 7,3% Zn with 1,4% as sphalerite, and 0,9% Pb with 0,40% as galena, led to good recoveries that (Table IV). [76]

Also Mangalam et. al. [76] proved efficiency of 8-hydroxyquinoline in chalcopyrite flotation.

Another example of chelating reagent in sulphide treatment is the flotation of cobaltite with α-nitroso- ß-naphtol [77].

It is known that α-nitroso- ß-naphtol (α NßN) is a chelating agent of cobalt. Fig.3 reports its formula and chelate structure. Flotation tests in the Hallimond tube on pure cobaltite

106

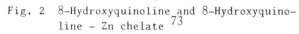

Fig. 2 8-Hydroxyquinoline and 8-Hydroxyquino-
line - Zn chelate [73]

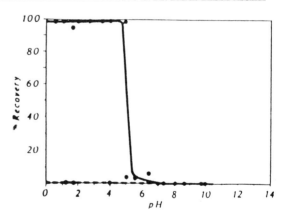

Fig. 3 α-Nitroso-βNapthol (αNßN) and
αNßN-Co chelate [77]

Table IV Flotation results of Zn-Pb ore with 8-hydroxyquinoline (oxine) [73]

Test	Product	Conditioning time, min.	Oxine g	Gasoline ml	% wt	Grade of float product		% Recovery	
						% Zn	% Pb	Zn	Pb
1	Concentrates	5	3-5	210	33.29	19.84	2.03	91.33	77.74
	Tailings				66.71	0.94	0.29	8.67	22.26
	Feed				100.00	7.22	0.87	100.00	100.00
2	Concentrates	10	3-5	210	34.65	19.92	2.09	94.29	80.48
	Tailings				65.35	0.64	0.27	5.71	19.52
	Feed				100.00	7.32	0.90	100.00	100.00
3	Concentrates	20	3-5	210	29.08	21.99	2.33	90.20	77.97
	Tailings				70.92	0.98	0.27	9.80	22.03
	Feed				100.00	7.09	0.87	100.00	100.00

with α-NßN and fuel oil as collector (Fig. 4)
gave a 100% recovery. Moreover, tests on
cobaltite/hematite and cobaltite/niccolite
gave good selectivity for cobaltite (Table V).
Rinelli and Marabini interpreted these results
on the basis of the greater affinity of α-NßN
for cobalt than nickel and copper. In fact,
the graph of conditional constants of α-NßN
complexes with Co, Ni and Cu (Fig. 5) shows
this selectivity.

At this stage, the results had confirmed
the soundness of the basic theoretical
premise. However problems had emerged that
made it difficult for the chelating agents
commercially available to be used on the plant
scale.

The main hindrages to the pratical use of
commercial chelating reagents with aromatic
structure are:
- lack of long-hydrocarbon chain in the
  molecule
- costitutional insolubility in water of most
  commercial chelating reagents
- high consumption needed to ensure

Fig. 4 Relationship between cobaltite recovery
and pH (fuel-oil concentration = 0.1
g/l) [77]

—————— with αNßN 0.025 g/l
----------- without αNßN

107

Table V Results of cobaltite-niccolite and cobaltite-hematite flotation tests (feed 50% cobaltite) [77]

Mixture	NβN concentration g/l	Fuel-oil concentration g/l	pH	Products	% Weight		% Recovery
					Cobaltite	Niccolite	
Cobaltite Niccolite	0.05	0.1	3	Float Non-float	42.36 54.64 100.00	80.26 19.73 100.00	12.81 87.19 100. 00
Cobaltite Hematite	0.03	0.05	3	Float Non-float	60.75 29.25 100.00	98.73 1.27 100.00	23.00 77.00 100.00

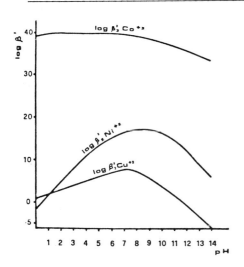

Fig. 5 Conditional constants of ꝺNβN complexes with Co - Ni and Cu as a function of pH [77]

satisfactory flotation results

The first factor generally makes the chelating agent alone incapable of rendering the mineral surface ufficiently aerophilic. To overcome this difficulty the chelating agent was provided artificially with a long-chain organic portion by introducing a neutral oil (fuel oil). But the introduction of a new liquid phase into the flotation pulp is damaging to the system as a whole and is not available on the industrial scale.

The insolubility in water and high consumption are both related to the chemical characteristics of the reagents commercially available, and particularly to the nature of their chelating groups.

Furthermore, too strong an affinity of chelation groups for mineral cations can lead to detachment of the chelate from the particle surfaces and consequential high consumption of

this type of reagent in comparison with classical collectors.

To conclude, for all practical purposes che chelating agent must have a long chain, be water-soluble and have proper chelating groups endowed with good but not too high an affinity for mineral cations and capable of acting selectively on a given mineral against an other.

At this point Marabini et al. orientated their research towards the design and synthesis of new compounds incorporating into a single molecule long-chain hydrocarbons and chelating groups satisfying the requirements for specific collectors.

Nevertheless, to envisage an ideal chelating collector, it was necessary:

- to state a method capable of evaluating in advance the efficiency of a ligand as selective collector towards a given mineral
- to fully understand the intimate mechanism of the interaction between ligands and mineral surfaces, studying the associated static and kinetic equilibria

CALCULATION METHOD FOR SELECTION OF CHELATING COLLECTORS

To solve the first problem, a method has been developed, with the help of a computerized program, for the selection of chelating reagents theoretically provided with a selective collecting power towards one given mineral rather than another, based on the calculation of conditional constants [78].

Conditional constants are so called because they depend on the experimental conditions such as metal-water and ligand-water reactions and consequently they vary as a function of pH. These constants can be calculated from published values of equilibrium constants. It can be assumed that a given ligand L acts as a selective collector for a mineral that contains the cation M_1 against another that contains the cation M_2 if it forms with M_1 a complex M_1L the conditional constant of which is

$$\log K'_{M_1L} \geqslant 6$$

and if the difference between the conditional constants of the complexes M_1L and M_2L is

$$\log K'_{M_1L} - \log K'_{M_2L} \geqslant 5$$

The first criterion characterizes the absolute chelating power toward the cation M_1. The second defines the difference in the relative chelating power of a ligand vis à vis two cations M_1 and M_2.
For a generic cation M the complexation reaction is
$$M + L \rightleftharpoons ML$$
the equilibrium constant of which is defined by the expression
$$K_{ML} = \frac{[ML]}{[M][L]}$$
where K_{ML} is the stability constant dependent on ionic strength and temperature, and $[M], [L]$ and $[ML]$ are the concentrations of free metal, free ligand and complex, mole l^{-1}, respectively.
From consideration of the influence of all the side reactions in which M and L are involved a new constant may be defined for the same equilibrium:
$$K'_{ML} = \frac{[ML]}{[M'][L']}$$
where K'_{ML} is the conditional constant and $[M']$ $[L']$ are concentrations of free and hydroxylated metal and of free and protonated ligand, mole l^{-1}, respectively.
From the dissociation coefficients
$$M = \frac{[M']}{[M]}$$
$$L = \frac{[L']}{[L]}$$
the conditional constant may be written
$$K'_{ML} = \frac{K_{ML}}{2_M 2_L}$$
and
$$\log K'_{ML} = \log K_{ML} - \log_M - \log_L$$
where $_M$ and $_L$ are calculated from mathematical relationships considering the equilibrium constants of the metal-water reactions. The validity of these criteria has been proved in many systems by flotation tests in the Hallimond tube and in laboratory cells.
The method has been applied both to the selection of collectors for separation of

rutile from iron oxide, and for the selection of ligands capable of separating Zn and Pb oxides from Fe-Mg-Ca oxides[79-80]. Two series of reagents containing chelating functional groups for Pb and Zn (Fig. 6) and an alkyl or alcoxyl chain linked to the aromatic ring have

MERCAPTOBENZOTHIAZOLE   AMINOTHIOPHENOL
    (MBT)   (ATP)

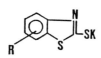

Where R   alkyl chain (12 C atoms)
          alcoxyl chain

Fig. 6   Mercapto Benzothiazole and Amino Thiophenol structure and chelating bond [81]

been prepared[81] and tested successfully on oxidized Pb-Zn ores.
Although obtained on oxidized ores, the results obtained with these two series of reagents of provide interesting pointers on the criteria to be followed in designing mixed aliphatic-aromatic chelate-type collector molecules. Studies are now being done on their application in sulphide ore treatment[85].

CHELATING COLLECTORS FOR SULPHIDE MINERALS

Examples of the use of chelating agents in sulphide flotation are known from studies on traditional thiol collectors. In fact, both dixanthogen (Fig. 7) and thionocarbamate (Fig. 8) act as chelating agents.[86]
Many researchers have studied the use of chelating reagents in metallic mineral flotation. Recently, a review of these studies has been published by Pradip.[87] He showed that most of the studies concern oxidized minerals there are very few applications to sulphides. However, having observed[86] that thionocarbamates have chelating properties, Ackerman et al.[88] began to investigate their action.

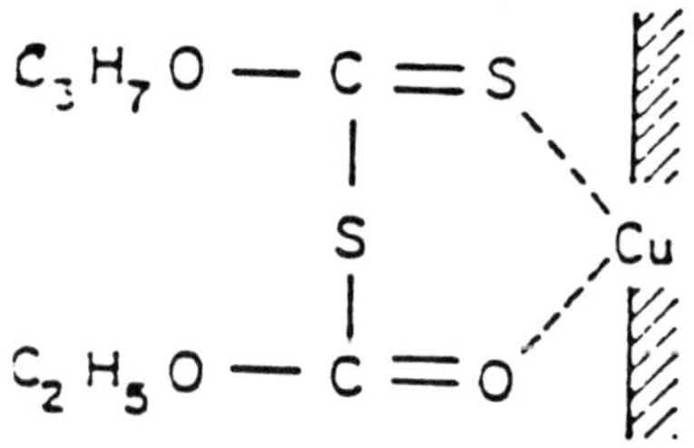

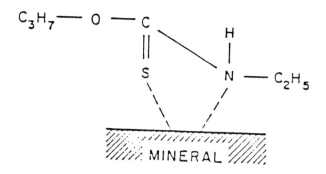

Fig. 7 Proposed attachment of xanthogen to the surface of a copper mineral [86]

They studied alkyl substituent effects on the performance of thionocarbamates as copper sulphide and pyrite collectors and found that O-alkyl or N-alkyl substituent better controls hydrophobicity, electron density and accessibility of the adsorbed collectors, with resulting influence on collector density. This is[89] in accordance with Somasundaran's studies on substituent effect on functional groups of a chelating agent.

Moreover, Ackerman et al.[90] proposed for the flotation of sulphide minerals a new commercial chelating agent, 1-hydroxy-ethyl-2-heptadecenyl glyoxalidine (amine 220), the structure and mechanism of surface attachment of which are reported in Fig. 9. They also found that the collecting power of glyoxalidine is superior to that of other non-sulphydryl collectors (Fig. 10).

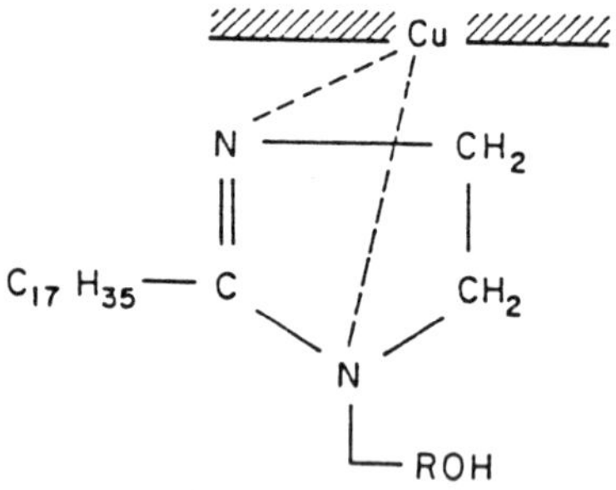

Fig. 8 Proposed attachment of thionocarbamate to a mineral surface (modified after Bogdanov et al., 1976) [86]

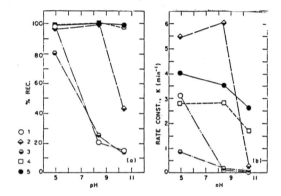

Fig. 10-a) Chalcopyrite flotation with various non-sulphydryl collectors. (a) % recovery vs. pH, (b) rate constant, K, vs. pH. 1=sodium laureate; 2=sodium oleate; 3=sodium dodecyl sulfate: 4=dodecyl amine acetate; 5=Amine 220 [90]

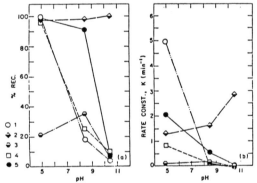

Fig. 10-b) Chalcocite flotation with various non-sulphydryl collectors. (a) % recovery vs. pH. (b) rate constant, K, vs. pH. 1= sodium laureate; 2=sodium oleate; 3=sodium dodecyl sulfate; 4=dodecyl amine acetate; 5= Amine 220[90]

Fig. 9 Proposed attachment of 1-hydroxyethyl-2-heptadecenyl glyoxalidine (Amine 220) to the mineral surface by forming a copper chelate [90]

Two classes of chelating collectors for sulphide mineral flotation[91] have been synthesized by Klimpel et al; the general molecular formula of the two classes are reported in Fig. 11. They also studied the

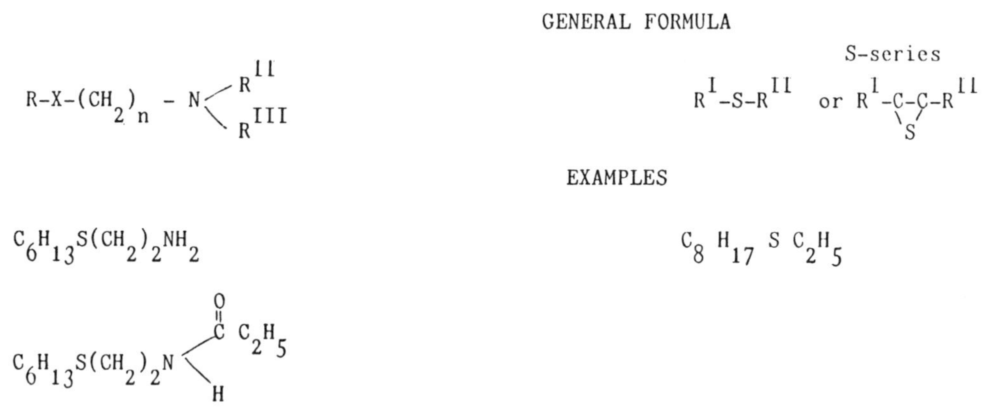

GENERAL FORMULA

$$R-X-(CH_2)_n - N \begin{smallmatrix} R^{II} \\ \\ R^{III} \end{smallmatrix}$$

S-series

$$R^I-S-R^{II} \quad \text{or} \quad R^1-\underset{\underset{S}{|}}{C}-\underset{S}{C}-R^{II}$$

EXAMPLES

$$C_6H_{13}S(CH_2)_2NH_2$$

$$C_8H_{17} S C_2H_5$$

$$C_6H_{13}S(CH_2)_2N \overset{\overset{O}{\overset{\|}{C}} C_2H_5}{\underset{H}{}}$$

Fig. 11 Chelating collectors for sulphide minerals [91]

effect of substituents and chain length of these reagents on copper recovery (Figs. 12 and 13) and chose the examples reported in (Fig. 11).

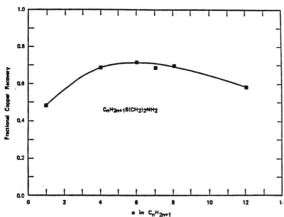

Fig. 12 The effect of chain length on the recovery of copper [91].

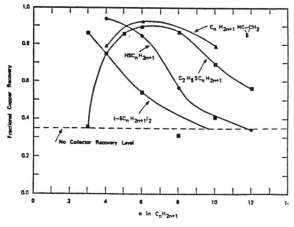

Fig. 13 The effect of chain length on the recovery of copper with various chemical collector structures [91]

Fig. 14 reports the selectivity of the three selected reagents over pyrite and copper/molybdenum recovery, and shows that F2

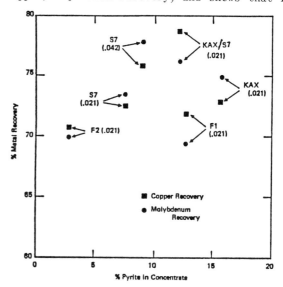

Fig. 14 Relationship of copper/molybdenum recovery to pyrite content. [91]
collector dosages in kg/metric ton indicated in ( ) MIBC frother at 0.015 kg/metric ton
F1, F2, and S7 are experimental collectors products of the Dow Chemical Co.

and S7 collectors increase recovery above commercial standards.

[92] In another work Klimpel and Hansen analysed all influences of controlled changes in chemical structure of this proposed series of sulphide mineral chelating collectors:

$$R^I-S-R^{II}-N \begin{smallmatrix} R^{III} \\ \\ R^{IV} \end{smallmatrix}$$

They confirmed that several members of this family have the potential of achieving usage on a commercial scale.

Recently, new classes of collectors with chelating properties for copper sulphide minerals have been presented by Nagaraj et al.[93]. They are alkoxycarbonyl alkyl thionocarbamates and thioureas, dialkyl (or diaryl) monothiophosphates and dialkyl monothiophosphinates. Their structure is reported in Fig. 15. FTIR – ATR and thermodynamic studies of these structures have shown[40] that alkoxycarbonyl alkyl thionocarbamates and thioureas adsorb on sulphide minerals via the formation of a six-membered chelate between metal cation and groups C=S and C=O. There are, however, differences between these two collectors in terms of flotability of the various sulphide minerals.

	Structure	pKa	Appearance	pH range of Stability
1. Alkoxycarbonyl Alkyl Thionocarbamates	$\overset{O}{R-O-\overset{\|\|}{C}}-NH-\overset{S}{\overset{\|\|}{C}}-O-R'$	10.5	Clear brown oily liquid (for R = C_2H_5; R' = $i\text{-}C_4H_9$)	3-11
2. Alkoxycarbonyl Alkyl Thioureas	$R-O-\overset{O}{\overset{\|\|}{C}}-NH-\overset{S}{\overset{\|\|}{C}}-NH-R'$	>12	Low melting solid with appreciable solubility in water, (R = C_2H_5; R' = $i\text{-}C_4H_9$)	3-12
3. Dialkyl or diaryl Monothiophosphates	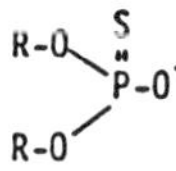 $\overset{R-O}{\underset{R-O}{}}\overset{S}{\overset{\|\|}{P}}-O^-$	1.55	Clear pale yellow liquid (R = $i\text{-}C_4H_9$)	1-12
3. Dialkyl Monothiophosphinates	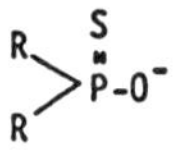 $\overset{R}{\underset{R}{}}\overset{S}{\overset{\|\|}{P}}-O^-$	3.17	Clear pale yellow liquid (R = $i\text{-}C_4H_9$)	1-12

Fig. 15 New Classes of Collectors[93]

In fact, as can be seen from Figs. 16 and 17 and 18, alkoxycarbonyl thionocarbamate floats copper-rich minerals such as bornite, covellite and chalcocite faster than chalcopyrite, whereas the modified thioureas float chalcopyrite very effectively. Such differences are observed not only with pure minerals but also with natural ores containing the various minerals.

As regards the phosphorus acids (Fig. 13) Nagaraj et al. have shown[93] that the monothiophosphates are acid-circuit collectors of sulphides, the monothiophosphinates are effective in neutral and mildly alkaline circuit and the dithiophosphates are effective at pH 9 (Fig. 19), and they proposed a mechanism to explain the differences in such collector activities between the phosphorus acids.

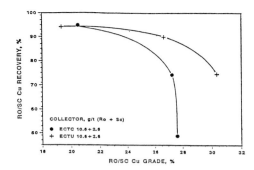

Fig. 16 The difference in collector activity
between Ethoxycarbonyl Thionocarbamate
and Ethoxycarbonyl Thiourea (Isobutyl
homologs) for a Western U.S. Copper ore
containing predominantly chaleopyrite
(pH 9.3; Cu recovery vs. grade; Cumula
tive times 1, 3 and 12 min.) [93]

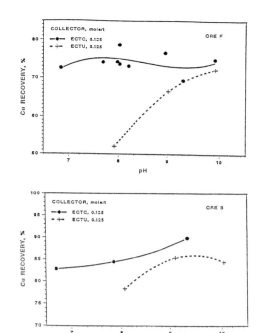

Fig. 18 The difference in Collector activity
between Ethoxycarbonyl Thionocarbamate
and Ethoxycarbonyl Thiourea (Isobutyl
homologs) as a function of pH for two
South American Cu ores containing predominantly copper-rich minerals (chalcocite, bornite and covellite). Flot.
Time 7 min. [93]

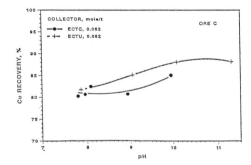

Fig. 17 The difference in Collector activity
between Ethoxycarbonyl Thionocarbamate
and Ethoxycarbonyl Thiourea as a function of pH for a Western U.S. Copper
ore containing predominantly chalcopyrite. (6 min. Flot.) [93]

CONCLUSIONS

There have been fewer studies on chelating
reagents as collectors for sulphide flotation
than on oxidized minerals. In fact, reagents
for flotation of oxidized Zn and Pb in the
pilot plant have already been proposed. This
opens up new prospectives in the application
of these reagents.

Applications on sulphides are presently only
on laboratory scale and on a limited number of
reagents, but further results obtained show
that this subject deserves attention. All
studies should consider not only the active

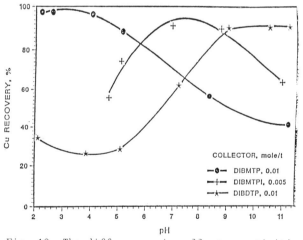

Fig. 19 The differences in collector activities
between monothiophosphate, monothio-
phosphinate and dithiophoshate (all
Isobutyl homologs) as a function of pH
for a Western U.S. Copper ore. Total
Flotation Time 7 min. [93]

functional group but also steric and thermodynamic factors.

On the one hand the mechanism of action of these reagents should be studied and, on the other criteria should be defined in order to project optimal chelating structures for a given mineral relative to its reticular structure and to its metallic active sites.

Moreover, efforts should be made towards reducing the reagent cost and reagent consumption in order to make these reagents economically advantageous in mineral processing application. However, as many problems in mineral and sulphide mineral processing not can be solved by conventional reagents, probably in the coming decades a number of reagents already tested in laboratory and at the pilot-plant scale will be commercialized and the mineral industry will collaborate with the production of tailor-made chelating collectors.

References

1) Healy T.W., Trahar W.J.. Challenges in mineral sulfide flotation. Challenges in mineral processing, Sastry K.V. ed., AIME, Colorado 1989.

2) Sutherland K.L. and Wark I.W.. Principles of flotation. Australasian Inst. of Mining metallurgy, Melbourne, 1855.

3) Rogers J.: Principles of sulfide mineral flotation. Fuerstenau M.C. ed, Froth Flotation, Gaudin A.M. memorial volume, AIME New York 1976.

4) Poling G.W.: Reactions between thiol reagent and sulfide minerals. Furestenau M.C. ed., Flotation, Gaudin A.M. memorial vol., AIME, New York 1976 pp. 334-363.

5) Ackerman P.K., Harris G.H., Klimpel R.R. and Aplan F.F.: Evaluation of flotation collectors for copper sulfides and pyrite I. Common Sulphydryl collectors. Int. Min. Proc., 21, 1987 pp. 105-127.

6) Granville R., Finkelstein N.P. and S.A. Allison. Review of the reactions in the flotation system Galena-Xanthate-Oxygen, Trans. IMM, B, 1972.

7) Hagihara H. Mono and Multilayer Adsorption of Aqueous Xanthate on Galena Surfaces. J. Phys. Chem., 56, 1952, 616.

8) Greenler R.G., An Infrared Investigation of Xanthate Adsorption by Lead Sulfide, J. Phys. Chem., 66, 1962, 879.

9) Leja J., Little L.H. and Poling G.W., Xanthate Adsorption Using Infra-Red Spectroscopy. 1-Oxidized and Sulphidized Copper Substrates. 2-Evaporated lead Sulphide, Galena and Metallic Lead Substrates., Trans. IMM, 72, 1963, 407.

10) Plante E.C. and Sutherland K.L., Effects of Oxidation of Sulphide Minerals on Their Flotation Properties., Trans. AIME, 183, 1949, 160.

11) Herd H.H. and Ure W., Surface Chemistry in the Flotation of Galena., J. Phys. Chem. 45, 1941, 93.

12) Taylor T.C. and Knoll A.F., Action of Alkali Xanthates on Galena., Trans. AIME, 112, 1934, 382.

13) Sparrow G., Pomianowski A. and Leja J., Soluble Copper Xanthate Complexes. Separation Science 12, 1977, 87.

14) Gaudin A.M. and Schuhmann R., The action of Potassium n-Amyl Xanthate on Chalcocite., J. Phys. Chem. 40, 1936, 257.

15) Heyes G.W. and Trahar W.J., Oxidation-Reduction Effects in the Flotation of Chalcocite and Cuprite., Int. J. of Min. Proc., 6, 1979, 229.

16) Kowal A. and Pomianowski A., Cyclic Voltammetry of Ethyl Kanthate on a Natural Copper Sulphide Electrode., Electroanal. Chem. and Interfac. Electrochem. 46, 1973, 411.

17) Richardson et al, Electrochemical Flotation of Sulfides: Chalcocite-Ethylxanthate Interaction. Int. J. of Min. Proc., 12, 1984, 87.

18) Walker G.W., Stout J.V. and Richardson P.E., Electrochemical Flotation of Sulfides: 1. Reactions of Chalcocite in Aqueous solutions. Int. J. Min. Proc. 12, 1984, 55.

19) Gaudin A.M. and Finkelstein N.P., Interaction in the System PbS-KEX-O_2., Nature, 201, 1965, 389.

20) Finkelstein N.P., Kinetic and Thermodynamic Aspects of the Interaction Between Potassium Ethyl Xanthate and Oxygen in Aqueous Systems., Trans IMM, 76 1967, C51.

21) Finkelstein N.P., Quantitive Aspects of the Role of Oxygen in the Interaction Between Xanthate and Galena., Separation Science 5, 1970, 227.

22) Gardener J.R. and Woods R., The Use of A Particulate Bed Electrode for the Electrochemical Investigation of Metal Sulfide Flotation., Aust. J. Chem., 26, 1973, 1635.

23) Healy T.W. and Moignard M.S.: A review of electrokinetic study of metal sulfides. Fuerstenau M.C. ed., Flotation, Gaudin A.M. memorial vol., AIME N.Y. 1976, p. 275.

24) Woods R.: Electrochemistry of sulphide flotation. ed. Flotation, Gaudin A.M. memorial vol., Fuerstenau M.C. AIME NY 1976.

25) Richardson P.E., Monst E.E. Surface stoichiometry of Galena in aqueous electrolytes and its effect on xanthate interactions. Fuerstenau M.C. ed., Flotation, A.M. Gaudin memorial vol., AIME N.Y. 1976.

26) Nowak P. et al. The applicability of Emf measurements to evaluation of thermodynamic properties of the Cu-S system., J. Electroanal. Chem. 171, 1984, pp. 335-358.

27) Nowak P., Krauss E., Pomianowski A.. The electrochemical characteristics of the galvanic corrosion of sulfide minerals in short-circuited model galvanic cells., Hydrometallurgy 12, 1984, pp. 95-110.

28) Barzyk W., Malysa K., and Pomianowski A.. The influence of surface oxidation of chalcocite on its floatability and ethylxanthate sorption., Int. J. Min. Proc., 8, 1981, pp. 17-28.

29) Chander S.: Oxidation reduction effects in depression of sulphide minerals. A review., Min. Met. Processing, 22, 1985, pp. 26-35.

30) Luttrell G.M., Joon R.M.. Surface studies of the collectorless flotation of chalcopyrite., Coll. Surf, 12, 1984, pp. 239-254.

31) Giesekke E.W.. A review of spectroscopic techniques applied to the study of interactions between minerals and reagents in flotation systems., Int. J. Miner. Process., 11, 1983, 19-56.

32) Kongolo M., Cases J.M., Burreau A. and Predali J.J.. Spectroscopic study of potassium amylxanthate adsorption on finely ground galena. M. Jones and R. Oblatt (Editors), Reagents in the Minerals Industry. IMM, Rome, 1984, pp. 79-87.

33) Walker G.W., Richardson P.E. and Buckley A.N.. Workshop on the flotation related surface chemistry of sulfide minerals., Int. J.M. Proc. 25, 1989, 155-158.

34) Little L.M., Poling G.W., Leja J.: IR spectra of xanthates compounds., Can. J. Chem., 39, 1989, 745-754.

35) Marabini A.M., Cozza C.. Determination of lead ethylxanthate on mineral surface by IR spectroscopy., Spectrochim. Acta, 388, 1983, p. 215.

36) Mielczarski J. and Leppinen J.: Infrared reflection-absorption spectroscopy study of adsorption of xanthates on copper., Surf. Sci, 187, 1987, pp. 526-538.

37) Mielczarski J., Nowak P., Strojek J.W. and Pomianoxski A.. Investigation of the products of ethylxanthate sorption on sulfides by IR-ATR spectroscopy, in J. Laskowski ed. Proc of the XIII Int. Min. Proc. Congress, Warszawa Part. A 1981, pp. 40-131, 1981.

38) Mielezarski J. In situ ATR-IR spectroscopic study of xanthate sorption an marcasite., Coll. Surf., 17, 1986, 235-248.

39) Leppinen J.O., Basilio C.I. and Yoon R.H.. FTIR study of thionocarbamate adsorption on sulfide minerals., Coll. Surf., 32, 1988, 113-125.

40) Leppinen J.O., Basilio C.I. and Yoon R.H. In situ FTIR study of ethylxanthate adsorption on sulfide minerals under conditions of controlled potential., Int. J. Min. Proc., 26, 1989, 259-274.

41) Johansson L.S., Juhanoja J., Laajalehto K., Suoninen E., Mielczarski J. XPS studies of xanthate adsorption on metals and sulfides., Surf. Interf. Anal., 9, pp. 1986, 501-505.

42) Termes S.C., Richardson P.E.. Application of FTIR spectroscopy for on situ studies of sphalerite with aqueous solutions of KETX and with diethyldixanthogen., Int. J. Min. Proc., 18, 1986, p. 167-178.

43) Mielczarski J., XPS study of ethyl xanthate adsorption on oxidized surface of cuprous sulphide., J. Colloid Interface Sci., 120, 1987, 201-209.

44) Laajalehto K., Johansson L.S., Mielczarski J., Anderson S. and Souninen E.. Electron spectroscopic studies of interaction of xanthate collectors with different substrates. In K.S.E. Forssberg (Editor), Proceedings of the XVI International Mineral Processing Congress, Stockholm, 1988, pp. 691-702.

45) Mielczarski J. and Minni E.. The adsorption of diethyldithiophasphate on cuprous sulphide., Surf. and Interf. Anal. 6, 1984, 221-226.

46) Mielczarski J. and Suoninen E.. XPS study of ethylxanthate adsorbed onto cuprous sulphide., Surf and Interf. Anal., 6, 1984, 221-226.

47) Mielczarski J. and Suoninen E.. XPS study of the oxidation of cuprous sulphide in aerated aqueous solutions., Coll. Surf, 33, 1988, p. 231-237.

48) Cécile J.S. , Bloise R., Barbéry G. Galena depression with chromate ions after flotation with xanthates. A kinetic and spectrometric study in Jones M. ed. Complex sulphide ores. IMM., London, 1980, pp. 159-170.

49) Predali K.C. et al. XPS study of xanthate adsaption on peprite mineral surfaces., J. Coll Interf. Sci, 103, 1985, n. 1, 1985.

50) Buckley A.M., Woods R. and Wauterlood H.J. An XPS investigation of the surface of natural sphalerites under flotation related conditions., Int. J. Min. Proc. 26, 1989, 29-49.

51) Pillai K.C., Young U.Y. and Bankris O.M.. XPS studies of xanthate adsorption auto galena surfaces., Appl. Surf. Sci., 16, 1983, pp. 322-344.

52) Page P.W. and Hazell L.B. X ray photoelectron spectroscopy (XPS) studies of potassium amylxanthate (KAX) adsorption on precipitated PbS related to galena flotation., Int. J. Min. Proc., 25, 1989, pp. 87-100.

53) Pillai K.C., Young V.Y. and Bockris Y.O.M. X-ray photoelectron spectroscopy studies of xanthate adsorption on pyrite mineral surfaces., J. Coll. Int. Sci, 103, 1985, pp. 145-153.

54) Predali J.J., Brion D., Harper J., Pelletier B.. Charactérisation par spectroscopie ESCA des états de surface de mineraux sulfurés fines de Pb-Zu-Cu-Fe lors de la flottation. Proc. of the XIII Int. Min. Proc. Congress, Warsaw part A, 1981, p. 55.

55) Mielczarski J., Suoninen E. Johansson L.S. and K. Laajalehto An XPS study of adsorption of methyl and amyl-xanthates an copper., Int. J. Min. Proc. 26, 1989, p. 181-191.

56) Bertil I. Palsson and K.S.E. Forssberg. Computer-assisted calculations of thermodynamic Equilibria in Sphalerite-xanthate system., Int. J. Min. Proc., 1989, 26, 223-258.

57) Bertil I. Palsson and Forssberg RSE. Computer-assisted calculations of thermodynamic Equilibria in the galena - ethylxantate system., Int. J. Min. Proc., 23, 1988, 93-121.

58) Partyka S., Arnaud M., Lindheimer M.. Adsorption of Ethyl-xanthate onto Galena at low surface coverages., Coll. Surf., 26, 1987, 141-153.

59) Maillot M., Cécile J.L., Bloise R.. Stability of ethylxanthate ion in neutral and weakly acidic media., Int. J. Min. Proc., 13, 1984, 193-210.

60) Shimoizaka J. et al. Depression of galena flotation by sulfite or chromate ion, in Fuerstenau M.C. ed., Flotation, Gaudin A.M. memorial vol., AIME, NY, 1976.

61) Rinelli G., Marabini A., Alesse A.. Depressing action of permanganate on pyrite and galena flotation. Jones M.J. ed. Complex sulphide ores. Rome IMM, London, 1981, pp. 199.

62) Popov S.S. et al.. Effect of the depressing agents FeSo4 and NaCN on the surface properties of Galena in the flotation system., Int. J. Min. Proc. 24, 1988, 111-123.

63) Eigillani D.A. and Fuerstenau M.C.. Mechanisams involved in cyanide depression of pyrite., Trans. AIME, 241, 1968.

64) Steiningen J.. The depression of sphalerite and pyrite by basic complexes of copper and sulphidryl flotation collectors., Trans. AIME, 1968, p. 34.

65) Finkelstein N.P. Allison S.A.. The chemistry of activation, deactivation and depression in the flotation of zinc sulphides: a review. Fuerstenau M.C. ed. Flotation, Gaudin A.M. memorial volume, AIME New York, 1976, pp. 414-457.

66) Ball B., Richard R.S.. The chemistry of pyrite activation and depression. Fuerstenau M.C. and Flotation, Gaudin M.A. memorial volume, AIME, New York, 1976.

67) Ghiani M., Satta F., Barbaro M. and Passariello B.. Flotation of sphalerite from pyrite by use of copper xanthate and sodium cyanide. Reagents in minerals industry, ed. Jones M. and Oblatt, IMM 1984.

68) Chander S. Inorganic depressants for sulfide minerals" Reagents in Mineral Technology, P. Somasndaran and B. Moudgil eds., Marcel Dekker, New York, pp. 428-468.

69) Gutzeit G. Chelate-forming organic compounds as flotation reagents., Trans-AIME 1946, v. 169, p. 272.

70) Martell A., Calvin M. Chemistry of the metal chelate compounds, Prentice Hall, 1952.

71) Usoni L., Rinelli G., Marabini A.M.. Chelating agents and fuel oil: A new way to flotation, AIME Centennial Annual Meeting, New York, February 26-March 3, 1971.

72) Marabini A.M.. Flottazione e flocculazione selettiva., Boll. Ass. Min. Subalpina XIX, n. 3-4, 1982.

73) Rinelli G., Marabini A.M.. Flotation of zinc and lead oxide-sulphide ores with chelating agents, 10th Int. Min. Proc. Congress, IMM, London, 1973.

74) Marabini A.M. New collector for cassiterite., Trans IMM, Sect. C. vol. 84, 1975, C 177.

75) Rinelli G., Marabini A.M., Alesse V. Flotation of cassiterite with salycilaldheide as a collector, Flotation, Gaudin A.M. Memorial volume Fuerstenau M.C.ed., vol. 1, 1976, p. 549.

76) Mangalam V. et al. Zeta potential and flotation studies of chalcopyrite fines with 8-hydroxyquinoline., Coll. Surf., 7, 1983, pp. 209-220.

77) Marabini A.M., Rinelli G. Flotation of cobalt minerals with chelating agents as collectors., ATB Metallurgie, XXII 1, 1982, p. 17-22.

78) Marabini A.M., Barbaro M., Ciriachi M. A calculation method for selection of

complexing collectors having selective acion an a cation., Trans IMM, Sec C, 92, 1983, CZO-C26.

79) Marabini A.M.. Study od adsorption of Salicylaldhiede on cassiterite., Trans. IMM, Sect. C., 87, 1978, C75.

80) Marabini A.M., Rinelli G.. Development of a specific reagent for rutile flotation., SME Trans., 274, 1983, p. 1822.

81) Marabini A.M., Alesse V., Barbaro M. New synthetic collectors for selective flotation of zinc and lead oxidized minerals., XVI Int. Min. Proc. Congress, Forssberg ed. Elsevier, Amsterdam Vol. 1988, pp. 1197-1208.

82) Marabini A.M., Barbaro M., Passariello B. Flotation of cerussite with a synthetic chelating collector., Int. J. Min. Proc., 25, 1989, p. 29.

83) Marabini A.M., Cases J. , Barbaro M., Chelating reagents as collectors and their adsorption mechanism, Challenges in Mineral Processing, Sastry K.V., ed AIME, Colorado, 1989.

84) Cozza C., Di Castro V., Polzonetti G., Marabini A.M.. An X-Ray Photoelectron Spectroscopy Study of the Interaction of Mercapto-Benzo-Thiazole with cerussite, in press on Int. J. of Min. Proc.

85) Bornengo G., Marabini A.M. and Alesse V., Italian Patent 22019 A/89, 1989.

86) Acherman P.K., Harris G.H., Klimpel R.R. and Aplan F.F., Evaluation of flotation collectors for copper sulfides and pyrite. III Effect of Xanthate chain length and branching., Int. J. Min. Proc., 21, 1987 4-156.P

87) Pradip: Application of chelating agents in mineral processing., Min. Met. Proc., 80; 1988.

88) Ackerman P.K., Harris G.H., Klimpel R. and Aplan F.F., Effect of alkyl substituents perfomance on thionocarbamates as copper sulphide and pyrite collectors. Reagents in mineral industry, ed. Jones M.J. and Oblatt, IMM, Lond, 1984.

89) Somasundaran P. and Nagaraj P.R., Chemistry and applications of chelating agents in flotation and flocculation, Reagents in mineral industry, ed. Jones M.J. and Oblatt R., IMM, London, 1984.

90) Ackerman P.K., Harris G.H., Klimpel R.R. and Aplan F.F., Evaluation of flotation collectors for copper sulfides and pyrite II. Non-sulphydryl collectors., Int. J. Min. Proc., 21, 1987, pp 129-140.

91) Klimpel R.R., Hansen R.D. and Fee B.S., New collector chemistries for sulfide mineral flotation; SME/AIME, Annual Meeting, Phoenix, Arizona, Jan 25-8, 1988.

92) Klimpel R.R., Hansen. Recent work on developing new sulfide mineral collectors based on chelation chemistry, SME/AIME Annual Meeting Las Vegas, Nevada, Febr. 27 - March 2, 1989.

93) Nagaraj D.R., Basilio C. and Yoon R.M.. The chemistry and structure-activity relationships for new sulfide collectors. 118th SME/AIME annual Meeting, Las Vegas, Nevada, Feb. 27 - March 2, 1989.

Mineralogy of and potential beneficiation process for the Molai complex sulphide orebody, Greece

M. Grossou-Valta
Department of Technology, Mineral Processing and Metallurgy, Institute of Geology and Mineral Exploration, Athens, Greece
K. Adam
Department of Research and Technology, Aegean Metallurgical Industries (METBA S.A.), Athens, Greece
D.C. Constantinides
Department of Economic Geology, Institute of Geology and Mineral Exploration, Athens, Greece
J.M. Prevosteau
Department of Analysis, Bureau de Recherches Géologiques et Minières, Orléans, France
E. Dimou
Department of Mineralogy and Petrology, Institute of Geology and Mineral Exploration, Athens, Greece

ABSTRACT

The Molai sulphide deposit situated in S.E.Peloponnese, Greece, was located after a systematic exploration programme which commenced in 1980. Geostatistical evaluation of the drilling campaign results indicated mineable ore reserves of 2.3 x 10^6 tons with an average estimated content of Zn: 9,0%, Pb: 1,8% and Ag: 50 g/t.

The geological and mineralogical features led to the conclusion that the Molai ore body belongs to the stratiform, syngenetic, volcanogenic massive sulphide deposits. Detailed mineralogical analysis of drill core and representative bulk ore samples indicated sphalerite as the main Zn bearing mineral, while smithsonite and monheimite were also observed. Galena and small quantities of anglesite and cerrusite were recorded as the Pb-bearing minerals. Silver is encountered in the tetrahedrite–tennantite mineral group mainly associated with sphalerite and in the Ag–Sb bearing minerals diaphorite, pyrargirite and stephanite observed as inclusions within galena.

The fine dissemination and association of base metal sulphides, combined with the presence of oxidised zinc and lead minerals adversely affect the beneficiation of the Molai sulphide deposit. The liberation size varies from 26μm for sphalerite to 12μm for galena and 2μm for anglesite. Prolonged grinding and high reagent consumption is required to improve zinc recoveries. Metal losses in slimes, early zinc activation in the galena circuit and silver distribution between the lead and zinc concentrate are recorded, as major problems encountered in the differential flotation of the Molai sulphide ore.

In this article mineralogical analysis data are employed to corroborate the prevailing theories on the Molai ore body genesis and formation as well as to assist in developing an economic beneficiation process for the treatment of this complex, finely disseminated sulphide mineralisation.

INTRODUCTION

The Molai polymetallic sulphide deposit is situated in the Lakonia county, S.E. Peloponnese (Fig.1). It was located after a systematic exploration programme carried out by the Institute of Geology and Mineral Exploration (IGME), which commenced in 1980 and included geological mapping in various scales (up to 1:1000), geochemical and geophysical surveys and an extended drilling campaign. The mineralisation was explored to a maximum distance of 1300m along strike and 200m across dip and was found to be developed along rather well defined zones and sub-zones. Geostatistical evaluation of the drilling campaign results from the surface (+165m) down to –50m level indicated mineable reserves of 2.3 x 10^6 tons with an average estimated grade of Zn: 9.0%, Pb: 1.8% and Ag: 50 g/t.[1] Above data will be reevaluated after the completion of ongoing research activities in the Molai area.[2,3] It is suggested that the sulphide ore body continues at depths below the levels already explored. Presently an exploratory gallery is being opened in the main ore body (Western zone) by Aegean Metallurgical Industries (METBA S.A), with the aim to investigate the optimal mining method and estimate the corresponding capital and operating costs.[2,3] During trial mining exploitation bulk ore material will be produced for subsequent beneficiation tests in laboratory and pilot plant scale.[4,5] Technical and economical data thus obtained will be employed in the final feasibility study for the potential exploitation of the Molai sulphide deposit.

GEOLOGICAL SETTING - SULPHIDE MINERALISATION

The geological environment of the deposit comprise an assemblage of clastic and volcanic rocks belonging to the Tyros beds. The base metal sulphides are hosted in the volcanic rocks consisting of tuffs, tuffites and other pyroclastics, as well as massive aphanitic and porphyritic lavas, which have an andesitic to basalt-andesitic composition and calc-alkaline character.

Intense alteration has affected both the volcanic rocks hosting the mineralisation and the adjoining rocks with an adverse impact on their mechanical properties.[2] The alteration is mainly represented by silicification, sericitisation and chloritisation. The Tyros beds have undergone regional metamorphism, which is placed on the boundary of Eocene to Oligocene. The metamorphic temperature is estimated to have reached 320-350°C with a pressure of about 2 Kb.[6]

Despite the existence of two fold systems in the ductile formations, the deformation in the more massive volcanics is manifested by the development of faults resulting in the displacement of the ore bodies and the variability in their thickness. Nevertheless, the latter as well as the ore grade variability is mainly attributed to the initial conditions prevailing during deposition.

The mineralisation is developed along several zones and subzones. The altered and mineralised zones exhibit a N-S trend and an eastern dip of 55° to 75°, Fig. 1. The main ore body of economic significance lies within the Western subzone whose reserves were given above. The average thickness of this subzone is 3.8m. Three other orebodies, namely : the Eastern zone, Western B-subzone and Western C-subzone of total reserves averaging 1.5×10^6 tons are considered as not mineable due to their limited thickness of 1.0-1.4m.[1]

The presence of different mineralised zones in the strata is attributed to the repetition of the ore genetic process and not to tectonic events. The overall characteristics of the above zones including the stratiform development, the type of alteration and the ore mineralogy are very similar.[6]

The spatial distribution of Zn, Ag, Cd, Fe and Ba obtained from the geostatistical analysis of drill-core samples assays (programme BECES[7]) is irregular and expressed by rich local concentrations. It has to be noted that the correlation coefficient between Zn and Ag (0.71) is higher than the corresponding value between Pb and Ag (0.41). There is a tendency for increased Ba concentrations towards the South where the ore body is observed to transcate.[8]

Chemical analysis of drill core samples indicated that across the ore body the oxidation degree of the mineralisation varied significantly. Average zinc oxidation in the Eastern and Western zone was estimated to 30-40% and 5-15%, respectively.

Two main types of mineralisation can be distinguished:
1) the disseminated and semi-massive fine-grained sulphides with both fragmental (lithic and sulphide fragments randomly distributed in the matrix) and bedded textural varieties.
2) the coarse-grained massive ore, which is recrystallised and remobilised. The latter type is minor and occurs occasionally. The stratiform development of the mineralisation is evident both in macroscopic and microscopic ore textures. Extensive brecciation is another main characteristic of the ore.

ORE GENESIS

The base metal mineralisation of the Molai deposit is closely associated with the permotriassic volcanic activity. The age of the ore was determined by seven lead isotopic analyses organised by IGME and conducted by N.Gale at Oxford University. The mean value was plotted on $^{206}Pb/^{204}Pb$ versus $^{207}Pb/^{204}Pb$ diagram shown in Fig.2 indicating an average age of 240 m.y. in agreement with that deduced from paleontological evidences on metasediment rocks.[10] These data support the theory for the syngenetic origin of the ore.

Furthermore, it is suggested that during the Molai ore genesis, volcanosedimentary conditions of deposition prevailed as evidenced by the following observations:
i) The ore consists of bedded sulphides that commonly display sedimentary features such as graded bedding or rhythmic layering (Fig. 3(a).[6]
ii) Soft sediment deformation of the semi-consolidated material resulted in slump folds and intraformational faults.
iii) The presence and form of imbricated mineralised breccias within the semi- massive fine grained ore, indicates that the above breccias have been transported from their original depositional sites, commonly lying within a short distance, Fig. 3(b).
iv) The existence of framboidal and colloidal textures within pyrite and sphalerite minerals as well as the presence of linear banded pyrite inclusions within sphalerite.[4,5]

Although further study is required for the final classification of the Molai ore body into a certain group of deposits, the features already described indicate that many similarities exist between the examined ore body and the deposits of the Kuroko type.[11]

ORE MINERALOGY

A detailed mineralogical stydy was conducted on representative drill-core and bulk ore samples originating from the main part of the ore body (Western zone). Microscopic studies indicated that the major sulphide minerals are: sphalerite (6-25 wt%), pyrite (3-7 wt%) and galena (1-3 wt%).[4]

The minor and rare metallic minerals observed, as well as the supergene and gangue minerals indentified are listed in Table I. Microanalysis results for the characteristic metallic minerals are presented in Table II.

Zinc is mainly associated with sphalerite, however an amount of approximately 5-15% is distributed within the supergene minerals, smithsonite-monheimite. Galena is reported as the major lead-bearing mineral, while small quantities of anglesite and cerrusite were also observed in the samples examined.

The ore minerals mainly occur as polycrystallic aggregates and rarely as unmixed concentrations or independent crystals, Fig. 4(a) & 4(b).

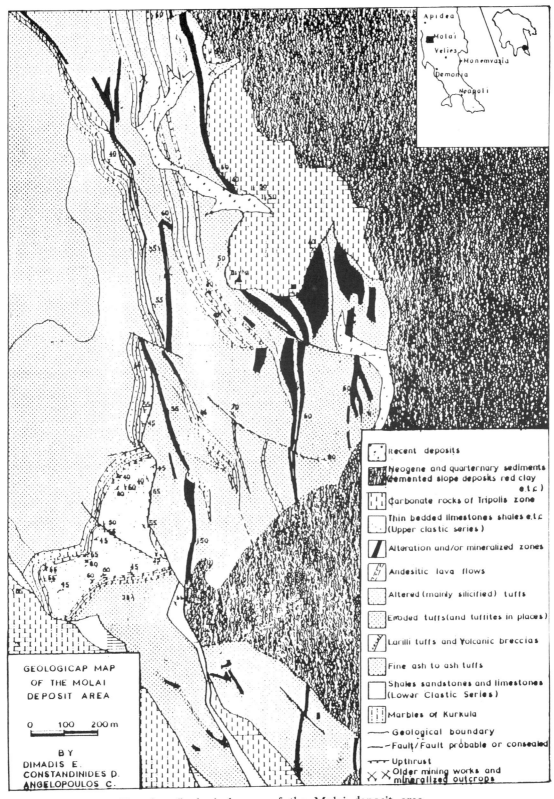

Within the figure, legend and labels:

Apidea
Molai
Velies
Monemvasia
Demonia
Neapoli

GEOLOGICAP MAP
OF THE MOLAI
DEPOSIT AREA

0   100   200 m

BY
DIMADIS E.
CONSTANDINIDES D.
ANGELOPOULOS C.

Recent deposits

Neogene and quarternary sediments
(Cemented slope deposits red clay
e.t.c)

Carbonate rocks of Tripolis zone

Thin bedded limestones shales e.t.c
(Upper clastic series)

Alteration and/or mineralized zones

Andesitic lava flows

Altered (mainly silicified) tuffs

Eroded tuffs(and tuffites in places)

Larilli tuffs and volcanic breccias

Fine ash to ash tuffs

Shales sandstones and limestones
(Lower Clastic Series)

Marbles of Kurkula

Geological boundary

Fault/Fault probable or consealed

Upthrust

Older mining works and
mineralized outcrops

Fig. 1 : Geological map of the Molai deposit area

121

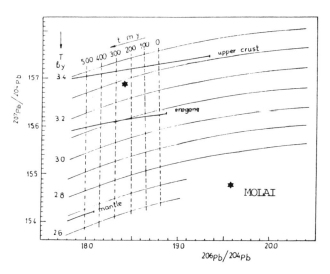

Fig. 2 : The mean value of the ratios $^{206}Pb/^{204}Pb$, $^{207}Pb/^{204}Pb$ plotted on the Amov diagram[9]

(a)   **(b)**

Fig. 3 : Macroscopic observations of the Molai sulphide ore:
  (a) Stratiform mineralisation of sphalerite (light grey) and tuff.
  (b) Massive ore, sphalerite (light grey) with imbricated partly
      mineralised breccias.
  Drill -core sample from the Western subzone. Scale: 1cm-1.5cm

Minerals of Molai ore body

A. Hypogene sulfides-sulfosalts-oxides minerals		Main	Minor	Rare
Sphalerite	ZnS	*		
Pyrite	FeS_2	*		
Galena	PbS	*		
Tetrahedrite-Tennantite	$Cu_{12}(As-Sb)_4S_{13}$		*	
Magnetite	Fe_3O_4		*	
Hematite	Fe_2O_3		*	
Geokronite	Pb_5SbAsS_8		*	
Freibergite	$(CuAgFe)_{12}Sb_4S_{13}$		*	
Chalcopyrite	$CuFeS_2$		*	
Marcasite	FeS_2		*	
Pyrrhotite	FeS		*	
Bornite	Cu_5FeS_4		*	
Enargite	Cu_3AsS_4			*
Famatinite	Cu_3SbS_4			*
Arsenopyrite	$FeAsS$			*
Boulangerite	$Pb_5Sb_4S_{11}$			*
Seligmannite (?)	$CuPbAsS_3(+Sb)$			*
Diaphorite	$Ag_3Pb_2Sb_3S_8$			*
Stephanite	Ag_5SbS_4			*
Pyrargirite	Ag_3SbS_3			*
Lazarevicite	$Cu_3(As,V)S_4$			*
				*

B. Supergene minerals				
Cerussite , Anglesite	$PbCO_3$, $PbSO_4$		*	
Smithsonite(monheimite)	$ZnCO_3,(Fe,Zn,Mn,Ca)CO_3$		*	
Geothite	$a-Fe_2O_3.H_2O(FeHO_2)$		*	
Chalcosite	Cu_2S		*	
Covellite	CuS		*	

C. Gangue minerals				
Quartz	SiO_2	*		
Calcite	$CaCO_3$	*		
Dolomite	$Ca(Mg,Fe)(CO_3)_2$		*	
Siderite	$FeCO_3$		*	
Kaolinite	$Al_2O_3.2SiO_2.2H_2O$		*	
Illite	$KAl_2(OH)_2AlSi_3(O.OH)_{10}$		*	
Chlorite	$Mg_5(Al,Fe)(OH)_8(Al,Si)_4O_{14}$		*	
Barite	$BaSO_4$		*	

TABLE II
Microanalysis of Molai metallic minerals

%	Sphalerite		Galena	Pyrite	Tetrahedrite Tennantite		Freibergite	Diaphorite
	ZnS		PbS	FeS$_2$	Cu$_{12}$(As-Sb)$_4$S$_{13}$		(CuAgFe)$_{12}$Sb$_4$S$_{13}$	Ag$_3$Pb$_2$Sb$_3$S$_8$
Ag	–	–	0.02	–	0.4	–	23.9	23.8
Sb	–	–	–	–	29.3	11.2	26.0	27.4
Zn	62.9	65.7	0.1	–	2.2	0.04	1.1	0.02
Cu	0.1	–	–	–	37.0	44.7	24.3	0.2
Fe	3.1	0.3	–	46.8	4.2	0.4	1.1	–
S	34.1	33.7	14.1	53.2	25.0	32.2	24.0	19.4
Pb	0.2	0.4	87.3	–	0.2	0.1	–	29.2
As	–	–	–	–	0.9	10.5	0.2	0.01
Cd	0.5	–	–	–	–	0.3	–	–
Total	100.9	100.1	101.5	100.0	99.5	99.1	100.6	100.1

Sphalerite occurs in two types. The first one is Fe free and consists of grains and aggregates ranging from 30 to 300μm in size. The second one, less frequently observed, contains up to 3% Fe (Table II) and consists of finer grains ranging from 20 to 200μm in size. As shown in Fig.4(c), sphalerite grains contain numerous and various inclusions - exsolutions covering their surface, such as: pyrite crystals from a few μm to 100μm in size, pyrrhotite, arsenopyrite, hematite, chalcopyrite, galena, anglesite tetrahedrite - freibergite - tennantite. In some cases these inclusions exhibit a linear-banded form into sphalerite, Fig. 4(d), implying that during the ore genesis sedimentary conditions of deposition prevailed. Also, in some sphalerite crystals the rare phenomenon of Fe exsolution in the form of magnetite can be observed, as illustrated in Fig. 4(e). It is anticipated that inclusions of iron bearing minerals within sphalerite grains would adversely affect the kinetics of zinc separation by flotation.[12,13]

Furthermore, on the boundaries of sphalerite grains galena was observed in a form that was difficult to liberate and would potentially result in the early activation of zinc in the lead circuit.[14] It has to be noted that sphalerite mineral grains contain approximately 0,5% Cd.[4,5,14]

Galena crystals are smaller in size ranging from 5 to 100μm, Fig.4(f). They contain very fine inclusions (<5μm) of anglesite, tetrahedrite-tennantite, boulangerite, and silver minerals. When galena is forming aggregates, pyrite is often included in a fine grained intergrowth texture. Microanalysis of galena grains indicated only traces of Ag.

Pyrite occurs as grains and aggregates ranging from 30 to 200μm in size, and as inclusions of a few microns within sphalerite and galena. In certain mineral specimens pyrite was also observed in the form of framboids.

Tetrahedrite, freibergite and tennantite are enclosed or intergrown mostly with sphalerite and to a lesser extend with galena.

The minor and rare minerals such as geokronite, seligmanite, bornite, famatinite, enargite and lazarevicite do not form intergrowth textures with the other metallic minerals but they occur as individual crystals.

The supergene minerals such as smithsonite, monheimite and anglesite are often observed in close assosiation with the base metal sulphides. Zinc carbonates also occur as fine inclusions (4-5μm) within the quartz and silicates.

Silver was located within the following mineralogical phases: in tetrahedrite-tennantite series, mainly in freibergite and in the three Ag-Sb bearing minerals, diaphorite, pyrargirite and stephanite. Freibergite occurs in the form of inclusions or intergrowths mainly associated with sphalerite while the other three Ag-Sb bearing minerals are exclusively found as microinclusions within galena.

The main characteristics of the Molai ore are the fine-grained texture, the intergrowths often observed between sphalerite – galena, galena – pyrite and the numerous inclusions present particularly in sphalerite grains. The above mineralogical features are expected to adversely affect the flotation response of lead and zinc minerals with conventional flotation reagents.

LIBERATION STUDY

The grain size of the metallic minerals significantly affect the selection of the subsequest benefication scheme. It is known that for selective flotation as mineral grains become finer, liberation and selectivity becomes less effective.

A detailed liberation study was conducted on a composite bulk sample from the Molai mine with the aim to identify the optimum grinding size for the differential and bulk flotation of lead and zinc minerals. The average chemical analysis of the examined sample is shown in Table III. Although the base metal assays are lower than the previously examined drill-core samples, the average mineralogical analysis of the bulk sample is reported to be the same.[4,5]

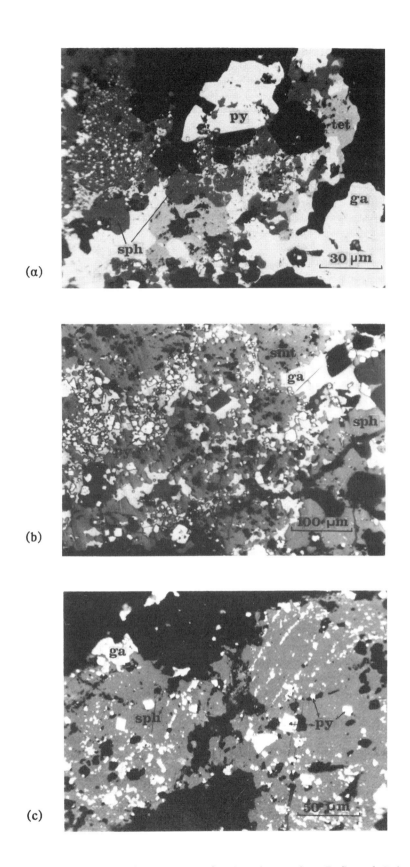

Fig. 4: Microscopic observations of mineral samples. Reflected light, // Nicols. (a) Polycrystallic aggregates: sphalerite(sph), galena(ga), pyrite(py), tetrahedrite(tet).(b) Aggregates and independent crystals of sphalerite, galena, pyrite into a smithsonite matrix(smt). (c) Pyrite, galena inclusions within sphalerite crystal.

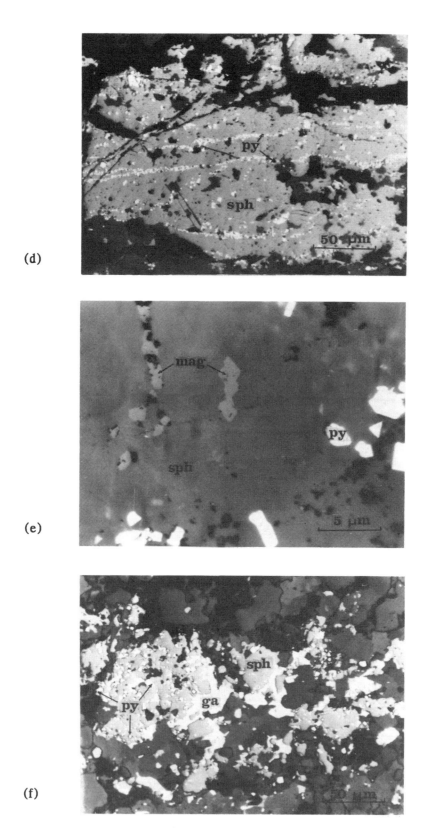

(d)

(e)

(f)

Fig. 4: Microscopic observations of mineral samples. Reflected light, // Nicols. (d) Linear banded pyrite inclusions within sphalerite crystal.
(e) Magnetite exsolution(mag) within sphalerite.
Oil immersion. (f) Fine grained galena–sphalerite intergrowths.

TABLE III
Chemical analyses of the Molai composite bulk sample

Element	%
Zn	6.3
Pb	1.0
Fe	4.7
S	5.9
$S(SO_4)$	0.6
Ag (ppm)	31.3
Cd (ppm)	332.0
SiO_2	52.0
Al_2O_3	12.0
K_2O	4.3
Na_2O	0.7
CaO	2.4
MgO	1.7
CO_2	3.5

The study was implemented with an Image Analyser Cambridge Quantimet 900, coupled with a Scanning Electron Microscope (S.E.M), Cambridge Stereoscan 250. The procedure followed is presented in detail elsewhere.[5]

It has to be noted that the size distribution obtained does not itself represent the actual mineral grain size, but the smallest dimension to which the ore must be ground in order to liberate the valuable minerals from the gangue material for bulk flotation, or to extract monomineral grains for differential flotation. For the purposes of this study, galena (PbS) and anglesite (PbSO$_4$) were defined as the lead bearing minerals, while sphalerite (ZnS) and zinc carbonate series of smithsonite ($ZnCO_3$) - monheimite (Fe,Zn,Mn,Ca)CO$_3$ were considered to be the zinc carrier minerals. Above phases are also presented in the backscattered electron images seen in Fig.5(a) and 5(b).

From analytical results in Table IV it is deduced that the Molai ore is classified in the fine grained minerals group.[12] To liberate the majority of sphalerite and galena as monomineral grains, the required grinding size is 26μm and 12μm respectively.

On the other hand for bulk flotation the liberation size of sphalerite estimated to 70μm, is significantly coarser than the corresponding values for zinc carbonates, galena and anglesite.

The majority of zinc carbonates is liberated at 20μm, however at this mesh of grind a fraction is still bound in the form of inclusions within silicates.[5] Galena and anglesite were the most difficult to liberate, the reason being their fine dissemination not only within zinc bearing minerals but within pyrite, quartz and other silicates.

Similar conclusions were reached after heavy liquid tests on drill-core samples from the same part of the ore body.[15] Tests conducted on particle size fractions ranging from −37+34μm to −13+10μm indicated that lead losses in the light fraction, varying from 5.0% to 21.0% were systematically higher than zinc ranging between 3.0% and 15.0%. For both zinc and lead significant portion of the metal values reporting in the tails were in the oxidised form.[15]

Thus, if bulk flotation is applied to the treatment of the Molai ore a grinding size of 60μm would be adequate, in order to liberate sphalerite and a considerable fraction of lead minerals from pyrite and gangue material. In that

TABLE IV
Liberation size of major Pb-Zn minerals of the Molai ore

Liberation size	Pb-minerals and sphalerite	Sphalerite	Oxidised Zn minerals	Galena and Anglesite
Differential Flotation	–	26*	–	12/2
Bulk Flotation	55	70**	20	7

* monomineral
** with metallic inclusions

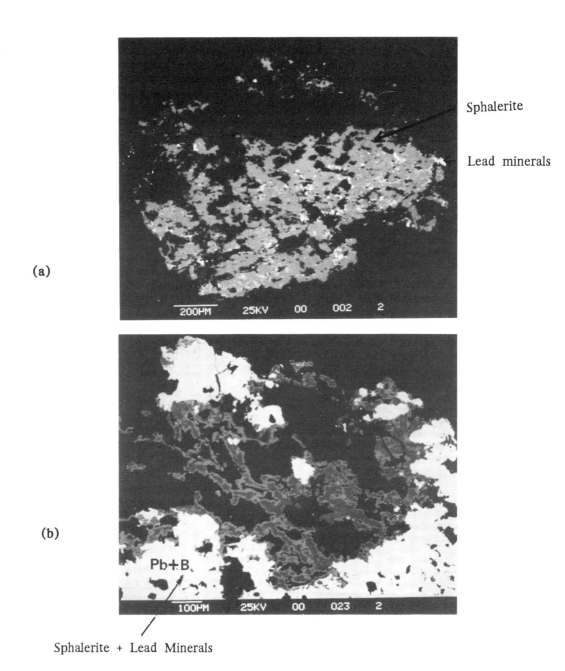

Sphalerite

Lead minerals

(a)

(b)

Pb+B.

Sphalerite + Lead Minerals

Fig. 5: Backscattered electron images with S.E.M.
  (a) Inclusions of lead bearing minerals within sphalerite matrix
  (b) Zonation within zinc carbonate minerals: Smithsonite (light grey) on
  monheimite grain boundaries (dark grey).

case, a percentage of valuable metals is not anticipated to report in the tails.

Above liberation study has also shown that due to the fine intergrowths between lead and zinc minerals and the numerous inclusions observed within sphalerite finer grinding of approximately 10μm is required for the separation of lead minerals. The effect of liberation on the overall performance of the subsequent beneficiation process is discussed in detail in the following section.

BENEFICIATION PROCESS

The beneficiation study of the Molai ore was initially conducted on characteristic drill-core samples originating from different parts of the ore body.[14,15] Above data formed the basis for an extended research programme, being currently in progress, with the aim to develop an economic process route for the treatment of the Molai sulphide minerals.[4,5] Testwork results presented in this section refer to composite drill-core samples from the main ore body (Western zone).[15] The average chemical analysis of the examined sample is given in Table V.

Due to the mineralogical features already decribed, separation by flotation was considered as the most appropriate method for the beneficiation of the Molai ore. Preconcentration for the upgrading of the flotation feed material was also examined. Laboratory and Pilot Plant testwork have indicated that separation with heavy media cyclone may be applied in order to scalp 10-25% of run-of-mine ore (r.o.m) with Pb and Zn losses in the tailing ranging from 2-10% to 1-5% respectively.[15] Thus, preconcentration has to be considered as a process option in view of the selected mining method and subsequent dilution of the r.o.m ore.

Differential Flotation

To investigate the sulphide mineral response under various flotation conditions, batch laboratory tests were initially conducted in order to produce separate lead and zinc concentrates. During the testwork the effect of process variables such as grinding time and size analysis, reagents combination and dosage, pH, conditioning and flotation time was evaluated.[15]

Flowsheet derived from the above tests is seen in Fig. 6. Fine grinding, 80% passing 40μm, was applied to liberate galena and sphalerite. Increased reagent consumption was observed for the depression and subsequent activation of sphalerite, as compared to published data on similar ores.[16] Metallurgical results of batch tests shown in Table V indicate that overall Zn and Pb recovery was of the order of 80% and 78% respectively. Based on laboratory test results marketable lead and zinc products can be recovered with subsequent cleaning stages.

The finely disseminated intergrowth of galena, sphalerite and pyrite combined with the presence of oxidised minerals were the main reasons for reduced selectivity of separation and low metal recoveries. As already described in the previous sections at this fine mesh of grind galena is not fully liberated. More than ten per cent of lead reports in the sphalerite concentrate in the form of locked particles (middlings). Fine intergrowth textures also resulted in valuable metal losses in the tails. High zinc assay in the galena concentrate was mainly attributed to early sphalerite activaction caused by the galena and anglesite slimes covering its surface. Mineralogical observation of the galena cleaner concentrate further support the above.[15]

The presence of oxidised minerals such as smithsonite and anglesite adversely affected the overall flotation recoveries. Despite the strong sulphidisation applied, more than 50% of the contained oxidised lead and 70% of oxidised zinc report in the tails, Table V.

Silver was equally distributed among the two rougher concentrates due to the association of Ag bearing minerals with both galena and sphalerite.

To further illustrate the above, Ag distribution versus the Pb and Zn recoveries in the galena and sphalerite cleaner concentrates is presented in Fig.7. The silver recovered in the zinc concentrate does not linearly correspond with the Pb contained in this product but follows zinc distribution as evidenced by mineralogical analysis, indicating the presence of silver bearing minerals such as freibergite within sphalerite grains.

TABLE V

Metallurgical balance of differential flotation tests on drill-core samples from the Molai ore

Product	Assay %					Recovery %				
	Pb	Pb_{ox}	Zn	Zn_{ox}	Ag ppm	Pb	Pb_{ox}	Zn	Zn_{ox}	Ag
PbS cleaner Concentrate	41.9	1.2	19.7	0.5	515	60	11	4	2	27
PbS cleaner Tails	8.1	1.1	26.2	0.8	188	18	16	8	4	15
ZnS cleaner Concentrate	1.0	0.2	59.3*	0.5	123	9	13	74	12	39
ZnS cleaner Tails	1.0	0.6	16.6	1.1	57	3	10	7	8	6
Tails	0.2	0.2	1.1	0.7	8	10	50	7	74	13
Head	1.7	0.3	11.6	0.7	46	100	100	100	100	100

*Cd: 0,35%

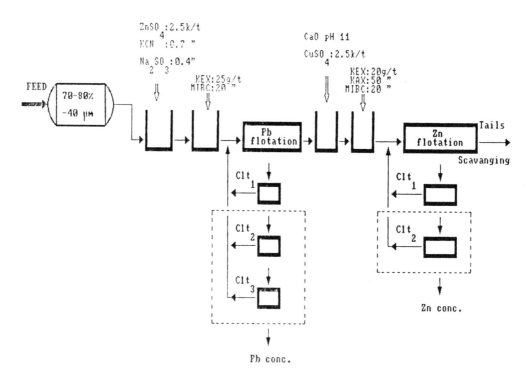

Fig. 6 : Flowsheet for differential flotation of Molai ore.

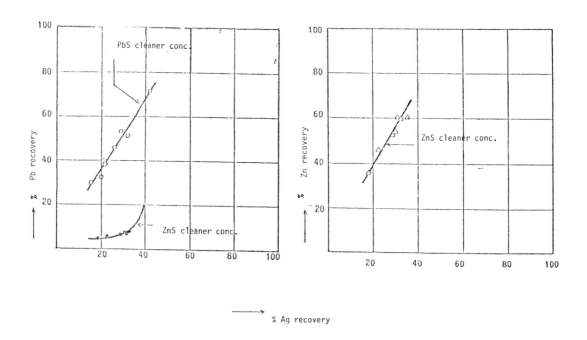

Fig. 7: Silver versus lead and zinc distribution in the galena and sphalerite cleaner concentrates.

Bulk Flotation

To mitigate the adverse effect of fine grinding on the overall flotation performance, an alternative process scheme was also examined in order to evaluate the recoveries attained when a bulk Pb-Zn concentrate is produced from the Molai ore. With the flow-sheet applied, Fig. 8, significant grinding energy and reagent consumption savings were observed.

From the corresponding metallurgical results shown in Table VI, it is deduced that with bulk flotation higher overall base metal recoveries were achieved. Due to the high Zn grade of the bulk concentrate, this product could be potentialy considered as a marketable zinc concentrate.[17]

To improve the grade of above concentrate various process schemes might be employed including regrinding followed by subsequent cleaning stages to separate pyrite and galena.[14] The optimum beneficiation scheme has to be selected in view of the subsequent metallurgical processing options to obtain the maximum economic benefit from the ore resource. As already stated an extended research project is currently in progress for the development of the Molai ore process route.[4,5] Technical and economical data thus obtained will be incorporated in the final feasibility study for the potential exploitation of the Molai sulphide deposit.

TABLE VI

Metallurgical balance of bulk flotation tests on drill-core samples from the Molai ore

Product	Assay %					Recovery %				
	Pb	Pb_{ox}	Zn	Zn_{ox}	Ag ppm	Pb	Pb_{ox}	Zn	Zn_{ox}	Ag
Bulk cleaner* Concentrate	7.1	0.5	48.0	0.1	180	76	32	75	6	70
Cleaner Tails	1.9	0.5	18.1	0.4	70	16	29	21	14	21
Tails	0.2	0.1	0.7	0.4	6	8	39	4	80	9
Head	1.7	0.3	11.6	0.7	46	100	100	100	100	100

*Cd: 0,3%

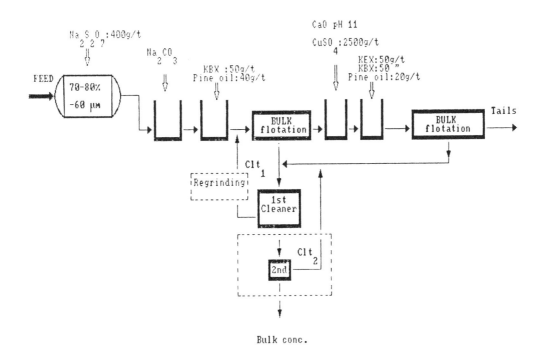

Fig. 8: Flowsheet for bulk flotation of Molai ore.

131

CONCLUSIONS

1. The Molai polymetallic sulphide deposit is situated in S.E.Peloponnese, Greece. To date evaluation of drilling campaign results indicated mineable ore reserves of 2.3×10^6 tons with an average estimated grade of Zn: 9.0%, Pb: 1.8% and Ag: 50 g/t. Presently an exploratory gallery is excavated in the main ore body to investigate the optimal mining method.

2. The Molai deposit belongs to the volcanic associated massive sulphides. The host rocks are mainly volcanics, such as tuffs, tuffites, massive arphanitic and porphyritic lavas. The ore predominantly comprise stratiform accumulation of base metal sulphide minerals which have been formed in a submarine volcanosedimentary environment by precipitation of hydrothermal fluids. Although the final classification of this deposit into a specific group is a matter of further investigation, the Molai ore deposit presents many similarities to the Kuroko type deposits.[11]

3. The genetic conditions have affected the mineralogical characteristics of the ore (fine grained textures, complex intergrowths between ore minerals, numerous inclusions particularly in sphalerite), the geometry of the ore bodies (thickness and average grade variation) and the rock mechanical properties. These features would subsequently affect the mining and mineral processing methods to be employed for the potential development of the Molai deposit.

4. Sphalerite is the main zinc bearing mineral in the Molai ore, although depending on the oxidation degree of the sulphide mineralisation, a fraction of contained zinc is distributed within the supergene minerals, smithsonite - monheimite. The occurrence of these two minerals would result in lower overall recoveries due to their poor response to conventional flotation reagents. Galena is the main Pb bearing mineral. Anglesite and cerrusite are also observed occasionally, on the boundaries of galena crystals. Silver is associated with the minerals of the tetrahedrite-tennantite group, mainly with freibergite. Furthermore, part of silver is also distributed within the Ag bearing minerals i.e. stephanite, diaphorite, pyrargyrite, exclusively reported as microinclusions in galena crystals.

5. As a result of the close association between lead and zinc minerals and the numerous microinclusions observed, the liberation sizes determined ranged from $28\mu m$ for sphalerite to $12\mu m$ for galena and $2\mu m$ for anglesite. However valuable minerals would be liberated from gangue material at coarser grinding sizes of approximately $60\mu m$.

6. For the preliminary beneficiation study of the Molai sulphide ore, laboratory differential and bulk flotation tests were conducted on composite drill core samples. Testwork results indicated that with differential flotation, after fine grinding and increased reagents consumption, marketable lead and zinc concentrates could be produced at recoveries lower than 80%. High zinc assay in the galena concentrate was mainly attributed to early sphalerite activation by galena and anglesite slimes. With bulk flotation lead, zinc and silver recoveries of the order of 90% were achieved with significant grinding energy and reagents savings. Due to the high zinc grade of the bulk concentrate this product could be potentially considered as a marketable sphalerite concentrate.

7. The optimum beneficiation scheme for the Molai sulphide ore has to be selected in view of the subsequent metallurgical processing options with the aim to maximise the economic benefit from the ore resource. An extended research programme is being currently in progress for the development of the above process route.

ACKNOWLEDGMENTS

The authors would like to acknowledge the financial support of the European Community, Contract No. MA1M-0051-C(AM), for part of the work presented in this article. They would also like to thank Mrs. R. Vasmoulakis for typing this manuscript.

REFERENCES

1. IGME: "Concise report on the Prefeasibility Study of the Molai massive sulphide deposit in Laconia". Institute of Geological and Mineralogical Exploration, Athens, 1985.

2. METBA: "Optimisation of the exploitation of thin vein polymetallic sulphide deposits through mathematical modelling and rock mechanics, an application to the Molai mine" CEC contract No. MA1M-0025-C(TT) 2nd Progress Report, Athens, December 1989.

3. METBA: "Molai Base metal massive sulphides" CEC Project GRE 02/09/002, 2nd Progress Report, Athens, April 1990.

4. IGME, BRGM, ITMI: "A process design study for a finely disseminated partially oxidized complex sulphide ore from the Molai area Southern Greece" CEC Contract No. MA1M-0025-C(TT). 2nd Progress Report, Athens, May 1989.

5. IGME, BRGM, IMM: "A process design study for a finely disseminated partially oxidized complex sulphide ore from the Molai area Southern Greece" CEC Contract No. MA1M-0025-C(TT). 3rd Progress Report, Athens, November 1989.

6. Angelopoulos C. and Constantinides, D.C.: "The Molai Zn-Pb-Ag deposit, Lakonia, S.E.Peloponnese", Bull Geol.Soc. Athens, Greece, Vol.XX/2, 1988, p.305-320.

7. Christakos, G.: "A computer program for the analysis and estimation of spatial functions", Intern. Report, Civil Eng.Dpt, MIT< Boston, Mass. 1980.

8. Angelopoulos, C. & Constantides, D.C: "The Molai zinc-silver-lead deposit, Lakonia/Greece", Intern. South Eur. Symp. on Expl. Geoch., Field trips guide book, Athens 1986, pp.61-67.

9. Amov, G.B: "Evolution of uranogenic and thorogenic lead. A dynamic model of continuous isotopic evolution". Earth and Planet Sci. Letters 65, Elsevier, Amsterdam, 1983, pp 61-74.

10. Brauer, R.: "Das Praneugen im Raum Molai Talanta/S.D.-Lakonien (Peloponnes, Griechenland)" Inang.- Dissert. zur Erlangung des Doctorgrades. Frankurt am, 1982.

11. Ohmoto, H and Skinner, J.P.: "The Kuroko and Related Volcanogenic Massive Sulphide deposits". Econ. Geol. Monogr. 5, 1983.

12. Norrgran, D.A. and Armstrong, R.O.: "Developing a selective flotation concentration technique for polymetallic ores" in Complex Sulphides, eds. A.D.Zunkel, R.S.Boorman,

A.E.Morris, R.J.Wesely, TMS, 1985, pp. 37-54.

13. Finkelstein, N.P. Allison, S.A.: "The Chemistry of Activation, Deactivation and Depression in the Flotation of Zinc Sulphide. A review". In Flotation, ed. M.G.Furestenau, Vol. 1, New York, 1976, pp. 414-457.

14. Ingerttila, K. Kaukkanen J., Hintikka, V.: "On Mineralogy and Flotation test of Molai ore", Technical Research Center of Finland, 1986.

15. Grossou-Valta, M. et al.: "Discussion on the possible treatment of Molai mixed sulphide ore after experimental investigation of typical samples". Metallurgical Research Reports No.37, Institute of Geology and Mineral Exploration, Athens, 1984.

16. Bulatovic, S.M., Wyslouzil, D.M.: "Selection of reagent scheme to treat massive sulphide ores" in Complex Sulphides ed. D.Zunkel, R.S. Boorman, A.E.Morries, R.J.Wesely, TMS. 1985, pp. 101-137.

17. Cases, J.M.: "Finely disseminated complex sulphide ores" in Complex Sulphide Ores, ed. M.J.Jones, IMM Conference Proceedings 1980, pp 234-247.

133

Studies of mineral liberation performance in sulphide comminution circuits

D.M. Weedon
T.J. Napier-Munn
Julius Kruttschnitt Mineral Research Centre, University of Queensland, Indooroopilly, Queensland, Australia
C.L. Evans
Mount Isa Mines Ltd., Mount Isa, Queensland, Australia (formerly, Julius Kruttschnitt Mineral Research Centre)

SYNOPSIS

This paper describes studies of two relatively coarse-grained sulphide ore grinding circuits (one copper and one lead/zinc), carried out to quantify the behaviour of the liberated minerals. The work formed part of a larger study whose objective is the modelling of mineral liberation for comminution circuit simulation.

The circuits were sampled and mass balances calculated. The mineralogy of the solids in each process stream was quantitatively determined using QEM*SEM, from which mass flows of liberated mineral were calculated. Samples of circuit feed were also broken in a laboratory pendulum apparatus, and the degree of liberation measured using QEM*SEM.

The plant results confirmed the strong effects of classification on liberation performance, and the differences in the behaviour of different classifier types. The effects are usually negative, as the denser liberated values tend to concentrate in the classifier underflow product, resulting in large circulating loads of liberated mineral which are reground unnecessarily, whereas the liberated gangue is preferentially selected for the downstream separation step (flotation in the case considered here). The data also demonstrated that the net liberation in secondary grinding can be quite small, and that the extent of liberation in products from the coarse (primary) grinding stage can be substantial.

The evidence in one case suggested that the proportion of liberation in a particular size interval is independent of comminution history, and that the propensity of a mineral to be liberated can be assessed using single particle breakage tests and QEM*SEM analysis.

These observations imply the need to consider aspects such as coarse flotation early in the grinding process, positive sizing devices to close the grinding circuit (eg screens rather than hydraulic classifiers), and whether conventional multiple grinding stages are always justified in practice.

INTRODUCTION

The principal purpose of any grinding circuit in a sulphide mineral processing plant is to break the ore to a size sufficiently small to release or liberate the value and gangue mineral grains, so as to permit the valuable component(s) to be recovered by downstream concentration processes, usually flotation. Penalties exist for undergrinding (inadequate liberation) and overgrinding (poor flotation response and excessive power costs). There is therefore considerable incentive to design or tune the grinding circuit to provide adequate liberation without overgrinding. Such optimisation is difficult in practice because:

1. The assessment of the extent of liberation in different parts of the circuit is time-consuming and expensive (to the extent that it is not a routine function in most concentrators).

2. The modelling of liberation in comminution is not yet sufficiently advanced to permit optimisation by simulation, although some progress has been made in recent years [1, 2].

3. The distribution of liberated species around the circuit, and the associated circulating loads, are strongly influenced by the characteristics of the comminution device itself, its associated classifier, and the properties of the different mineral phases. This introduces complex interactions and renders an analysis of liberation performance difficult.

The Julius Kruttschnitt Mineral Research Centre is currently investigating ways of incorporating liberation information into established mathematical models of comminution processes, to permit more comprehensive design and optimisation of grinding circuits. In order to provide a base level of data for this programme, and to advance the understanding of the liberation process in conventional sulphide concentrators, the grinding circuits in two well known such concentrators in Australia were surveyed to determine their liberation performance. This paper presents a preliminary analysis of the results of this exercise, and draws conclusions relating to the incidence and behaviour of liberated minerals in the circuits.

EXPERIMENTAL DETAILS

Circuits

Two plants were surveyed. The first was the ZC Mine Concentrator of Pasminco Mining at Broken Hill, formerly the Zinc Corporation, in which the valuable minerals are galena and sphalerite; it was surveyed in September 1987. The ZC plant contained three parallel grinding circuits, all of which were sampled. Circuits 1 and 2 each incorporated an open circuit rod mill followed by a closed circuit ball mill, lead rougher flotation, and a second closed circuit ball mill treating the lead rougher tailing. In Circuit 1 the ball mill was closed by a rake classifier, whereas in Circuit 2 classification was by hydrocyclones. Circuit 3 was configured similarly to Circuit 2, but without the lead roughing and secondary ball mill stages; the feedrate to this circuit was also about half that of the other two. Flowsheets of the three circuits are shown in Figures 1 - 3.

The second plant was the No. 4 Copper Concentrator of Mount Isa Mines, in which the principal valuable mineral is chalcopyrite; it was surveyed in October 1986. This concentrator contained two parallel grinding lines, one of which was sampled. An open circuit rod mill was followed by two stages of closed circuit ball milling, the classifiers being hydrocyclones in each case. The flowsheet is shown in Figure 4.

Surveys and Sample Analysis

Each stream of each circuit was sampled incrementally over a period of about 1-2 hours while the circuit was as far as possible in steady state, as determined by the stability of plant instrument readings. A variety of plant instrumentation was available, but for the purposes of this exercise only the mass flow of ore to the rod mill in each circuit was determined instrumentally. Fresh water additions were also measured by plant instrumentation. All other mass flows were calculated by mass balancing.

The samples were processed to determine solids size distribution (by sieving to 38 µm), solids concentration (by drying and weighing), and metal assay. Sub-samples were taken for liberation analysis using QEM*SEM (see below); these were pre-screened into individual size fractions using conventional woven wire sieves and microsieves.

The mass flow, solids concentration, assay and particle size distribution data were then mass balanced to produce self-consistent estimates of flows and size distributions in each stream, subject to the constraint that input = output, both at each process node and over the whole circuit. In nearly every case, the data adjustments required to meet this criterion were small, confirming that the sampling and analysis procedures were satisfactory. A summary of the mass balanced data is given for each circuit on the circuit flowsheets (Figures 1-4).

Determination of Mineral Liberation by QEM*SEM

Sub-samples of the solids in all process streams were sized and submitted to the CSIRO Division of Mineral and Process Engineering for analysis by QEM*SEM, to determine the size-by-size incidence of composite and liberated particles in each stream.

"QEM*SEM" is an acronym for "Quantitative Evaluation of Materials (Minerals) using Scanning Electron Microscopy". It is an automated image analysis system developed by the CSIRO and capable of scanning large numbers of particles to determine their mineralogical composition. It analyses scanning electron microscope images of samples using back scattered electrons and X-ray emissions, and is a useful tool in the estimation of liberation in mineral processing circuits, its results being quantitative and reproducible [3].

QEM*SEM measures mineral samples in polished sections. A size fraction from a circuit stream sample is mounted in an epoxy resin briquette before cutting flat and polishing with diamond paste. Great care is taken to present the particles in the resin in a random manner to ensure no bias exists. The briquette is mounted in a motorised stage which is computer controlled in the x - y plane, and the surface is then systematically scanned by an electron beam. The back scattered electrons and emitted X-rays are recorded and interpreted to provide details of particle numbers, mineral phases and associations with other minerals.

Two modes of analysis were used by QEM*SEM in this work:

1. Bulk Mineralogical analysis.
2. Particle Analysis

The first technique involves linear scans of a large number of particles in polished section. The X-ray spectrum and back scattered electron intensity for each measurement point are collected. Each mineral phase is identified and the volume of mineral present is estimated statistically. The weight percent is calculated by multiplying the volume fraction by the density of each phase and normalising the results. Elemental comparisons are calculated using the weight percent of each phase and the known mineral stoichiometry, which is supplied to the system prior to analysis [4].

The second technique gives a quantitative measure of the degree of liberation and intergrowth of individual particles for up to six nominated phases or groupings of phases. All particles measured are classified into percentile categories which represent the modal abundance of the reference material in each particle. Also shown are the number and percentage of particles in each of the ten liberation classes, the area percent of the particles in that class and the grain size of the phase of interest.

There are limitations to the size of particle which can be measured by QEM*SEM. The minimum size is 10-20 µm because of difficulties in obtaining

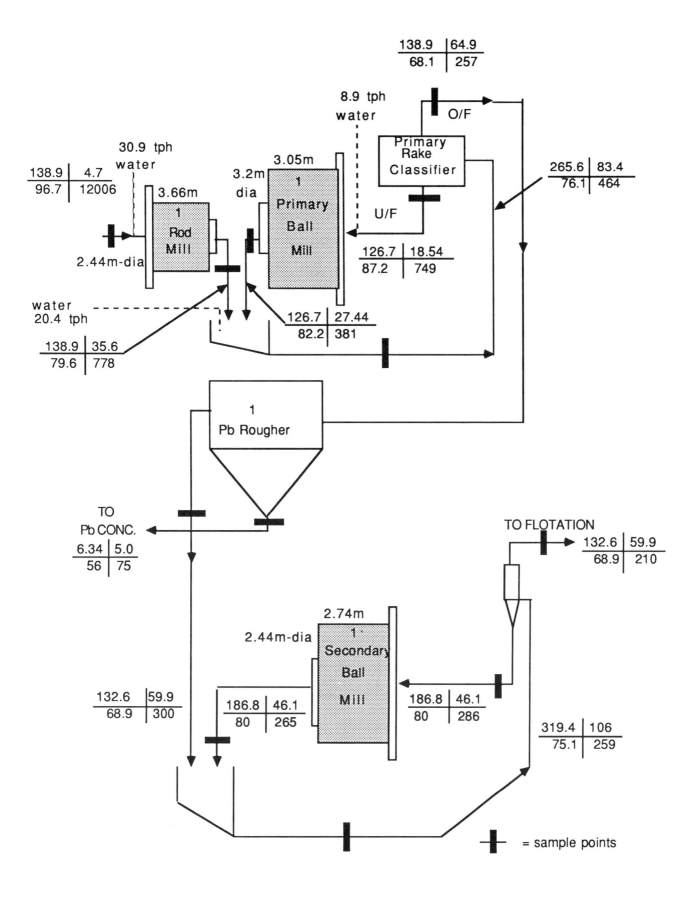

138.9	64.9
68.1	257

8.9 tph
water

Primary
Rake
Classifier

O/F

U/F

138.9	4.7
96.7	12006

30.9 tph
water

3.66m

1
Rod
Mill

2.44m-dia

3.2m
dia

3.05m

1
Primary
Ball
Mill

265.6	83.4
76.1	464

126.7	18.54
87.2	749

water
20.4 tph

126.7	27.44
82.2	381

138.9	35.6
79.6	778

1
Pb Rougher

TO
Pb CONC.

6.34	5.0
56	75

TO FLOTATION

132.6	59.9
68.9	210

2.74m

2.44m-dia

1
Secondary
Ball
Mill

132.6	59.9
68.9	300

186.8	46.1
80	265

186.8	46.1
80	286

319.4	106
75.1	259

= sample points

No 1 CIRCUIT

KEY

Solids tph	Water tph
% Solids	P 80 um

Figure1. Flowsheet of ZC Mine grinding circuit No1 showing mass
balance data and sample points.   137

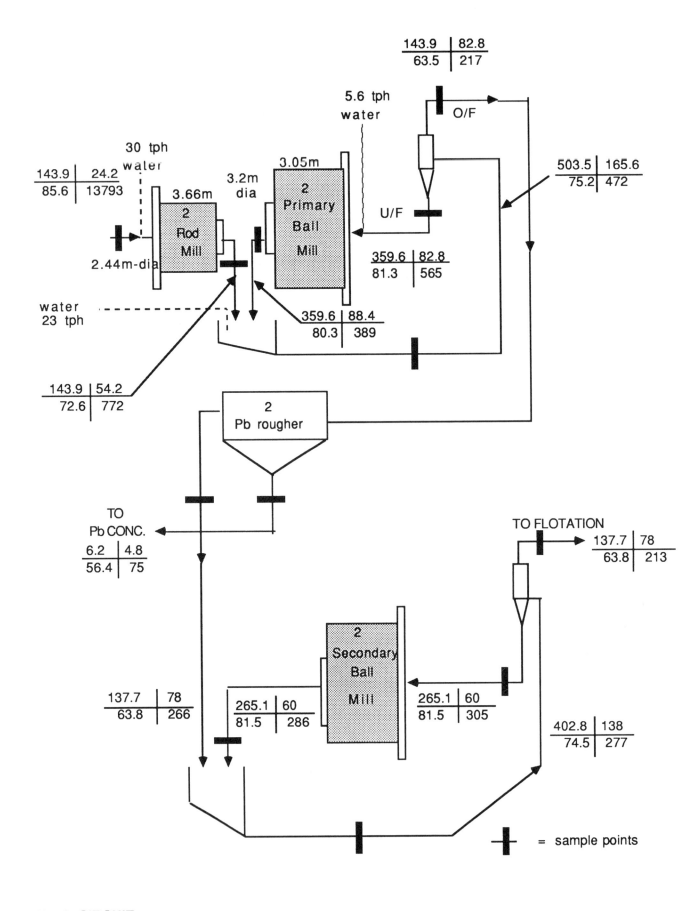

143.9	82.8
63.5	217

O/F

5.6 tph
water

30 tph
water

143.9	24.2
85.6	13793

3.05m

3.66m

3.2m
dia

2
Rod
Mill

2
Primary
Ball
Mill

503.5	165.6
75.2	472

2.44m-dia

U/F

359.6	82.8
81.3	565

water
23 tph

359.6	88.4
80.3	389

143.9	54.2
72.6	772

2
Pb rougher

TO
Pb CONC.

6.2	4.8
56.4	75

TO FLOTATION

137.7	78
63.8	213

2
Secondary
Ball
Mill

137.7	78
63.8	266

265.1	60
81.5	286

265.1	60
81.5	305

402.8	138
74.5	277

= sample points

No 2 CIRCUIT

KEY

Solids tph	Water tph
% Solids	P 80 um

Figure 2. Flowsheet of ZC Mine grinding circuit No2 showing mass balance data and sample points.

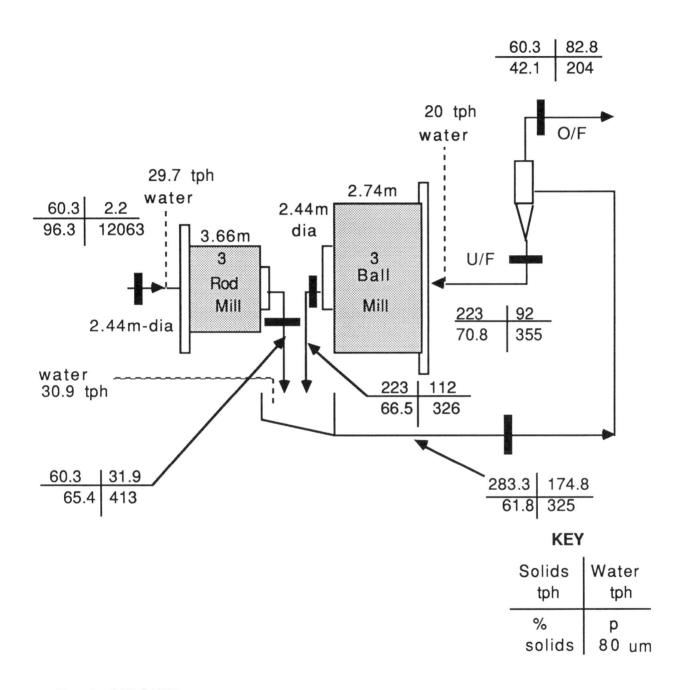

No 3 CIRCUIT

Figure 3. Flowsheet of ZC Mine grinding circuit No3 showing mass balance data and sample points.

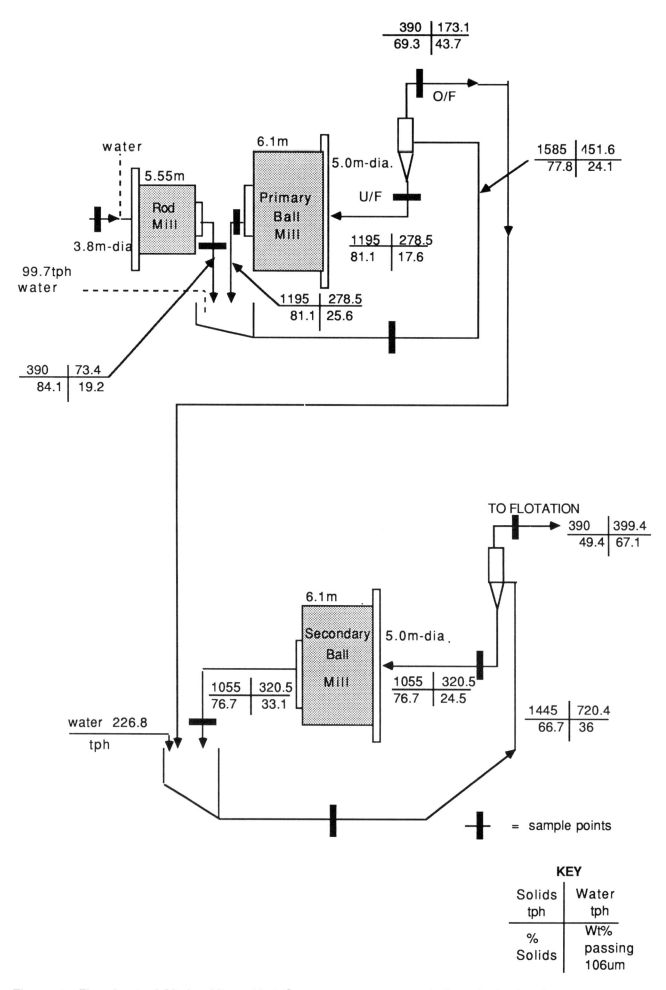

$\frac{390}{69.3}$ | $\frac{173.1}{43.7}$

O/F

$\frac{1585}{77.8}$ | $\frac{451.6}{24.1}$

water

6.1m

5.55m

5.0m-dia.

Rod
Mill

Primary
Ball
Mill

U/F

3.8m-dia

$\frac{1195}{81.1}$ | $\frac{278.5}{17.6}$

99.7tph
water

$\frac{1195}{81.1}$ | $\frac{278.5}{25.6}$

$\frac{390}{84.1}$ | $\frac{73.4}{19.2}$

TO FLOTATION

$\frac{390}{49.4}$ | $\frac{399.4}{67.1}$

6.1m

Secondary
Ball
Mill

5.0m-dia.

$\frac{1055}{76.7}$ | $\frac{320.5}{33.1}$

$\frac{1055}{76.7}$ | $\frac{320.5}{24.5}$

$\frac{1445}{66.7}$ | $\frac{720.4}{36}$

water 226.8
tph

✚ = sample points

KEY

Solids tph	Water tph
% Solids	Wt% passing 106um

Figure 4   Flowsheet of Mt Isa Mines No4 Copper concentrator grinding cicuit showing mass balance data and sample points.

unambiguous X-ray signals at finer resolutions. The maximum size is about 300 μm due to time and cost constraints in measuring the many particles necessary to obtain a statistically significant sample. For the present work it has been arbitrarily assumed that all the minerals in the fine sizes (-28 μm in this case) are liberated and all particles above 300 μm are unliberated. The former assumption is reasonable (by extrapolation from coarser sizes), but the latter is open to some doubt, and the mass flows of liberated mineral may therefore be underestimated in streams containing significant proportions of +300 μm material.

Stereological bias is an unavoidable problem in the current state of image analysis technology. It arises due to the two dimensional representation of three dimensional particles. In simple terms, sectioning the particle will result in some composites being seen by measurement systems such as QEM*SEM (incorrectly) as monomineralic, whereas truly liberated particles will always be seen (correctly) as monomineralic. As a result, image analysis by measurement systems such as QEM*SEM will overestimate liberation by a proportion inversely related to the apparent degree of liberation.

There is also the question of defining a liberated particle. Even though the proportion of truly liberated particles (those which are monomineralic) in a particular fraction might be quite small, many of the composites will be sufficiently liberated for their recovery kinetics not to be limited by the degree of liberation. In this work, a liberated particle was arbitrarily defined as one in which the plane section surface area was composed of at least 80% of the phase of interest. Barbery [5] has estimated that the error in the estimate of degree of liberation caused by stereological bias inherent in such a definition might be up to 15% absolute.

Once the proportion of "liberated" particles in a particular stream/size interval had been determined, the mass flow of liberated mineral in each size/stream was calculated by multiplying this proportion by the total solids mass flow in that size interval (estimated from the mass balancing step), and then re-balancing the mineral flows.

DISCUSSION OF RESULTS

Distribution of liberated mineral in the circuits

The production and behaviour of liberated mineral becomes very apparent when the mass flows are tracked through the circuit. Figures 5-8 map the size-by-size mass flows of four liberated phases through the ZC Nos. 1-3 Circuits, and Figure 9 presents similar data for MIM. The plots are constructed such that material flows through the circuit approximately from left to right. The more obvious features of such plots include the following:

1.   The relatively high degree of apparent liberation achieved early in the comminution process, as evidenced by the tonnages of liberated mineral in rod mill discharge compared with final classifier overflow.

2.   The substantial extent to which liberated mineral is returned by the classifiers for regrinding, and the variation in the performance of different classifiers.

3.   The different behaviour of the individual mineral phases in the circuit.

4.   The very limited amount of additional apparent liberation achieved in some of the second stage grinding circuits.

These aspects are discussed in more detail below.

Production of liberated mineral

The incidence of liberated mineral in the early stages of grinding was relatively high in both sites. For example at ZC the rod mill discharge contained a mass flow of liberated galena and sphalerite which was at least 60-70% of the amount of liberated mineral reporting to lead roughing (i.e. after one further stage of grinding). At MIM, the rod mill produced nearly 50% of the tonnage of liberated chalcopyrite reporting to the final circuit product (i.e. after two further grinding stages). This incidence of liberated mineral after primary grinding suggests that in some cases it may be appropriate to consider a coarse particle concentration stage early in the comminution process. The benefits would be both in improved mineral recovery by minimising overgrinding, and in a reduction in downstream milling tonnages, by reduction in the considerable circulating loads of liberated values occurring in most circuits (see below).

The creation of additional liberated mineral in the final grinding stages was variable and mineral-dependant. In the MIM case, it was interesting to note that the secondary ball mill (drawing about 2.5 MW) increased the tonnage of apparently liberated chalcopyrite by less than 2%, which is negligible in the context of the prevailing experimental error. This conclusion should however be viewed in the context of the stereological bias mentioned earlier. Espinosa-Gomez [12] has estimated (after applying the 2D/3D correction to the 90-100% liberation class) that the increase in liberation may actually approach 12%, and this was supported by laboratory flotation tests which suggested that the flotation of primary cyclone overflow operated on a lower grade recovery curve than that of the secondary cyclone overflow. This illustrates the difficulty of interpreting liberation data of this kind.

For the ZC Circuit 1, the tonnage of liberated galena apparently declined slightly after regrinding of the lead rougher tailing. This is an artefact of the data due to uncertainties in the incidence of (unmeasured) + 212 μm and -28 μm mineral and the vagaries of the definition of "liberation" mentioned earlier. However the additional liberation achieved, if any, was clearly small in this case also (though it should be stated that the sphalerite liberation was substantial). In some situations, therefore, there is justification in considering whether the expense of multiple grinding stages is warranted in view of the limited additional liberation sometimes achieved.

The relative behaviour of the different sizes of liberated mineral in grinding, as observed in Figures 5-9,

Figure 5a. GALENA No1 Circuit

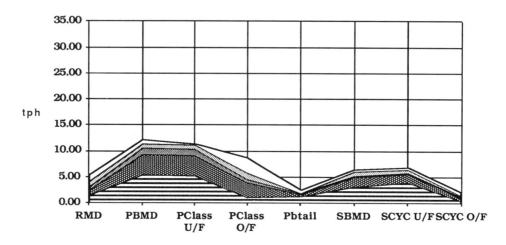

Figure 5b. GALENA No2 Circuit

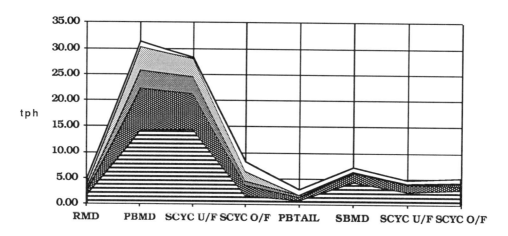

Figure 5c.   GALENA No3 Circuit

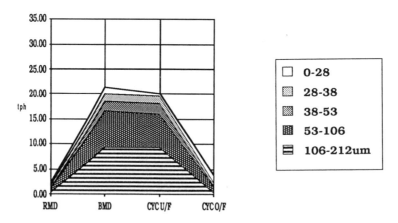

Figure 5.   The tph of liberated galena flowing through circuits 1, 2 and 3 at ZC Mine.

KEY

PCLASS U/F	=primary rake classifier underflow.
PBMD	=primary ball mill discharge.
SCYC O/F	=secondary cyclone overflow.
RMD	=rod mill discharge.

Figure 6a. PYRITE No1 Circuit

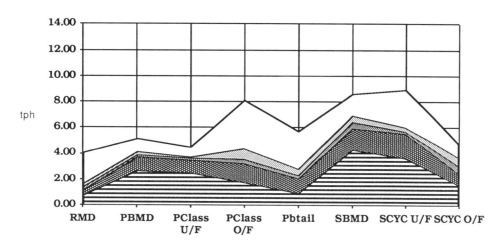

Figure 6b. PYRITE No2 Circuit

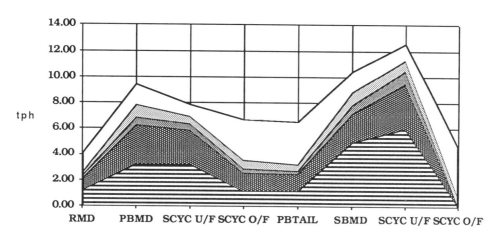

Figure 6c. PYRITE No3 Circuit

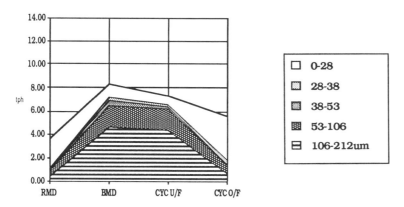

Figure 6. The tph of liberated pyrite flowing through circuits 1, 2 and 3 at ZC Mines.

KEY

PCLASS U/F	=primary rake classifier underflow.
PBMD	=primary ball mill discharge.
SCYC O/F	=secondary cyclone overflow.
RMD	=rod mill discharge.

Figure 7a. SPHALERITE No1 Circuit

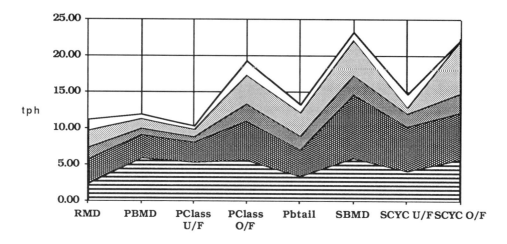

Figure 7b. SPHALERITE No2 Circuit

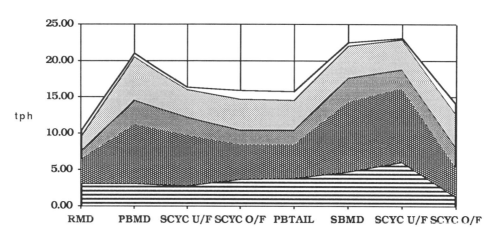

Figure 7c. SPHALERITE No3 Circuit

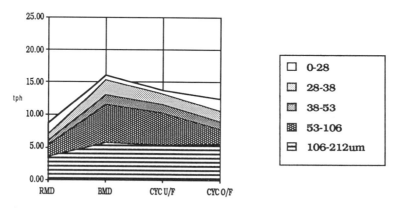

Figure 7. The tph of liberated sphalerite flowing through circuits 1, 2 and 3 at ZC Mine.

KEY

PCLASS U/F	=primary rake classifier underflow.
PBMD	=primary ball mill discharge.
SCYC O/F	=secondary cyclone overflow.
RMD	=rod mill discharge.

144

Figure 8a. NON-SULPHIDE GANGUE No1 Circuit

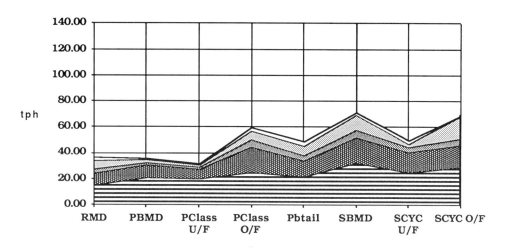

Figure 8b. NON-SULPHIDE GANGUE No2 Circuit

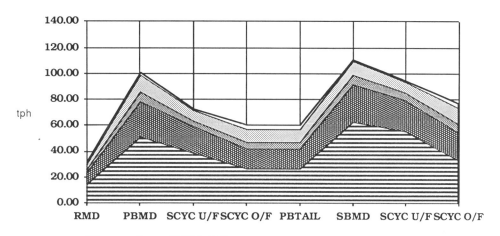

Figure 8c. NON-SULPHIDE GANGUE No3 Circuit.

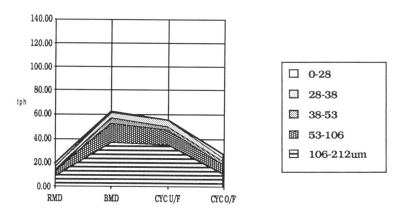

Figure 8. The tph of liberated non-sulphide gangue flowing through circuits 1, 2 and 3 at ZC Mine.

KEY

PCLASS U/F	=primary rake classifier underflow.
PBMD	=primary ball mill discharge.
SCYC O/F	=secondary cyclone overflow.
RMD	=rod mill discharge.

Figure 9a. CHALCOPYRITE

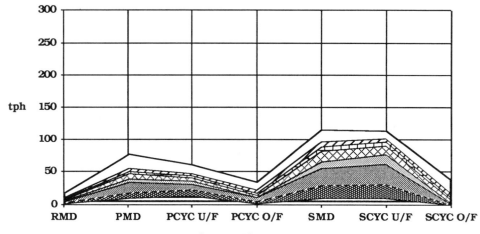

Figure 9b. QUARTZ

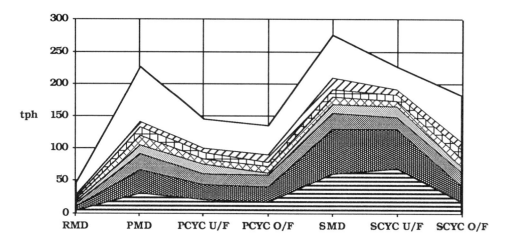

Figure 9c. PYRITE

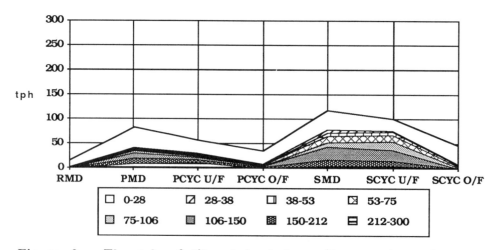

KEY: □ 0-28   ▨ 28-38   ▢ 38-53   ⊠ 53-75   ▨ 75-106   ▨ 106-150   ▨ 150-212   ⊟ 212-300

Figure 9.   The tph of liberated chalcopyrite, quartz and pyrite flowing through the No4 Copper concentrator circuit, Mt Isa Mines.

KEY

PMD	Primary mill discharge.
PCYC U/F	Primary cyclone underflow.
SCYC O/F	Secondary cyclone overflow.
SMD	Secondary mill discharge

is a function of two mechanisms which control the appearance of liberated mineral into, and removal from, each size interval: the size reduction of existing liberated grains, and the creation of newly liberated grains from (coarser) composites. With the data presented in this paper it is not possible to decouple these effects, but attention will be given to this problem at a later stage of the investigation, in the context of the modelling of the liberation process. For the present it is sufficient to observe that in most of the comminution stages illustrated in Figures 5-9, the size reduction mechanism is dominant. This is best seen in the behaviour of liberated galena in the ZC Circuit 2 primary ball mill (Figure 5b). Reading the size-by-size tonnage from the ball mill feed (primary cyclone underflow) to ball mill discharge (i.e. from right to left), the production of additional liberated galena is seen to increase with decrease in particle size (i.e. the gradient of the line increases). The increases in the size-by-size tonnages of liberated galena resulting from the primary ball milling stage were:

Size Interval (μm)	% increase in liberated galena
106-212	-1
53-106	16
38-53	11
28-38	28
0-28	327

(The negative increase in the coarsest size implies that breakage out of that size interval exceeded breakage into it from coarser composites).

A similar trend is seen in many of the other ball mill stages. The MIM secondary ball mill illustrates the phenomenon well (Figure 9a). Figure 10 shows the size distribution of liberated chalcopyrite entering and leaving the mill, the tonnages of which were apparently nearly identical. It suggests that the coarse mineral remained unground, whereas the finer mineral (below about 150μm) was reduced in size.

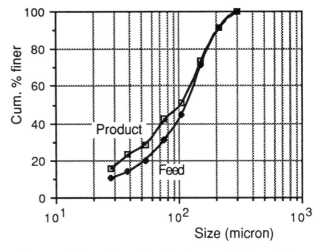

Figure 10 - Size distributions of liberated chalcopyrite, MIM secondary ball mill

When designing a sulphide grinding circuit, therefore, a distinction should if possible be made between the roles of size reduction *per se*, to render an appropriate size distribution for flotation, and size reduction to achieve liberation. In practice, the appropriate conditions are usually assessed through batch grinding and flotation

Table 1- Ratio of underflow to overflow mass flows of liberated mineral, for all classifiers and minerals

Mineral	Galena	Pyrite	Chalcopyrite	Sphalerite	NSG**	Quartz
Nominal SG*	7.5	5.0	4.2	4.0	2.7 ?	2.65
Hardness (Moh)	2.5	6-6.5	3.5-4.0	3.5-4.0	6-7	7
Classifier ***						
ZC 1 PRAKE	1.30	0.42	-	0.55	0.48	-
ZC 1 SCYC	3.26	1.88	-	0.66	0.73	-
ZC 2 PCYC	3.38	1.17	-	1.03	1.23	-
ZC 2 SCYC	0.93	2.77	-	1.65	1.23	-
ZC 3 PCYC	5.25	1.30	-	1.09	2.02	-
MIM PCYC	-	1.64	1.72	-	-	1.08
MIM SCYC	-	2.12	3.03	-	-	1.24

*   Average SG of pure mineral (from literature). In the present work, significant deviations from this figure could occur for "liberated" grains due to the definition of liberation and the stereological bias mentioned earlier.

**   Non-sulphide gangue, mostly quartz; variable SG.

***   For identification of classifier codes, see Appendix.

testing, but the present work has demonstrated that the interaction between these roles is complex, presumably due to the different liberation sizes and grinding rates of the different minerals.

Circulating Loads

In dealing with liberated mineral, the term "circulating load" (referring to the flow of liberated mineral returning to the mill) is not strictly valid in view of the creation of new liberated mineral in the mill. However it is a helpful concept, and will be used here. In all the circuits studied, the circulating load of liberated mineral varied systematically with mineral density and size distribution, as would be expected from the hydrodynamic principles of classification, and as has been observed by other workers [6, 7, 8]. In Figures 5-9 the effect is seen in terms of the difference in mass flow of classifier underflow and overflow, and the systematic change in this difference with mineral density. The difference in the underflow-overflow distribution for the ZC rake classifiers and cyclones is particularly striking. Table 1 shows the ratio of classifier underflow to overflow mass flow for each mineral circuit, as a measure of circulating load.

In the ZC data, the circulating load (ratio) generally follows mineral density, galena usually having the largest circulating load and the non-sulphide gangue the smallest. There are some exceptions. The galena ratio for the No. 2 circuit secondary cyclone was anomalously low, and it is not known why this should be so. In three cases (Circuit 1 primary classifier, Circuit 2 primary cyclone, and Circuit 3 primary cyclone) the NSG ratio exceeded that of the pyrite. Anomalous pyrite behaviour can be attributed to its relatively fine size (see Table 2 below) and the uncertainty in the density of "liberated" minerals is such that the true densities of sulphide (dense phases) and non-sulphide (light phases) will be closer to each other than indicated in Table 1; for example in extreme cases the density of a "liberated" pyrite grain could be as low as 4.2 and that of a "liberated" NSG grain as high as 4.3. Some inconsistencies in the trends seen in Table 1 can therefore be expected.

The same trends are apparent in the MIM data. The low density quartz had the lowest circulating load (as expected) but the chalcopyrite consistently exhibited a greater circulating load than the denser pyrite. Again, this is attributable to the finer size of the pyrite (Table 2), and the relatively small difference in densities.

The key conclusion apparent from Table 1 and the mass flow maps of Figures 5-9 is that a large proportion of liberated mineral is returned by each classifier for further grinding, and that this proportion is greatest for the denser and thus the valuable minerals. The effect was particularly severe at ZC where the hydrocyclones returned about 75% of the liberated galena to their associated ball mill, compared with only 53% of the NSG. Since the NSG was presumably also harder than the galena, this ensures that the galena reporting to lead roughing was significantly finer than the NSG - an average of 45% - 38μm, compared with 21% - 38 μm for the NSG. The same applies to the MIM situation: the liberated chalcopyrite reporting to flotation (secondary cyclone overflow) was 66% - 38 μm, whereas the associated quartz was only 46% - 38 μm.

It is probable that some grinding capacity is taken up with the breakage of mineral grains which are already liberated, and although in some cases it may be necessary to achieve a size reduction to optimise flotation recovery, it seems likely that both overgrinding of liberated mineral and unnecessary consumption of power and grinding media do occur. There would therefore seem to be some incentive in sulphide comminution circuits to optimise the liberation/size reduction conditions for each valuable mineral. However, such benefits are often difficult to realise in practice. A project some years ago to reduce the preferential recirculation of liberated galena at the North Broken Hill Concentrator, through modifications to the classifiers (including the installation of a sieve bend), was not successful [13].

The Influence of the Classifier

Classifiers behave according to hydrodynamic principles, and their performance has a profound influence on the associated grinding circuit. The present case is no exception.

Table 2 - Fineness of Selected Classifier Feeds

Mineral	Pyrite	Chalcopyrite	NSG	Quartz
	% - 38 μm	% - 38 μm	% - 38 μm	% - 38 μm
Classifier				
ZC 1 PRAKE	49.0	-	18.0	-
ZC 2 PCYC	36.5	-	17.1	-
ZC 3 PCYC	37.7	-	12.1	-
MIM PCYC	58.7	34.8	-	38.7
MIM SCYC	45.0	27.1	-	31.0

Table 3 - Classification parameters for all classifiers (total material)

Classifier	α	d$_{50c}$ (μm)	R$_f$
ZC1 PRAKE	1.59	362	0.198
ZC 1 SCYC	<0.5	334	0.421
ZC 2 PCYC	0.81	233	0.493
ZC 2 SCYC	<0.5	193	0.426
ZC 3 PCYC	0.47	122	0.507
MIM PCYC	2.48	565	0.582
MIM SCYC	0.94	122	0.411

In general, the performance of the hydrocyclones in all circuits was relatively poor, with high recoveries of water to underflow, high corrected cut-points and low efficiency values. Table 3 summarises the classification parameters for all the hydrocyclones and for the rake classifier in the ZC Circuit 1. The parameters are those of the well-known Whiten efficiency curve [7]:

$$Ec = \frac{e^{\alpha x} - 1}{e^{\alpha x} + e^x - 2} \qquad (1)$$

where
	x	=	d/d$_{50c}$
	d	=	particle size (μm)
	d$_{50c}$	=	corrected cut-point (μm)
	α	=	efficiency parameter

and Ec is the efficiency (or partition number) corrected for by-pass according to the expression

$$Ec = \frac{Ea - R_f}{1 - R_f} \qquad (2)$$

where Ea = actual efficiency (calculated from mass balanced size distributions), and R$_f$ = proportion of water reporting to underflow.

The parameters in Equation 1 were estimated by non-linear fitting procedures.

A comparison of the ZC primary rake classifier in Circuit 1 and the corresponding primary cyclones in Circuit 2 is particularly instructive. Figure 11a shows the experimental points and fitted uncorrected efficiency curves for these two units. It is clear that the rake classifier cut coarser and recovered much less water than the cyclones; its classification efficiency was also higher (Table 3). As a consequence, the circulating load of liberated mineral associated with the rake classifier was much lower than that of the hydrocyclones (Figures 5a/b, 6a/b, 7a/b; Table 1). Another consequence was that the

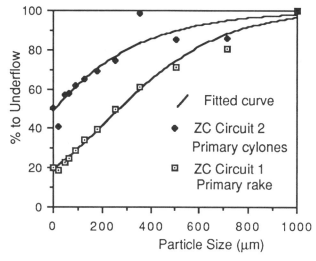

Figure 11a - ZC Ccts.1&2 primary classifier uncorrected efficiency curves

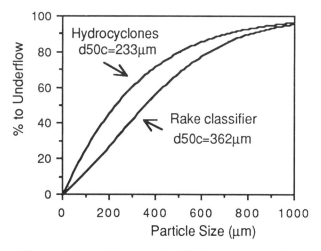

Figure 11b - Corrected efficiency curves for ZC rake and cyclones

total product from the rake classifier circuit was coarser than that of the hydrocyclones, even though the liberated galena size distribution was much the same. The lead rougher flotation performance in both cases was similar, most of the losses of liberated lead reporting to the secondary grinding circuit being in the coarse sizes (Figures 5a and 6a).

Figure 11b shows the corrected efficiency curves for the two units, confirming that even after by-pass is accounted for, the rake classifier cuts coarser and somewhat more sharply than the hydrocyclones.

The data from all the classification units confirm that the various liberated minerals separated according to their densities, although the effect was often masked by the relatively poor overall performance of the classifiers, particularly the high water recoveries. The effect is best seen in the MIM data. Table 4 gives the fitted classification parameters for the MIM primary and secondary hydrocyclones, for liberated pyrite, chalcopyrite, quartz and the bulk material, and Figure 12 shows the fitted corrected efficiency curves for these species.

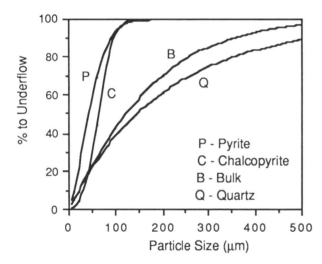

Figure 12 - Corrected efficiency curves for MIM liberated minerals

Clearly the corrected cut-points increase monotonically with decrease in density, as expected and as has been observed by other workers [6, 8, 9]. The bulk material (incorporating all liberated and composite

species) cut finer than the least dense phase (liberated quartz) but coarser than the denser phases. Again, this is to be expected from the distribution of the phases in the hydrocyclone feeds.

All other things being equal, the ratio of the cut-points of any two phases would be expected to be defined by:

$$\frac{d_{50c(1)}}{d_{50c(2)}} = \left(\frac{\delta_2 - \rho}{\delta_1 - \rho}\right)^n \qquad (3)$$

where δ = mineral density
  ρ = water density
  n = hydrodynamic exponent (0.5 for laminar flow and 1.0 for turbulent flow).

Lynch (1977) presented the results of Marlow's work [9] to show that Equation 3 was valid, with n = 1, for ore treated by the ZC Mine Concentrator (formerly NBHC); this was based on lead and zinc assays and the assumption that these metals were present exclusively as liberated mineral. The present data do not quantitatively fit Equation 3, even though the trends evident in Table 4 and Figure 12 are as expected. This is almost certainly because of the large uncertainties in the estimates of α and d_{50c} due both to the spectrum of densities inherent in the definition of liberation being used (as mentioned earlier), and to the relatively poor classification efficiencies which themselves generate large standard deviations in the parameter estimates.

The relationship between the extent of liberation achieved in a mill and the circulating load ratio of its associated classifier is illustrated for all minerals and circuits in Figure 13. The increase in liberation is defined as the ratio of the tph of liberated mineral leaving the mill to the tph of liberated mineral entering the mill. The circulating load ratio is defined as previously (Table 1). The inverse trend is clear and merely confirms that as the circulating load (of liberated mineral) increases, the increase in liberation tends to unity (i.e. no increase) until in the limit all the mineral returned to the mill is already liberated, and no additional liberation can by definition occur. In such cases the mill acts only to reduce the mineral size distribution, an example of this being the MIM secondary mill (Figure 10). In simple terms, at low

Table 4 - Classification parameters for different minerals, MIM primary and secondary circuits

Mineral	Primary Hydrocyclones		Secondary Hydrocyclones	
(notional SG)	d_{50c} (μm)	α	d_{50c} (μm)	α
Pyrite (SG 5.0)	166	4.50	42.5	1.79
Chalcopyrite (SG 4.2)	183	2.08	63.3	4.17
Quartz (SG 2.65)	1020	<0.5	145	0.60
Bulk	565	2.48	122	0.94

circulating loads the liberation mechanism is dominant, whereas at high circulating loads the size reduction mechanism is dominant. Figure 13 also emphasises the relatively high incidence of large circulating loads and associated minimal liberation.

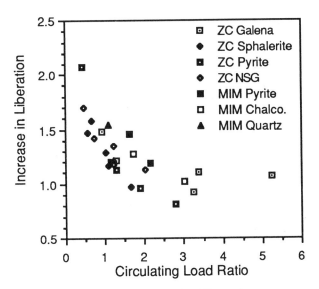

Figure 13 - Increase in liberation vs circulating load, for all minerals

In summary, it is clear from the data that the liberated minerals separated according to density, and that the circulating loads apparent in Figures 5-9 (and Table 1) were a direct consequence of this. The ZC rake classifier cut coarser and more efficiently than the corresponding hydrocyclones. The large water recoveries (40-60%) experienced by the hydrocyclones in all circuits also contributed to the circulating loads, particularly of the fine minerals. High circulating loads of liberated mineral returned by the classifier will always be associated with low or zero liberation in the mill.

Particle Breakage and Liberation Trends

It is self-evident that in general terms the proportion of a particular mineral that is liberated will increase with decreasing size. Two interesting questions then arise, of some process significance:

1. Does the proportion liberated increase linearly with decrease in size, or is the trend non-linear, liberation increasing rapidly below a certain critical "liberation size"?

2. Does the proportion liberated in a given size interval depend on breakage history, i.e. on the number and type of comminution machines that have acted on the particle?

Figure 14 summarises the answers to these questions for the MIM data, relating to the liberation of chalcopyrite. The graph shows the proportion of chalcopyrite liberated in each of seven size intervals (represented by their geometric mean sizes), for three mill products: the rod mill discharge (RMD), primary ball mill

discharge (PBMD) and the secondary ball mill discharge (SBMD).

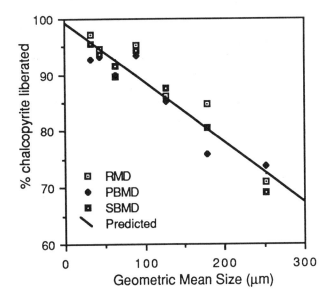

Figure 14 - Proportion liberated vs size for different mill products, MIM chalcopyrite

In this case it does seem that the relationship is linear, there being no end-point in liberation until the theoretical limit of zero size is reached (the finest mean size plotted is 32.6 μm, so the resolution of extrapolation is limited). Indeed, the correlation can be represented adequately by the very simple relation

$$\% \text{ liberated} = 100 - 0.1 \text{ (size)} \qquad (4)$$

(It should be remembered that the stereological bias is not accounted for in Figure 14; since the bias is an inverse function of liberation, it is unlikely that the relationship for <u>true</u> liberation is as given in Equation 4).

In addition, the data for all three mills fall naturally on the same line with no mill-specific trend being evident. This implies that, for this ore, mineral and circuit, a given size interval will always exhibit the same degree of liberation, regardless of the number or type of grinding stages which those particles have experienced.

A long term objective of this work is to model the liberation process so that it can be incorporated into comminution models to predict both the size distribution and composition of circuit products. To do this, the liberation behaviour of each ore must be characterised. In the present study, a twin pendulum single particle breakage apparatus was used to break particles of ZC rod mill feed to assess their liberation behaviour. The pendulum apparatus, which has been described in detail elsewhere [10], was used to break individual particles with a range of known initial sizes at a range of known specific energies (a procedure used to determine the appearance function of an ore for use in comminution models).

The broken product was sized and analysed by QEM*SEM to establish the incidence of liberated mineral in the various size fractions. Figures 15a and 15b summarise the liberation behaviour of <u>one</u> initial size of particle (-5.6 + 4.75mm) at <u>one</u> input energy (1.3 kWh/t), for the ZC ore, and they illustrate some interesting features of the liberation behaviour of this material. Figure 15a displays the liberation information obtained directly from the pendulum test. It records the distribution of liberation of each mineral with particle size. Here "distribution" for a particular size means the proportion of the <u>total</u> mineral present in the starting material which exists in a liberated form in <u>that</u> size interval, accumulated over all size intervals from fine to coarse, for the particular conditions of the test (initial size and energy level). Such plots therefore reflect three characteristics of the ore:

- The size distribution to which it breaks (under the specific test conditions).

- The inherent size distribution of the mineral phases in question.

- The phase-specific texture of the ore, and the mineral hardnesses.

These plots increase from fine to coarse sizes, reaching a maximum equal to the overall proportion of the mineral liberated in the breakage event (not reached in the present data). Over the range illustrated in Figures 15a, the plots are approximately linear.

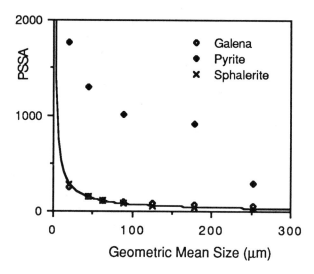

Figure 15b - Phase specific surface area for ZC sulphides
Pendulum breakage: 4.75-5.6mm, 1.3 kWh/t

There could be several reasons for this. One likely reason is evident in Figure 15b, which is a plot of the phase specific surface area (PSSA) against mean size, for each mineral. The PSSA is a parameter calculated by QEM*SEM which measures the surface area of the phase in question as a proportion of the total volume of that phase present [11]. Such plots are approximately horizontal at coarse sizes, the value of PSSA in this region (corresponding to composite particles) giving an indication of the relative inherent grain size of the different minerals. At lower sizes, the value of PSSA increases rapidly as the mineral is liberated and then broken further, since the breakage of liberated grains increases surface area without increasing total volume.

Figure 15b shows clearly that the pyrite had a significantly finer inherent grain size than the other sulphides. The liberated pyrite would therefore tend to be finer than the other liberated minerals, and this is clearly seen in the mass flow maps of Figures 5-9 and in Table 2.

The inherent grain sizes of galena and sphalerite appear very similar (almost indistinguishable in Figure 15b), even though their actual liberation behaviour appears different in Figure 15a. This is probably because the texture, mineral association or hardnesses are such that the galena of a given size liberates more freely after fracture than the sphalerite of the same size.

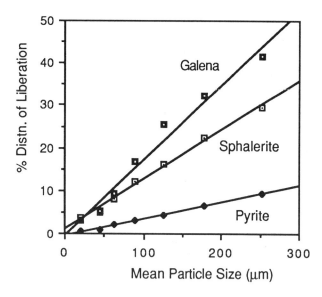

Figure 15a - Liberation distribution from pendulum test, for ZC sulphides

It is clear that the galena liberates most easily, and the pyrite least easily. The differences are quite substantial - over 30% of the galena is liberated in the -212 μm fraction, whereas only about 6% of the pyrite is liberated. (The absolute liberation values should be treated with caution, as they are a function of test conditions, particularly the input energy. The full picture is only apparent when considering the complete range of initial particle sizes and input energies).

SUMMARY AND CONCLUSIONS

Clearly the incidence of liberated mineral in a particular process stream in a comminution circuit is a function of the properties of the ore and the characteristics of the comminution and classification machinery. The investigation which is the subject of this paper seeks to quantify and decouple these effects so that they can be incorporated into models which can predict both the size distribution and composition of the product from a particular comminution circuit. The early conclusions of the investigation relating to the ZC and MIM surveys can be summarised as follows:

1. Significant amounts of valuable mineral are liberated early in the comminution process. Where practicable, this may justify the use of coarse particle concentration processes in cases where there are benefits to be gained thereby from reducing circulating loads and overgrinding.

2. The circulating load of liberated mineral (that reporting to classifier underflow) can be large. Its magnitude is a function of mineral size and density and the classifier characteristics. The circulating load is unnecessarily inflated when the classifier performance is poor, particularly in regard to a high water recovery to underflow. There may therefore be benefits in maximising classifier performance through correct selection of dimensions, feed solids concentration and feedrate, and use of automatic control.

3. The ZC rake classifier cut coarser and operated more efficiently than the corresponding hydrocyclones.

4. All the classifiers separated the liberated minerals as expected from hydrodynamic principles; dense minerals cut finer than light minerals, and thus generated larger circulating loads. The physical process of flotation is also hydrodynamic in nature and classification is therefore often an appropriate method of preparing feed for flotation. However in some cases there may be benefits in seeking to promote the sizing rather than the classification function in order to limit the circulating load of dense (valuable) mineral. The use of fine screening devices should be considered, where practicable.

5. In some grinding stages, minimal liberation was apparently achieved, and even size reduction of the liberated mineral was sometimes limited. It is important to establish the preferred role of each grinding stage (liberation, or size reduction of liberated grains, or both), and to maximise that role where possible.

6. There is some evidence (for the MIM copper ore) that liberation is independent of grinding history, i.e. that a given size interval will always contain the same proportion of liberated mineral, regardless of how many breakage events it has experienced (at least over the range of sizes studied here).

7. By utilising QEM*SEM to analyse the products of laboratory breakage tests, it is possible to draw conclusions about the propensity of each mineral to liberate following a breakage event. This propensity is a function of the inherent mineral grain size, the size distribution of the product of the breakage event, the texture of the ore and the mineral hardness. Of the sulphides in the ZC case, pyrite showed the lowest liberation, probably due to its finer grain size. Galena showed the highest liberation, possibly due to a greater tendency to break along grain boundaries than the sphalerite, even though these two minerals exhibited similar grain size distributions.

It is to be hoped that the trends reviewed in this paper will provide some helpful guidelines for the designers and operators of sulphide mineral comminution circuits, and expose the issues which must be addressed in maximising the metallurgical performance of such circuits. The challenge for the future is to decouple and model the liberation process quantitatively for use in circuit design and optimisation by simulation. An essential component of this work will be the refinement of the procedures by which QEM*SEM data are interpreted (including the definition of liberation), the extension of QEM*SEM's capabilities to finer sizes, and the development of valid methods for the correction of stereological bias.

Acknowledgements

The authors would like to thank the Management and Metallurgical Staff of the ZC Mine Concentrator of Pasminco Mining (formerly ZC Mines), and the Mount Isa Mines Limited Copper Concentrator, for permitting access to their plants and for the extensive help which they gave in sampling and sample analysis. Both companies also contributed materially to the cost of conducting the QEM*SEM analysis. The work was funded by a research grant from the Australian Mineral Industries Research Association for a collaborative JKMRC project in mineral processing, of which both companies were sponsors.

We wish to thank Dr. David Sutherland, Paul Gottlieb and Greg Wilkie of the CSIRO Division of Mineral and Process Engineering QEM*SEM Project for many useful discussions and for providing the QEM*SEM analysis service. We also thank our colleague Dr. Toni Kojovic for fitting the cyclone parameters quoted in the paper.

Finally, we wish to express our great sadness at the recent death of Professor Gilles Barbery of Laval University, with whom we had some very helpful discussions and who had agreed to participate in the analysis of the data generated in this work. Professor Barbery's wide-ranging talents were well known, but his work on the complex problem of stereology in mineral liberation will be particularly missed.

References

1. Barbery G. and Leroux D., 1988. Prediction of particle composition after fragmentation of heterogeneous materials. *Int. J. Mineral Proc.*, 22, 9-24.

2. King R.P., 1979. A model for quantitative estimation of mineral liberation by grinding. *Int. J. Mineral Proc.*, 6, 207-220.

3. Sutherland D.N. et al, 1988. The development and applications of QEM*SEM. *Chemeca 87*, Melbourne, 2, 106.1 - 106.6.

4. Miller P.R., Reid A.F. and Zuiderwyck M.A., 1982. QEM*SEM image analysis in the determination of model analysis, mineral associations and mineral liberation. *Proc XIV Int. Mineral Proc. Cong.*, Toronto. CIM.

5. Barbery G., 1989. Private communication.

6. Finch J.A. and Matwijenko O., 1977. Individual mineral behaviour in a closed grinding circuit. *CIM Bull.*, Nov.

7. Lynch A.J., 1977. Mineral crushing and grinding circuits. *Elsevier.*

8. Manlapig E.V., Drinkwater D.J., Munro P.D., Johnson N.W., and Watsford R.M.S., 1985. Optimisation of grinding circuits at the lead/zinc concentrator, Mount Isa Mines Limited. *Proc IFAC Symp.* "Automation for Mineral Resource Development", Brisbane. Aus. Inst. Min. Metall., 265-274.

9. Marlow D., 1973. A mathematical analysis of hydrocyclone data. *MSc thesis,* University of Queensland (JKMRC).

10. Narayanan S.S. and Whiten W.J., 1988. Determination of comminution characteristics of ores from single particle breakage tests, and its application to ball mill scale-up. *Trans. Inst. Min. Metall., Section C,* 97, 115-124.

11. Jackson B.R., Gottlieb P. and Sutherland D.N., 1988. A method for measuring and comparing the mineral sizes of ores from different origins. *Third Mill Operators Conf.,* Cobar, May, 61-65 Aus. Inst. Min. Metall.

12. Espinosa-Gomez R., 1989. Private Communication. (Mount Isa Mines).

13. Watters T.J., 1990. Private Communication. (Pasminco Mining, Broken Hill).

APPENDIX

Codes for Classifiers, Tables 1 - 3

ZC 1 PRAKE — ZC Circuit 1, primary rake classifier
ZC 1 SCYC — ZC Circuit 1, secondary hydrocyclones
ZC 2 PCYC — ZC Circuit 2, primary hydrocyclones
ZC 2 SCYC — ZC Circuit 2, secondary hydrocyclones
ZC 3 PCYC — ZC Circuit 3, primary hydrocyclones
MIM PCYC — MIM primary hydrocyclones
MIM SCYC — MIM secondary hydrocyclones

Methods of recovering platinum-group metals from Stillwater Complex ore

E.G. Baglin

Reno Research Center, Bureau of Mines, U.S. Department of the Interior, Reno, Nevada, U.S.A.

SYNOPSIS

As part of its strategic and critical minerals program, the U.S. Bureau of Mines investigated methods for beneficiating and recovering platinum-group metals (PGM) from ores found in the Stillwater Complex, the only primary resource of platinum-group metals located in the United States. Because the ore is sulfidic in nature, it was concentrated by froth flotation. A 45 kg/h process development unit (PDU) was constructed and 85 to 95% of the platinum and palladium were recovered from the ore by using mercaptobenzothiazole or sodium isobutyl xanthate as collectors. Bench-scale research showed that conventional matte smelting and leaching methods could be used to recover the precious metals from the flotation concentrate. More than 95% of the precious metals reported to the matte phase during smelting. Several leaching schemes were evaluated that selectively extracted the base metals from the matte and allowed recovery of high-grade PGM residues that were suitable as feed material for a precious metals refinery. Oxidative leaching of roasted flotation concentrate to selectively extract precious metal values was also investigated as an alternative to conventional matte smelting-leaching strategies.

Ongoing research is evaluating the effect of bioleaching using Thiobacillus ferrooxidans on sulfide oxidation and liberation of precious metals present in Stillwater Complex flotation concentrate.

INTRODUCTION

Exploration work in the Stillwater Complex, located in south-central Montana began in 1967. By 1976, four zones with high platinum-group metal values had been identified. At that time, the Bureau of Mines initiated a research program to define methods for upgrading and processing the ore to recover the precious metal values. Because the most significant PGM mineralization in the Stillwater Complex occurs in conjunction with iron, nickel, and copper sulfides, conventional froth flotation was selected as the most logical means for concentrating the ore. This paper summarizes development of the flotation process from bench to PDU-scale and also describes several methods for extracting the precious metals from the flotation concentrate.

Currently Bureau research is focussing on bioleaching of Stillwater flotation concentrate with acidophilic chemolithotropic bacteria, such as Thiobacillus ferrooxidans, to destroy the sulfide minerals and liberate the PGM. A process that can oxidize sulfide minerals without generating SO_2, and, at the same time, liberate the precious metals so that they are amenable to subsequent chemical leaching, may provide an inexpensive and environmentally preferable alternative to smelting for processing the sulfidic concentrate.

SAMPLE DESCRIPTION

Ore Samples

The most significant PGM mineralization in Stillwater Complex ores occurs in conjunction with iron, nickel, and copper sulfides, which are found in mineralized layers of the Banded Zone of the Complex. The deposit being mined is known as the J-M reef, which is continuous for about 45 km in the rugged Beartooth Mountains, near Nye, MT. Early Bureau work utilized ore from the West Fork exploration adit, but most samples originated in the Minneapolis adit, which is part of the current mine workings. Average grade of the Stillwater ore is about 26 g combined platinum and palladium per tonne with a 1:3.4 ratio of platinum to palladium. Geological, stratigraphical, petrographical, and mineralogical descriptions of the zones of PGM concentration are discussed in references 1-4.

Flotation Concentrates

Flotation concentrates were obtained from several sources: (1) bench-scale flotation tests conducted in Bureau laboratories[5], (2) the 45 kg/h PDU operated by the Bureau[6-7], (3) a pilot mill operated by Anaconda Minerals Co., and (4) Stillwater Mining Co.'s production mill[8].

Microprobe examination showed that the concentrates were composed primarily of sulfide minerals in siliceous gangue. The principal accessory minerals were pentlandite [$(Fe,Ni)_9S_8$], pyrite (FeS_2), and chalcopyrite ($CuFeS_2$), with minor to trace amounts of heazlewoodite (Ni_3S_2), galena (PbS), sphalerite (ZnS), and millerite (NiS). The PGM were distributed in the form of platinoid minerals (Pt-Fe alloys, PGM sulfides) and in solid solution in the pentlandite, where

155

palladium partially replaced nickel in the crystal lattice. The other principal PGM minerals observed in the samples were vysotskite, (Pd,Ni,Pt)S, and braggite, (Pt,Pd,Ni)S. The gangue minerals were Al, Ca, Fe, and Mg silicates.

RESULTS AND DISCUSSION

Flotation Studies

The Bureau began research on Stillwater Complex ores in the mid 1970s with a bench-scale study employing froth flotation to recover platinum and palladium values in a sulfide concentrate[5]. The first samples were obtained from the Johns-Manville West Fork exploration adit. The platinum-palladium mineralization in the West Fork ore occurs in sulfide-bearing unaltered anorthositic rocks.

Recovery of 80 to 90% of the Pt-Pd values was achieved in laboratory flotation units from an acidified pulp with a mercaptobenzothiazole (AERO 404)-polypropylene glycol methylether (Dowfroth 250) reagent suite.* Typical rougher concentrates contained 345 g PGM per tonne. Cleaner concentrates that analyzed 860 to 1,030 g PGM per tonne were obtained by refloating the rougher concentrate.

The West Fork adit was abandoned in 1976 because water under high pressure was encountered. The Bureau work continued with ore samples received from Anaconda Minerals Co. Anaconda's major focus of exploration was the development of the Minneapolis adit, which is located in an area east of the West Fork Adit. Ore from the Minneapolis adit was separated into two fractions: (1) fresh anorthositic gabbro containing lesser quantities of serpentinized and sericitized rocks, the products of alteration, and (2) material that was predominately serpentinized and sericitized rocks with some anorthositic gabbro.

A PDU flotation mill was constructed to concentrate the unaltered anorthositic ore[6]. A 5-day campaign processed an average of 48 kg/h of ore through the mill. The reagent suite used for floating the ore was 200 g AERO 404, 15 kg H_2SO_4, and 5 g Dowfroth 250 per tonne of ore. The solids content of the flotation slurry was 36 wt% and the acidity was maintained at pH 4.

Results are summarized in Table I. Although approximately 90% of the platinum and palladium were recovered in the concentrate, the overall grade was only 38 g Pt and 120 g Pd per tonne. During the last few hours of the campaign, samples of the flotation froth were taken simultaneously from each of the six flotation cells. The analytical results indicated that the grade decreased from the first to the last cell. If the froth products in the last four cells had been upgraded in a cleaner circuit, the overall grade of the final concentrate would have been considerably higher.

---

*Reference to specific brand names is for identification only and does not imply endorsement by the Bureau of Mines.

Because of acid-consuming minerals in the ore, the flotation scheme devised for the anorthositic ore was not satisfactory for the serpentinized ore[7]. Bench-scale tests were conducted to screen collectors for the serpentinized ore, and satisfactory results were obtained with xanthates combined with normal dodecyl mercaptan.

Three 32-h campaigns in the PDU were run to determine the effect of potassium amyl xanthate (AERO 350), sodium amyl xanthate (AERO 353), and sodium isobutyl xanthate (AERO 317), each in combination with normal dodecyl mercaptan (Pennfloat 3). During the three campaigns, an average of 39 kg of ore per hour was fed to the mill. Reagent dosages were 150 g xanthate, 100 g Pennfloat 3, and 7.5 g Dowfroth 250 per tonne of ore. Flotation was conducted at 16 wt% solids and a natural pH of 8.2. Best results were obtained with AERO 317, and results are outlined in Table II. Recoveries were acceptable, and the concentrate averaged 220 g/tonne PGM.

A 68-kg sample of the rougher concentrate was cleaned in the bank of six pilot mill cells. Cleaning upgraded the concentrate approximately two-fold to an average grade of 463 g/tonne PGM.

Table II Pilot mill flotation of serpentinized ore with AERO 317-Pennfloat 3

	Concentrate	Tailing	Composite
Solids....kg/h..	2.4	37	39
Analysis:			
Pt...g/tonne..	58	0.17	3.8
Pd...g/tonne..	161	1.71	11.3
Au...g/tonne..	3.4	<0.03	0.34
Cu.......wt%..	0.31	0.009	0.03
Ni.......wt%..	0.45	0.030	0.06
Distribution, %:			
Pt............	96	4	100
Pd............	86	14	100
Au............	95	5	100
Cu............	69	31	100
Ni............	49	51	100

Table I Pilot mill flotation of anorthositic ore with AERO 404-Dowfroth 250

	Concentrate	Tailing	Composite
Solids....kg/h..	4.1	44	48
Analysis:			
Pt...g/tonne..	38	0.34	3.8
Pd...g/tonne..	120	1.71	12.0
Au...g/tonne..	2.1	0.07	0.24
Cu.......wt%..	0.24	0.01	0.03
Ni.......wt%..	0.35	0.03	0.06
Distribution, %:			
Pt............	91	9	100
Pd............	87	13	100
Au............	74	26	100
Cu............	69	31	100
Ni............	52	48	100

Table III Analysis of Stillwater flotation concentrates
and smelting products

	Weight, g	g/tonne			wt%			
		Pt	Pd	Au	Cu	Ni	Fe	S
Flotation con- centrate.....	500	127	278	21	1.4	2.1	10.6	7.2
Matte........	93	521	1,423	93	7.5	10.5	53.0	31.5
Slag..........	438	5.5	12	.7	.1	.1	1.2	1.0

Flux: CaO 37 g, CaF$_2$ 18.3 g, SiO$_2$ 13.3 g.

Table IV Analysis of products from
first-stage leaching of
Stillwater matte

Products	Analysis	Distribution, %
Solution, g/l:		
Cu............	<0.001	<0.01
Ni............	12.1	99.1
Fe............	56	99.8
Pb............	<0.0015	<1.9
Residue:		
Cu......wt%..	66	>99.9
Ni......wt%..	0.9	0.9
Fe......wt%..	1.0	0.2
Pb......wt%..	3.2	>98.1
S........wt%..	18.9	NAp
Pt..kg/tonne..	5.59	>99.9
Pd..kg/tonne..	13.03	>99.9
Au..kg/tonne..	0.79	>99.9

NAp Not applicable.
Conditions: 275 g minus 200-mesh
matte, 2.58 l 2.1M H$_2$SO$_4$ (H$_2$SO$_4$-to-
matte ratio = 1.9), 4 h, 95°C.

Matte Smelting and Leaching

Matte smelting followed by leaching of the
metal values is the conventional metallurgical
processing sequence for treating sulfide
flotation concentrates containing copper,
nickel, iron, and PGM. The processing
sequence includes: (1) smelting to remove
gangue material and to produce an iron-nickel-
copper-PGM green matte, (2) converting by
blowing air through the green matte to oxidize
iron sulfide and to transfer the oxidized iron
to the slag as fayalite (2FeO·SiO$_2$), (3)
leaching the resultant copper-nickel-PGM white
matte to dissolve the copper and nickel and to
obtain a PGM residue, and (4) refining the
residue for recovery of PGM. Converting is
accompanied by evolution of large quantities
of SO$_2$ gas.

Direct leaching of the green matte to extract
metal values without SO$_2$ evolution was
investigated as an alternative to converting
and leaching of white matte[9]. Oxygen pressure
leaching of green mattes has been employed by
industry to avoid SO$_2$ generation[10-11], but
high-pressure autoclaves are required. In the
Bureau's investigation, green mattes were
prepared and leached under ambient pressure
conditions. The objective of this research
was to selectively extract the base metals
from the mattes and to recover the PGM in a
high-grade residue.

To prepare the mattes, 500-g charges of
Stillwater flotation concentrate were smelted
in an inductively heated graphite crucible
with a flux consisting of 37 g CaO, 18.3 g
CaF$_2$, and 18.3 g SiO$_2$. An argon blanket was
maintained over the charge to partially
exclude oxygen, to prolong crucible life, and
to minimize the oxidation of the sulfidic
charge. After melting, the charge was
maintained between 1,400° and 1,600°C for
approximately 1/2 h. The molten charge was
poured into a graphite mold. After cooling,
the matte and slag were separated and weighed.
More than 94% of the copper, nickel, and
precious metals reported to the matte.

Analyses of a Stillwater flotation concentrate
and the matte and slag obtained from smelting
are presented in Table III. Volatiles and
material remaining in the crucible account for
the 42-g weight loss during smelting. The
principal identifiable mineral phase in the
matte was troilite (FeS).

Several two-stage leaching schemes were
developed that removed the base metals from
the matte. All began with a sulfuric acid
leach, which selectively removed more than 99%
of the iron and nickel.

The following conditions gave the best results
in the first-stage H$_2$SO$_4$ leaching:

o Particle size: minus 200 mesh
o H$_2$SO$_4$ concentration: 2.1M
o H$_2$SO$_4$-to-matte ratio: 1.9
o Leaching temperature: 80°-95°C
o Leaching time: 4 h

Analyses of a typical pregnant liquor and
residue from first-stage leaching of
Stillwater matte are summarized in Table IV.
Distribution data demonstrate how completely
the iron and nickel were separated from the
copper, PGM, and gold. The small amount of
lead in the matte reported to the residue.
The major mineral in the first-stage residue
was djurleite (Cu$_{1.93}$S). The residue was
leached in a second stage to solubilize
copper.

Ferric chloride and ferric sulfate were
selected as second-stage lixiviants because of
their low cost and high capacity for
extracting copper from first-stage residues.
Ferric iron oxidized the copper and sulfide in
the djurleite and solubilized copper, leaving
the PGM in the residue. The reaction between
ferric iron and the djurleite was
stoichiometric, according to the following
equation:

$$Cu_2S + 4Fe^{3+} \rightarrow 2Cu^{2+} + 4Fe^{2+} + S°. \quad (1)$$

Products	Ferric chloride		Ferric sulfate	
	Analysis	Distribution, %	Analysis	Distribution, %
Solution:				
Cu.......g/l..	5.0	98.7	5.2	99.4
Ni.......g/l..	0.21	98.4	0.21	97.4
Fe.......g/l..	8.7	>99.9	9.2	>99.9
Pb......mg/l..	202	99.4	<7.5	<1.0
Pt......mg/l..	<0.02	<0.06	<0.02	<0.05
Pd......mg/l..	<0.02	<0.02	<0.02	<0.02
Au......mg/l..	<0.004	<0.07	<0.004	<0.06
Residue:				
Cu.......wt%..	4.9	1.3	1.7	0.6
Ni.......wt%..	0.26	1.6	0.32	2.6
Fe.......wt%..	0.15	0.02	0.06	0.01
Pb.......wt%..	0.09	0.6	10.5	>99.0
S........wt%..	79.8	NAp	75.8	NAp
Pt..kg/tonne..	24.1	>99.9	21.2	>99.9
Pd..kg/tonne..	63.8	>99.9	56.2	>99.9
Au....oz/ton..	3.9	>99.9	3.5	>99.9

NAp Not applicable.
Conditions: 5 g first-stage residue, 600 cm^3 0.15M HCl or
H_2SO_4 containing stoichiometric quantity of ferric chloride or
sulfate, respectively (pH 1.0), 6 h, 70°C.

The second-stage residue contained less than 5 wt% copper and more than 75 kg/tonne PGM (Table V).

The following conditions gave the best results in the ferric chloride and ferric sulfate leaching:

o Ferric salt - stoichiometric amount according to equation 1
o pH 1.0 (adjusted with HCl or H_2SO_4)
o Leaching temperature: 70°C
o Leaching time: 1 h, $Fe_2(SO_4)_3$; 4 h, $FeCl_3$

After ferric sulfate leaching, the principal species in the residue were elemental sulfur and lead sulfate. The residue was upgraded two-fold by dissolving the elemental sulfur in carbon disulfide. Lead sulfate would be removed during refining. After ferric chloride leaching, the second-stage residue contained sulfur and a noncrystalline, unidentifiable phase. The lead content was 0.09 wt% compared with 10.5 wt% in the sulfate leached residue. Ferric chloride solubilized the lead, while the ferric sulfate precipitated it.

Advantages of the sulfate system over the chloride system are (1) the second-stage extraction time is shorter, 1 h at 70°C for ferric sulfate compared with 4 h for ferric chloride leaching, (2) ferric sulfate and sulfuric acid are less expensive than ferric chloride and hydrochloric acid, (3) sulfate is less corrosive and less volatile than chloride, (4) sulfate solution is more suitable for electrolytic deposition of copper, and (5) lead sulfate in the residue can be easily removed during refining, whereas lead in the second-stage chloride liquor requires additional processing steps for removal.

Figure 1 shows a flowsheet summarizing the two-stage procedure developed for leaching Stillwater matte. The PGM content was upgraded almost 500-fold, from approximately 410 g/tonne in the flotation concentrate to more than 200 kg/tonne in the residue. The final residue contained more than 99% of the PGM in the matte.

Roasting was investigated as an alternative to oxidative second-stage leaching. Roasting oxidized the djurleite to the leachable copper species, Cu_2O, CuO, $CuSO_4$, and $CuO·CuSO_4$.

Dilute sulfuric acid solubilized more than 98% of the copper from first-stage residues when the roasting temperature was at least 360°C. One mole of sulfuric acid was required per mole of copper solubilized, and the PGM remained in the solids. Roasting between 360° and 500°C was the best range for separating copper and PGM. Roasting at temperatures below 360°C resulted in decreased copper extraction, while PGM values were lost to the leaching liquor when roasting was conducted at 500°C or above.

Analyses of the products of the roasting-leaching procedure are reported in Table VI. Copper extraction was complete in 1/2 h at 25°C with 0.2M H_2SO_4. The copper content of the pregnant liquor was increased to 48 g/l, which is satisfactory for copper electrowinning, by increasing the slurry solids in leaching from 23 to 100 g/l. The final residue contained 300 kg precious metals per tonne. A flow diagram for the roasting-leaching procedure is shown in Figure 2.

Leaching of Roasted Concentrate

A procedure was investigated for selectively leaching PGM and gold directly from the flotation concentrate, which would eliminate the need for smelting and significantly decrease processing time compared to conventional treatment of PGM sulfide concentrates[12].

The concentrate used was a cleaner concentrate prepared from Minneapolis adit anorthositic ore in the Bureau's PDU using an acidified pulp and AERO 404 as the collector. The

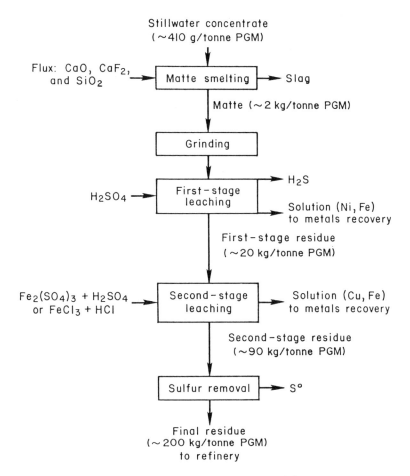

Fig. 1 Two-stage matte leaching.

Table VI Analyses of products from leaching of roasted first-stage residue from Stillwater matte

Products	Analysis
Solution, g/l:	
Cu............	12.3
Ni............	0.04
Fe............	0.04
Pt............	$<2\times10^{-5}$
Pd............	$<2\times10^{-5}$
Residue:	
Cu......wt%..	10.2
Ni......wt%..	0.19
Fe......wt%..	0.44
Pb......wt%..	7.0
Pt..kg/tonne..	67.2
Pd..kg/tonne..	224.6
Au..kg/tonne..	7.0

Conditions: 4.0 g first-stage residue roasted at 350° to 400°C, 175 cm^3 0.2M H_2SO_4 (H_2SO_4-to-Cu molar ratio = 1:1), 0.5 h, 25°C.

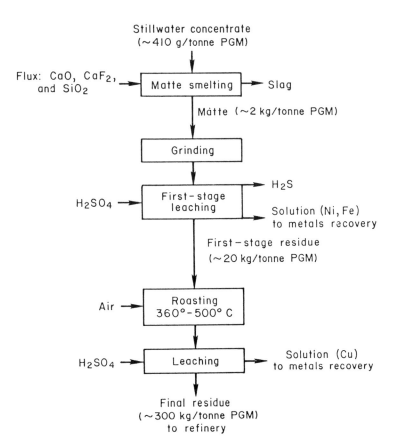

Fig. 2 Matte roasting-leaching procedure.

Table VII Analysis of flotation
          concentrates

	Concentrate sample			
	A	B	C	D
Analysis:				
Pt.....g/tonne..	301	300	105	342
Pd.....g/tonne..	556	1,139	296	1,175
Au.....g/tonne..	15	14	6	19
Cu........wt%..	1.4	2.0	0.6	1.4
Ni........wt%..	0.8	2.5	0.7	3.2
Fe........wt%..	8.2	12.0	7.4	11.6
S.........wt%..	5.9	8.7	2.8	7.0
Al$_2$O$_3$......wt%..	7.1	4.3	4.9	5.2
CaO.......wt%..	3.3	3.0	5.0	3.2
MgO.......wt%..	20.3	20.8	24.1	18.5
SiO$_2$.......wt%..	46.7	40.7	48.1	43.0

analysis of this concentrate is shown in Table VII (sample A). Experiments were conducted using 25 to 300-g samples.

Preliminary studies indicated that roasting the concentrate at temperatures above 800°C made the platinum, palladium, and gold amenable to oxidative leaching in strong HCl solutions. Effective oxidants included hydrogen peroxide, chlorine, sodium hypochlorite, nitric acid, and persulfate salts. Leaching was effective only when solution potentials were above 1.1 V, Eh. Most of the experimental data were generated using hydrogen peroxide, which is easy to handle and yields water as the reduction product. On the basis of cost and selectivity, sodium hypochlorite, chlorine, and hydrogen peroxide are the most favorable reagents. If sodium hypochlorite were used for leaching, control of sodium ion buildup in the pregnant solution would be necessary. Peroxide and chlorine add no contaminating ions to the strong HCl leaching solution.

Ninety percent or more of the platinum, palladium, and gold were extracted by hot H$_2$O$_2$-6M HCl solution when the roasting temperature was 800°C or higher. However, appreciable quantities of base metals and gangue minerals were also solubilized. Hydrogen peroxide-strong HCl solutions effectively solubilize precious metals from roasted concentrate because (1) the high acid concentration increases the half-reaction potential for H$_2$O$_2$ acting as an oxidizing agent,

$$H_2O_2 + 2H^+ + 2e^- \rightarrow 2H_2O \qquad Eh° = 1.78 \text{ V}, \quad (2)$$

and (2) high chloride ion concentrations (>6M) promote the oxidation of the precious metals by complexing and stabilizing dissolved ions.

Further investigation showed that leaching temperature had a significant effect on the selectivity of metal extraction. The data in Table VIII show that leaching at ambient temperature (approximately 25°C) preferentially extracted the precious metals from the concentrate and left almost all of the base metals in the residue. When the leaching temperature was increased to 105°C, PGM extraction increased 8 to 20%, while coextraction of base metals and gangue minerals increased several-fold. Utilizing preferential extraction of the precious metals at 25°C simplifies processing of the roasted concentrate by eliminating steps for base metal recovery. Loss of the copper and nickel in the tailings represents less than 5% of the monetary value of the concentrate, and their recovery could be undertaken if the economics were favorable.

Table IX shows the effect of roasting temperature on the extraction of precious metals during ambient temperature leaching. Extraction increased with temperature up to 1,050°C and decreased at higher temperatures owing to fusing of the charge. Maximum platinum, palladium, and gold extractions of 95, 90, and 99%, respectively, occurred after roasting at 1,050°C. Hydrogen peroxide additions totaling 0.005 to 0.03M were sufficient to maintain the oxidation potential during ambient-temperature leaching of the roasted concentrates.

X-ray diffraction examination of concentrates roasted at temperatures greater than 800°C showed a ferrite-type major phase, MFe$_2$O$_4$ (M = Cu and/or Ni), possibly mixed with maghemite (α-Fe$_2$O$_3$), which has a very similar crystal structure. Microprobe studies identified a silicate phase and a number of nickel, iron, and nickel-iron oxidic phases. Some PGM were converted from sulfide minerals to their elemental states by the roasting process. Some very small palladium-rich particles existed as relatively pure inclusions in the nickel-iron oxide phase. Larger discrete grains of Pt°-Pd° particles, which ranged in size up to 15 μm, were also observed. These particles exhibited varying ratios of palladium to platinum. Thus, both locked and liberated grains of palladium occurred in the roasted concentrate, but platinum existed primarily in liberated form.

Table VIII Effect of leaching temperature on H$_2$O$_2$-
           6M HCl leaching of 800°C roasted
           concentrate

Leaching temp, °C	Extraction, %									
	Pt	Pd	Au	Cu	Ni	Fe	Al$_2$O$_3$	CaO	MgO	SiO$_2$
25[1]....	87	72	96	3.6	2.5	2.8	32	42	11	0.6
70.....	98	82	>99	18	50	37	55	57	22	1.3
105.....	95	92	>99	47	91	84	81	69	52	.2

Conditions: 50 g roasted concentrate A, 840 cm^3
0.7M H$_2$O$_2$-6M HCl leaching solution, 6 h.
[1]22 h.

The lower solubility of palladium compared to platinum and gold can be attributed to locked palladium and to the fact that PtO_2, which can form during roasting, decomposes to metal and oxygen above 650°C, while PdO, the most stable oxide of palladium, can exist at red heat and is insoluble in mineral acids. Higher roasting temperatures favor decomposition of PdO to metal and result in increased palladium extraction during leaching.

Roasting temperature also affected base metal and gangue mineral attack. Coextraction of base metals after roasting at 1,050°C was less than after roasting at lower temperatures, and magnesium extraction decreased to approximately 1%. Since MgO comprises about 20 wt% of the concentrate, any decrease in solubilization is significant to the overall process. The principal contaminants of the pregnant solution were Al^{3+} and Ca^{2+} ions.

Cursory roasting-leaching experiments were undertaken with three other concentrates prepared from Minneapolis adit ore. Analyses are summarized in Table VII and leaching results are given in Table X. Concentrate B came from flotation of a mixture of anorthositic and serpentinite ores using a natural pH-xanthate circuit that included three cleaning cycles. This concentrate was received from a pilot mill operated by Anaconda Minerals Co. in Tucson, AZ.

Concentrate C was prepared by twice cleaning rougher concentrates recovered from serpentinized ore in the Bureau's pilot mill. An AERO 317-Pennfloat 3 collector combination was used at a natural pH of 8.2. Concentrate D is production mill concentrate obtained from Stillwater Mining Co. The SMC flotation circuit employs AERO 350 as the collector at pH 7.8 to 8.2.

None of the concentrates leached as well as the original sample and palladium was consistently less soluble than platinum and gold. Extraction from concentrate D peaked at a roasting temperature of 990°C, with 94% of the platinum, 77% of the palladium, and 97% of the gold being solubilized. SEM examination of the residue showed palladium-rich inclusions locked in the ferrite phase, as well as a small percentage locked in the silicate phase. It is believed that by judicious choice of grind size, roasting temperature, and leaching conditions, the palladium extraction can be increased to above the 77% obtained in these tests.

Incorporating a H_2SO_4 preleaching stage into the processing scheme significantly decreased contamination of the pregnant solution because easily soluble gangue minerals were removed prior to oxidative leaching. Preleaching was conducted by heating a pulp containing 10 wt% solids in 1M H_2SO_4 at 70°C for 4 h.

Table IX Effect of roasting temperature on H_2O_2-6M HCl leaching of Stillwater concentrate at ambient temperature

Roasting temp, °C	Eh, V	Extraction, %									
		Pt	Pd	Au	Cu	Ni	Fe	Al_2O_3	CaO	MgO	SiO_2
NR[1].....	1.12	1.2	8.9	40	12	14	21	42	30	7.4	1.8
750...	1.14	54	52	85	20	8.9	6.3	36	47	22	.8
800...	1.18	90	73	99	3.3	<.1	1.8	28	42	11	1.2
880...	1.18	92	75	99	3.0	3.5	2.0	33	40	7.6	1.6
960...	1.18	92	81	98	1.3	<.1	.8	32	40	3.3	1.2
1,050...	1.17	95	90	99	.9	<.1	1.1	38	40	1.0	1.0
1,110...	1.20	83	74	78	2.3	1.8	3.7	20	21	.7	.5
1,130...	1.20	86	78	82	6.9	6.9	7.0	3.2	3.3	2.0	.1

NR Not roasted.
Conditions: 25 g roasted concentrate A, 420 cm^3 0.03M H_2O_2-6M HCl leaching solution, 25° C, 6 h.
[1]0.2M H_2O_2-6M HCl.

Table X Effect of roasting temperature on H_2O_2-6M HCl leaching on different Stillwater concentrates

Concentrate	Roasting temp, °C	Extraction, %									
		Pt	Pd	Au	Cu	Ni	Fe	Al_2O_3	CaO	MgO	SiO_2
A.........	800	90	73	99	3.3	<0.1	1.8	28	42	11	1.2
	960	92	81	98	1.3	<.1	.8	32	40	3.3	1.2
	1,050	95	90	99	.9	<.1	1.1	38	40	1.0	1.0
B.........	800	76	45	92	5.9	4.0	4.2	20	53	20	2.0
	960	68	60	91	1.3	.8	.6	18	20	2.6	.9
	1,040	89	75	96	.8	.6	.5	20	19	1.4	1.0
C.........	800	89	48	99	7.8	5.3	5.5	29	60	9.8	1.6
	960	85	72	99	<.2	<.2	1.2	22	36	2.7	.7
	1,050	90	82	98	.1	1.0	.2	26	27	2.5	1.7
D.........	950	64	72	84	11.0	11.0	6.7	46	50	9.6	4.4
	990	94	77	97	8.4	9.0	5.5	43	48	5.7	3.8
	1,050	82	78	97	6.0	5.8	4.6	39	42	3.7	3.7
	1,070	74	77	92	9.7	8.8	7.4	46	51	5.4	3.9

Conditions: 25 g roasted concentrate, 420 cm^3 0.03M H_2O_2-6M HCl leaching solution, 25°C, 6 h.

Table XI shows the selectivity of two-stage leaching, and Table XII compares analyses of the pregnant solutions and tailings obtained by oxidative leaching with and without H_2SO_4 pretreatment. The precious metals content exceeded 60 mg/l in both pregnant solutions, but contamination by base metals and gangue minerals was decreased from a total of 3.2 g/l to 0.1 g/l when the concentrate was preleached prior to oxidative leaching.

Sulfuric acid consumption during preleaching depends on the nature of the gangue minerals in the concentrate and the duration and temperature of the preleaching operation. Acid consumption for the samples tested ranged from 0.04 to 0.22 g H_2SO_4 per gram of concentrate. Dilute H_2SO_4 for preleaching could be produced from the SO_2 in the roaster offgas.

Platinum, palladium, and gold were effectively recovered from the pregnant solution by three techniques: (1) precipitation as sulfides by sparging the solution with hydrogen sulfide, (2) cementation on nickel granules, and (3) collection on a column of activated coconut charcoal.

Based on the experimental results described, a flow diagram for the roasting-leaching procedure was devised and is presented in Figure 3. Flotation concentrate is roasted at 1,050°C and preleached in H_2SO_4 to dissolve impurity elements. Oxidative leaching in 6M HCl is conducted on the solids to extract the PGM and gold. The precious metals are recovered from solution for treatment at a refinery.

Bioleaching of Stillwater Flotation Concentrate

New research has been initiated by the Bureau to biologically oxidize the sulfide minerals in Stillwater flotation concentrate with Thiobacillus ferrooxidans. Bioleaching avoids the generation of SO_2 by oxidizing sulfide to sulfuric acid and is considered to be a more environmentally acceptable way of removing sulfide than processes that include high-temperature smelting or roasting steps. The concept is to decompose the sulfide minerals and liberate the PGM so that they become amenable to subsequent chemical leaching. Results of this research will be published in the near future.

SUMMARY

Bureau of Mines research showed the feasibility of upgrading Stillwater Complex ore by flotation concentration. In addition, procedures were developed for recovering platinum-group metals and gold from the concentrate by matte smelting and leaching and by direct leaching of preroasted concentrate. Current Bureau research efforts are directed towards liberation of the precious metals from Stillwater Complex flotation concentrate by bioleaching, to provide an alternative to high-temperature processes that include smelting or roasting steps.

ACKNOWLEDGMENT

The author gratefully acknowledges the assistance of Anaconda Minerals Co. (Atlantic Richfield Corp.), Johns-Manville Corp., and Stillwater Mining Co. for providing ore and concentrate samples used in this investigation.

Table XI Two stage leaching of 1,050°C roasted Concentrate

	Extraction, %	
	Stage 1 (1M H_2SO_4)	Stage 2 (H_2O_2-6M HCl)
Pt......	0.2	97
Pd......	.3	92
Au......	<.5	99
Cu......	5.1	.06
Ni......	3.2	.13
Fe......	3.9	.62
Al_2O_3...	49	.17
CaO.....	56	1.6
MgO.....	4.0	.20
SiO_2....	4.6	.01

Conditions: Stage 1 - 300 g roasted concentrate A, 2.8 l 1M H_2SO_4, 70°C, 4 h. Stage 2 - 150 g preleached concentrate, 2.5 l 0.03M H_2O_2-6M HCl leaching solution, 27°C, 6 h.

TABLE XII Analysis of products from single-and two-stage leaching of 1,050°C roasted concentrate

	Pregnant solution analysis, mg/l		Tailing analysis, wt%	
	Single-stage	Two-stage	Single-stage	Two stage
Pt.....	21.8	24.6	[1]20	[1]12
Pd.....	37.5	42.2	[1]79	[1]62
Au.....	1.1	1.2	[1].14	[1].14
Cu.....	11	<.5	2.0	1.3
Ni.....	<.7	<.7	1.1	.9
Fe.....	72	40	11	11
Al_2O_3..	1,800	20	5.3	4.5
CaO....	880	16	2.3	1.6
MgO....	140	28	23	22
SiO_2...	300	<2	51	52

Conditions: 150 g roasted concentrate A, 2.5 l 0.03M H_2O_2-6M HCl leaching solution, 27°C, 6 h.
[1]g/tonne.

162

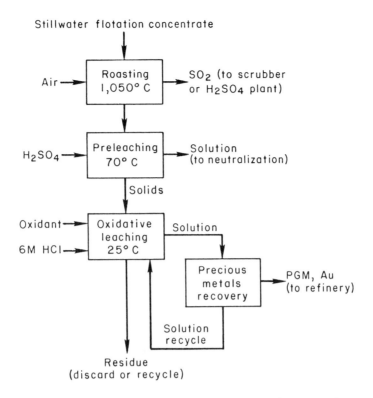

Stillwater flotation concentrate

Air → Roasting 1,050° C → SO₂ (to scrubber or H₂SO₄ plant)

H₂SO₄ → Preleaching 70° C → Solution (to neutralization)

Solids

Oxidant → / 6M HCl → Oxidative leaching 25° C → Solution

Precious metals recovery → PGM, Au (to refinery)

Solution recycle

Residue (discard or recycle)

Fig. 3 Flotation concentrate roasting-leaching procedure.

REFERENCES

1. Todd, S. G., Keith, D. W., and LeRoy, L. W. The J-M Platinum-Palladium Reef of the Stillwater Complex, Montana: I. Stratigraphy and Petrology. Economic Geology, vol. 77, 1982, pp. 1454-1480.

2. Bow, C., Wolfgram, D., Turner, A., Barnes, S., Evans, J., Zdepski, M., and Boundreau, A. Investigations of the Howland Reef of the Stillwater Complex, Minneapolis Adit Area: Stratigraphy, Structure, and Mineralization. Economic Geology, vol. 77, 1982, pp. 1481-1492.

3. Cabri, L. J., editor. Platinum Group Elements: Mineralogy, Geology, Recovery. Canadian Institute of Mining & Metallurgy Special Volume 23, 1981, 267 pp.

4. Kennedy, Alan. Stillwater Platinum-Palladium Mine. Engineering & Mining Journal, Nov. 1987, pp. 418-427.

5. Bennetts, J., Morrice, E., and Wong, M. M. Preparation of Platinum-Palladium Flotation Concentrate From Stillwater Complex Ore. Bureau of Mines Report of Investigations 8500, 1981, 18 pp.

6. Morrice, E. Pilot Mill Flotation of Anorthositic Platinum-Palladium Ore From the Stillwater Complex. Bureau of Mines Report of Investigations 8763, 1983, 8 pp.

7. Morrice, E., Walkiewicz, J. W., and Casale, G. Pilot Mill Flotation of Serpentinized Platinum-Palladium Ore From the Stillwater Complex. Bureau of Mines Report of Investigations 8885, 1984, 11 pp.

8. Hodges, G. J., and Clifford, R. K. Recovering Platinum and Palladium at Stillwater. J. Metals, vol. 40, No. 6, 1988, pp. 32-35.

9. Baglin, E. G., Gomes, J. M., and Wong, M. M. Recovery of Platinum-Group Metals From Stillwater Complex, Mont., Flotation Concentrates by Matte Smelting and Leaching. Bureau of Mines Report of Investigations 8717, 1982, 15 pp.

10. Plasket, R. P., and Romanchuk, S. Recovery of Nickel and Copper From High-Grade Matte at Impala Platinum by the Sherritt Process. Hydrometallurgy, vol. 3, No. 2, 1978, pp. 135-151.

11. Pearce, R. F., Warner, J. P., and Mackiw, U. N. A New Method of Matte Refining...by Pressure Leaching and Hydrogen Reduction. Journal of Metals, vol. 12, No. 1, 1960, pp. 28-32.

12. Baglin, E. G., Gomes, J. M., Carnahan, T. G., and Snider, J. M. Recovery of Platinum, Palladium, and Gold From Stillwater Complex Flotation Concentrate by a Roasting-Leaching Procedure. Bureau of Mines Report of Investigations 8970, 1985, 12 pp.

163

China's sulphide deposits — their occurrence and treatment

Yu Xingyuan
The Nonferrous Metals Society of China, Beijing, China
Li Fenglou
Beijing Central Research Institute for Mining and Metallurgy, Beijing, China
Huang Kaiguo
Department of Mineral Engineering, Central South University of Technology, China

IN CHINA SULPHIDE DEPOSITS ARE THE major source of lead, zinc, copper, nickel, molybdenum, antimony, mercury and sulphur. Nearly 40% of the production of gold, almost all of that of silver and virtually all of that of the platinum group metals come from sulphide deposits. Lead, zinc and copper sulphide deposits are widely distributed throughout China in, geologically, a great variety of occurrences. In contrast, antimony and mercury sulphide deposits are found in the provinces of Hunan, Guizhou and Guangdong.

Large nickel–copper sulphide deposits occur only in Gansu and Xinjiang provinces in north-west China. China is also endowed with a wide variety of unique sulphide-bearing tungsten–tin deposits, which occur mainly along the great tectonic belt that stretches from southeast to southwest China bordering Thailand and Burma.

The exploration and exploitation of sulphide deposits in China underwent a nation-wide and rapid development only after the founding of The People's Republic of China, virtually no adequate facilities and establishments able to undertake research and development on the proper utilization of China's resources of sulphides having existed previously. Moreover, it should be pointed out that flotation technology, the most effective beneficiation method for sulphides, had experienced worldwide industrial application and rapid development after the second world war. Due to the complex and/or refractory character of most of China's polymetallic sulphide deposits, the main trends in the treatment strategy for these sulphides are noted below.

(1) Full utilization of all the valuable minerals in the ore, coupled with awareness of its environmental aspects. Hence, considerable efforts have been made to introduce innovative treatment methods and technology aimed at the realization in industrial/commercial operation of 'no-waste' and/or 'zero-discharge' treatment systems.

(2) Integration of modern beneficiation methods with innovative chemical treatment of 'dirty' concentrates or polymetallic middlings. The aim is to obtain the maximum amount of clean concentrates as quickly as possible and to regroup 'dirty' concentrates and/or middlings in a rational manner for chemical treatment.

These trends in treatment strategy are described and illustrated by a representative example for each base metal or group of base metals in which significant achievements in modern and innovative beneficiation and chemical extraction methods are apparent. Most of the technologies that are cited in the paper have been in industrial use since the late 1970s.

CHINA'S SULPHIDE DEPOSITS

China's sulphide deposits are the principal source of such base metals as copper, lead, zinc, nickel, molybdenum, antimony and mercury. Some 40% of the supply of gold and virtually all of that of silver and the platinum group metals come from sulphide deposits in which the precious metals are recovered as by-products during the processing and extraction of the base-metal sulphides.

Lead, zinc and copper sulphide deposits are widely distributed in China. In contrast, antimony and mercury sulphide deposits are confined to Hunan, Guizhou and Guangdong.

Large nickel–copper sulphide deposits have not been explored systematically. At present only Gansu and Xinjiang provinces are known to have this kind of deposit. It is also a major source of platinum group metals and cobalt.

China is also endowed with a variety of unique tungsten and tin deposits with significant amounts of base-metal sulphides. These deposits occur mainly along the huge tectonic belt that stretches from southeast to southwest China bordering Thailand and Burma.

Lead–zinc deposits

Lead–zinc deposits are well distributed throughout China, large deposits and industrial bases being in Guangdong, Hunan, Yunnan and Shaanxi provinces. They are not only the main source of lead and zinc but also a significant source of silver, which is intimately associated with lead minerals—galena, in particular.

There are three main types of lead–zinc deposit—(1) strata-bound (for the most part associated with Proterozoic and Upper Palaeozoic carbonate rocks or Lower Palaeozoic metamorphosed rocks (marble)), (2) hydrothermal vein (in Proterozoic weakly metamorphosed rocks or related to Mesozoic magmatic activity) and (3) porphyritic type deposits in Yanshanian Granite (a huge deposit of this type occurs in Shaanxi province).

In terms of lead and zinc content these ores are of medium or low grade. Generally, a significant part of the sulphide grains is fine to ultrafine and intimately intergrown, the consequential beneficiation and metallurgical problems depending on the degree of fineness and textures of intergrowth. The lead–zinc deposits sometimes contain minor amounts of copper and, as such, can be considered as a Cu–Pb–Zn complex ore.

Copper deposits

Copper sulphide deposits are also well distributed. Current mining of large deposits takes place in Jiangsu, Yunnan and Gansu provinces.

Copper sulphide ores are also a major source of molybdenum and precious metals, including platinum-group metals. In China about 40% of the total production of gold and silver is closely associated with copper sulphide ores.

The main source of platinum-group metals and cobalt is massive pyritic copper–nickel deposits that are intimately associated with basic–ultrabasic rocks.

The five main types of copper deposit are (1) Yanshanian skarn deposits in contact with Upper Palaeozoic rocks, (2) Yanshanian porphyritic deposits in granodiorite porphyry, (3) strata-bound deposits in the Carboniferous series or in weakly metamorphosed Kunyang Group Proterozoic rocks, (4) volcanogenic-type deposits in Lower Palaeozoic volcano-sedimentary metamorphosed rocks and (5) pyritic copper–nickel deposits closely associated with basic–ultrabasic rocks.

The porphyritic-type deposits are generally

low-grade, but pose no major metallurgical problems as the grain sizes of the copper minerals are rather coarse to medium and textures are simple. The other types have a high to medium ore grade, but they are polymetallic in character and the sulphide grains are fine to ultrafine and, generally, intergrown with large amounts of pyrite, pyrrhotite, etc.—in particular, the copper–nickel deposits associated with basic and ultrabasic rocks.

Antimony deposits

The occurrences of antimony sulphide deposits are concentrated in central southern and south China, especially in Hunan and Guangxi provinces. Most of these deposits are small in size, but are generally of high grade and amenable to conventional beneficiation and extraction methods. The most important type of deposit is of the strata-bound type, occurring in Devonian Series or associated with Proterozoic weakly metamorphosed rocks.

Sulphide-bearing tungsten–tin deposits

Sulphide-bearing tungsten–tin deposits are unique in China in that they mostly occur along the tectonic belt from southeast to southwest China. Generally, from east to west the ratio of the content of tungsten and tin in the ore changes: at the eastern end are the tungsten deposits (tungsten content dominant) and at the western end the tin deposits (tin content dominant). In between the ratio is such that they could be considered as bimetallic tungsten–tin deposits. A feature of these deposits is the occurrence of base-metal sulphides together with oxide minerals of tungsten and tin. The ratio of base-metal sulphides also increases generally with the increase of the tin content from east to west. At the present time these sulphide-bearing tungsten–tin deposits are the major sources of those metals.

The W–Sn ore grade ranges from high to medium. The grain size distribution of the valuable minerals is rather wide—from coarse to fine and ultrafine. Most of the typical tungsten deposits belong to the wolframite–quartz vein deposits in Devonian metamorphosed rocks in the vicinity of the granitic contact zone. There are also large- to medium-size scheelite skarn-type deposits. The main tin deposits are principally sulphide-bearing cassiterite deposits of pyritic type, generally distributed in the outer contact zone of a medium acidic granite intrusion; the host rocks are limestone, dolomitic limestone or sandy shale.

EXTRACTION FEATURES OF CHINA'S SULPHIDE DEPOSITS

From the point of view of process mineralogy it should be noted that most of China's large sulphide deposits are polymetallic in character, the content of each valuable mineral is, on average, low and the sulphides are frequently fine to ultrafine and intimately intergrown.

For complex tin or tungsten deposits the base metals occur as sulphides, whereas tin, tungsten, beryllium, rare earths and others occur as oxides. Although they are polymetallic and contain many metal values, only the grade of tin or tungsten reaches or exceeds the economically exploitable level.

To obtain the optimal utilization of all values in the ore, following some years of research and development and industrial practice, a more or less comprehensive and complex flowsheet has been generally adopted that has the main features noted below.

Multistage grinding in closed circuit with multistage separation and concentration up to a mesh of grind of −200 to 300 mesh.

Before the founding of the People's Republic of China the beneficiation of ores had depended for the greater part on traditional gravity concentration. The most effective method for sulphides, i.e. flotation, was at that time still in its infancy. Nowadays, however, flowsheets for concentrators treating polymetallic ores are far more comprehensive and sophisticated, incorporating both the main concentration methods (gravity concentration and flotation) and other physical methods, such as a variety of magnetic electrical separators and sorting machines.

Innovative flotation circuits, such as a multistage flotation system of the same mineral or group of minerals according to their slight differences in floatability, were established to ensure the recovery and grade of the concentrated products; multifeed flotation circuits leading to less reagent consumption and collectorless flotation of sulphides were introduced at the commercial scale. The beneficial effects of these circuits were intensified by the use of newly developed powerful and selective flotation reagents.

Ultrafine particle separation techniques, such as carrier flotation, centrifugal gravity concentration, the multicarved deck slime table, the rocking–shaking slime vanner, etc., have been operated industrially for many years.

The introduction of various chemical treatments in dealing with refractory middlings or for upgrading 'dirty' concentrates enhanced overall recovery and utilization of the valuable constituents in the ore.

The following gives some details of these features as applied to specific ores

Treatment features of lead–zinc complex ores[2]

Most of the polysulphide ores in China with large reserves belong to lead–zinc type deposits in which the ratio of copper, lead and zinc is in the order of Cu(0–1):Pb(5–10):Zn(10–20).

Up to the late 1960s the flotation circuit that was generally adopted was selective flotation and/or bulk flotation with subsequent selective flotation. The reagent regime commonly used is the conventional cyanide–dichromate system in alkaline to high alkaline medium size. The overall mesh of grind is often of coarse to medium

size. This is still the practice in treating rich coarse-grained ores from small mines. However, gradually, the major part of the production of lead and zinc has come from lead–zinc ores characterized by an extreme fineness of the natural grains of the sulphides and their intimate intergrowth. With this type of ore the cyanide flotation system appeared in most cases unable to produce good-grade concentrates at reasonably high recoveries.

Other reasons for the need to switch to a cyanide-free flotation technology are environmental hazards caused by cyanide-bearing liquid and solid wastes, the presence of secondary copper sulphide minerals in prohibitive amounts (26–40% of the content of total copper) that can seriously affect the efficiency of the cyanide–dichromate flotation system, and, last but not least, the partial loss of precious metals caused by cyanide complexing: the high alkalinity of the cyanide flotation process with its viscous froth is not able to produce good-quality concentrates at reasonable recoveries.

On the other hand, the cyanide-free flotation system, combined with multistage grinding based on a relatively more precise process mineralogy in the quantitative determination of the degree of fines and liberation of the natural grains of the sulphides, offers great possibilities to cope successfully with the metallurgical problems mentioned above.

The cyanide-free flotation system can be divided into two sub-systems—bulk flotation of Cu–Pb or Pb–Zn and separation of Cu and Pb in the bulk concentrate.

The most important reagent regime for bulk flotation, currently applied commercially, is the lime–(sodium sulphide)–sulphite or bisulphite-(zinc sulphate) system. Zinc sulphate is optional and only used for strongly depressed sphalerite activated by copper ions from secondary copper minerals. Sodium sulphide is applied when large amounts of secondary copper sulphides are present in the ore relative to the total amount of copper.

A recent development of the cyanide-free system for the bulk flotation of lead–zinc ores containing only 0.1 wt% Cu is the Na_2CO_3–$ZnSO_4$–sodium thiosulphate system followed by Cu–Pb separation with dichromate. A good-quality copper concentrate (25 wt% Cu) was obtained from an ore containing only 0.1 wt% Cu.

There are many variants for Cu–Pb separation tailored to the process mineralogical characteristics of the bulk concentrates—the dichromate–sulphite system, the hydrogen peroxide–ferrosulphate–sulphite system, the 2-carboxylic cellulose–sulphite system in which 2-carboxylic cellulose is used as the lead depressor instead of dichromate (in this case a reduction of two-thirds of the reagent cost could be obtained as compared with the dichromate system), and a simple sulphite system with an increased temperature of flotation ($T = 60$–$70°C$). This can also be applied successfully to Pb–Zn separation after activating the sphalerite particles.

The overall beneficial effects of the free-cyanide flotation system can be summarized thus: (1) capability to recover fine to ultrafine values with high efficiency, (2) production of commercial-grade copper, zinc and lead concentrates at high overall recovery and (3) increased recovery of precious metals with increased recovery of the precious-metal carriers (copper sulphide and galena).

Some examples from practice substantiate these finds.

Xiaotieshan mine works a complex polysulphide ore that contains 0.4–0.6 wt% Cu, 3–4 wt% Pb and 5–6 wt% Zn; 47% and 29% of the natural grains of galena belong, respectively, to the −40 μm or −20 μm sizes. The natural grains of sphalerite and pyrite are somewhat coarser, but of the same order. The metallurgical results are shown in Table 1.

An ultrafine-grained pyritic lead–zinc ore is treated at Fankou mine,[3] which is the largest

Table 1 Metallurgical results, Xiaotieshan concentrator

Metal	Feed, wt%	Concentrate, wt%	Recovery, %
Copper	0.4–0.6	25–26	85–87
Zinc	5–6	50–53	92.93
Lead	3–4	56–58	75.80

lead–zinc mine in China. The ore contains 4.5–5 wt% Pb and 10–12 wt% Zn, but the natural grain sizes of galena, sphalerite and pyrite are ultrafine and extremely intergrown with one another. Conventional methods of selective or bulk flotation all failed to produce commercial-grade concentrates of lead and zinc.

Lengthy R&D showed that the best way to utilize the ore from Fankou mine was to produce a mixed concentrate of lead and zinc for subsequent treatment of the product by the Imperial Smelting Process.

By the conventional bulk flotation circuit and the introduction of various reagent regimes it was not possible to upgrade the Pb–Zn bulk concentrate in accordance with the requirements of the subsequent ISP treatment because of the intolerable contamination of pyrite impurities. The grade of the Pb–Zn mixed concentrate never exceeded 39 wt%. But, after the introduction of a unique differential bulk flotation circuit with the corresponding reagent regime, a substantial increase in the Pb–Zn concentrate from 39 wt% to 60 wt% total metal content was obtained with a high recovery of lead and zinc (90% and 98%, respectively). As galena is the carrier of silver, this also means the highest possible recovery of silver (the ore contains 75–80 g Ag/t).

The essential feature of the differential bulk flotation circuit is that the roughing flotation is divided into two stages, each stage differing in the method used to float lead and zinc. In the first stage the main purpose is to obtain the maximum recovery of lead and to strongly depress pyrite. During this stage locked particles of galena and sphalerite are also floated. This is done by using a selective and powerful collector for galena, such as ammonium dibutyl dithiophosphate and benzene ammonium dithiophosphate—also beneficial to precious-metals recovery in the galena–pyrite separation. No activator is used. Flotation takes place in a weakly alkaline medium (pH 8–8.5).

The second stage is to ensure the maximum recovery of sphalerite while strongly depressing pyrite. The xanthate–copper sulphate system in a high alkaline medium is applied (pH is controlled at 11.5–12). The principal flowsheet is shown in Fig. 1.

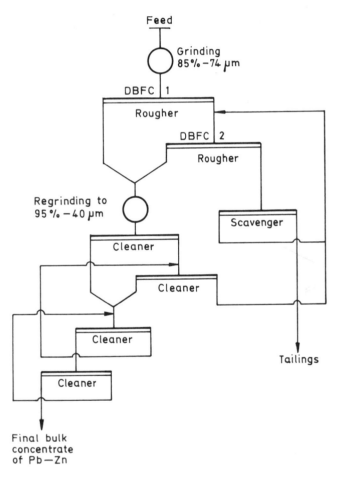

Fig. 1 Differential bulk flotation circuit (DBFC) for lead–zinc complex ore, Fankou mine

The ultrafine grind is a prerequisite to the upgrading of the final bulk lead–zinc concentrate.

The differential bulk flotation system for the treatment of fine- to ultrafine-grained complex sulphides has been widely applied.

The ISP pyrometallurgical method of treating Pb–Zn bulk concentrates from Fankou mine yields the following metallurgical performance: recovery of lead and zinc, 94%; acid utilization, 91–92%; recovery of by-products, 80–85% for Ag, 87% for germanium and 48% for mercury.

Treatment features of sulphide-bearing tungsten–tin ores [4,5]

As described above, there is a great variety of this type of ores in which wolframite and cassiterite occur together with base-metal sulphides. They are defined arbitrarily as tungsten ores when wolframite is the main economic mineral and as tin ores when cassiterite is predominant. The common treatment features of these complex tungsten and tin ores are noted below.

(1) Multistage grinding and classification closely combined with corresponding multistage gravity concentration to recover the liberated heavy minerals particles as rapidly and to the maximum extent possible after each stage of grinding and classification.

(2) In the main gravity concentration minor circuits are incorporated that comprise other separating operations for mixed concentrates of heavy minerals, e.g. removal of sulphides by flotation, separation of wolframite and cassiterite by magnetic means, separation of scheelite and cassiterite by corona-type electrical separators or flotation (for finer particles), etc.

(3) All slimes (primary and secondary slimes of −200 mesh) are collected and undergo a special treatment in a separate slimes plant. Various home-made gravity concentrators, such as the centrifugal concentrator,[7] the rocking–shaking belt vanner,[6] the multicarved deck slime table,[8] combined with fine to ultra-fine particle flotation to recover wolframite, scheelite, cassiterite slimes, etc., are employed.

(4) The upgrading of low-grade 'dirty' concentrates and the treatment of refractory middlings by chemical treatment with further

grinding to liberate the locked particles and subsequent treatment by physical separation techniques fail to produce the final concentrates. Current industrial applications of chemical treatment for these products include the following:

(i) The fuming process for the treatment of rich tin middlings (3–5 wt% Sn)

(ii) Chloridizing roasting of polymetallic tin middlings (1–3 wt% Sn) and subsequent hydrometallurgical treatment for separation and recovery of the metal values in the enriched dust product containing Sn, Pb, Zn, Cu, In, Cd, etc.

(iii) Treatment of anode slimes from the tin leaching– electrowinning plant containing lead and silver by the LPF process (acid leaching–precipitation–flotation of silver)

(iv) Separation of tungsten–cassiterite mixed products containing 10–20 wt% Sn and 20–40 wt% WO_3 by a variety of hydrometallurgical methods, producing tungsten trioxide powders and tin concentrates.

(v) Removal of phosphorus impurities from tungsten concentrates to produce standard high-grade concentrates by acid leaching

(vi) Production of ammonium paratungstate from 'dirty' tungsten concentrate from the slime treatment plant by use of the alkaline leaching–solvent extraction–ammonium conversion process

Most tungsten ores occur in vein-like deposits in which the main tungsten mineral, wolframite, is in close association with the quartz veins. During mining operations dilution of the barren host rock is more or less unavoidable. Hence, most tungsten ores are subjected to preconcentration by a home-made sorting machine, the magneto-optical sorting machine,[5] or dense media separation techniques to discard a significant amount of barren rocks (40–60% of run of mine).

Owing to the fineness, texture and disseminated mineralization of complex tin ores, preconcentration is often not so effective as that in tungsten ores.

Generally speaking, not only are the degree of fineness and intergrowth of the valuable minerals higher in tin ores than in tungsten ores but the amount of sulphides in the ore is also higher and more pyritic in character. All this leads to some differences in the set-up of the framework for the overall mineral processing circuit and the incorporation and combination of minor separating devices into the head circuit (see Figs. 2 and 3).

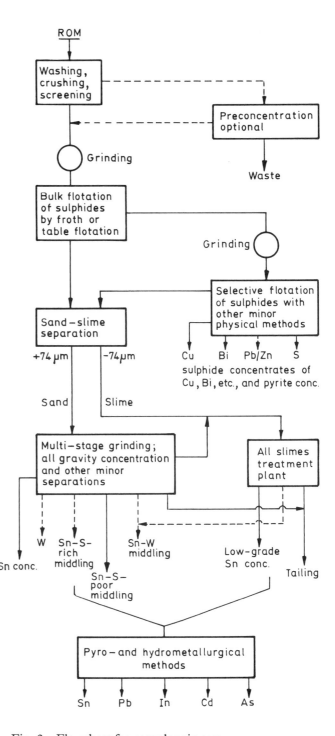

Fig. 2   Flowsheet for complex tin ores

171

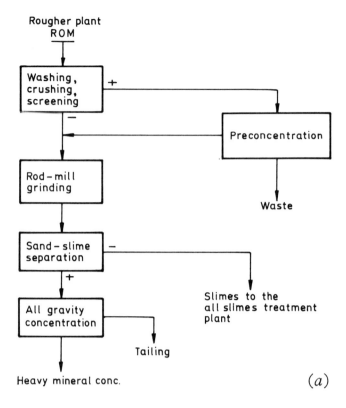

(a)

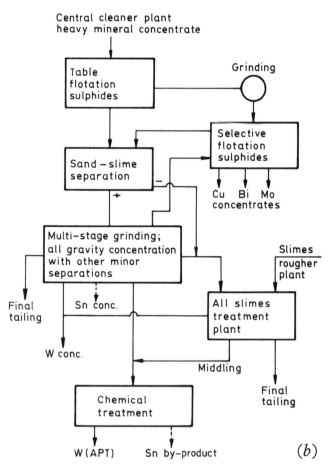

(b)

Fig. 3   Flowsheets for complex tungsten ores

In the principal flowsheet for complex tin ores the main gravity concentration circuit is often preceded by a bulk flotation of the sulphides, whereas in the case of tungsten ores the principal flowsheet can be divided into two parts—the primary concentration of heavy minerals, including sulphides, and the subsequent more complex secondary cleaning concentration incorporating various kinds of physical separations and even chemical treatments of semi-products.

The cleaning concentrator is often designed as a large centralized concentrator to treat a variety of rougher concentrates produced by small mines. It is therefore characterized by high flexibility of operations in solving metallurgical problems arising from variations of the mineralogical composition of the incoming rougher concentrates, such as the provision of blending operations, multifeed system and its divertors, surging–storage tanks, etc.

Before the introduction of the comprehensive and rather sophisticated flowsheet for complex tin and tungsten ores, the pre-war flowsheet was all-gravity concentration without separate slimes treatment. Consequently, the final concentrates of tin and tungsten were low-grade and recoveries were very low—less than 50% for tin and less than 60% for tungsten ores (mainly due to losses of Sn and W in the discarded slimes and the inability of the process to deal with refractory polymetallic middlings). With the introduction of the new flowsheet tin and tungsten recoveries of up to 80% could be obtained with standard-grade concentrates. Additionally, by-products of metals (Cu, Bi, Mo, etc., and some precious metals) were also produced. Optimal utilization of all values in the sulphide-bearing polymetallic tin and tungsten ores was thus realized.

Main features in treatment of copper sulphide ores

China's copper sulphide deposits are generally monometallic in character. Only copper is of economic grade, other base-metal sulphides being effectively below the cutoff grade, with the exception of a few copper–zinc deposits. The natural grains of copper minerals are, on average, of medium size, but in some pyritic copper sulphide ores copper minerals are ultrafine and intergrown in the pyrite matrix.

Generally speaking, most of the copper sulphide ores are therefore relatively more amenable to treatment by conventional sulphide flotation. However, important R&D in China led to new proven techniques to increase both the grade and recover of the copper concentrate, among which may be cited the following.

(1) Replacement of conventional xanthate and Aerofloat collectors by new powerful and selective collectors for the recovery of fine particles and precious metals and the selective separation of copper sulphides from pyrite. Industrial practice in Chinese mines has shown that among these newly developed collectors ammonium dibutyl dithiophosphate and cyanoethyl diethyl dithiocarbamate appear to have excellent properties. Industrial applications on skarn-type copper ores and pyritic copper ores of different grades and natural grain sizes have given good metallurgical performances. With a feed grade of 0.5–3 wt% Cu the copper concentrates produced have a grade of 25–28 wt% Cu with 92–94% recoveries. Recovery of precious metals also showed a definite increase.

(2) Innovative development of the multifeed flotation circuit.[9] The concept of this newly developed circuit is best illustrated by reference to Fig. 4.

Flotation circuit feed is divided into two parts. Froth from the first part is wholly or partially combined with the feed from the second part for flotation. This circuit is especially suited to the treatment of very low-grade and fine-grained ore. After each stage of multifeeding the grade is increased by the addition of froth product.

The main advantages of the multifeed circuit are the reduction of reagent consumption by 20–30%, the better metallurgical performance due to a thicker froth in the second feed flotation circuit, resulting in better drainage of impurities

and detrimental ions in the pulp, the autogenous *carrier effect* of coarse particles of the first froth on fine and ultrafine particle flotation in the second feed,[11] and the ease and speed of switching from a conventional flotation circuit to a multifeed flotation circuit or back to the original circuit make plant testing easy to carry out and require only a few hours' downtime.

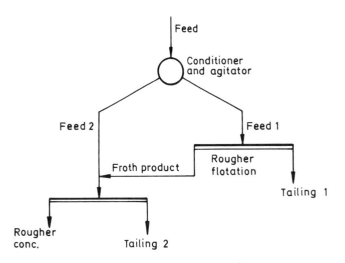

Fig. 4   Principle of multifeed flotation circuit

A further development of the multifeed flotation circuit, also called the serial flow multifeed circuit, as applied in the largest copper feed concentrator treating low-grade porphyritic copper sulphide ore of the Dexing copper mine complex, is shown in Fig. 5.

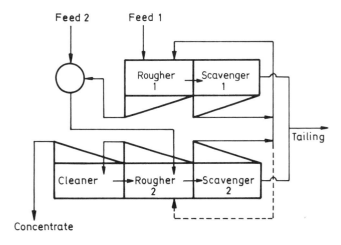

Fig. 5 Application of the serial flow multifeed circuit

(3) The differential-speed flotation system is based on the idea that the same mineral species show different flotation rates, i.e. the flotation

time needed is different. Hence, total flotation time should be subdivided into different periods and the corresponding flotation froth of each subdivision skimmed off separately, as shown in Fig. 6.

Applications and testing of the multifeed circuit and differential speed flotation circuit have been undertaken with sulphide ores of antimony, copper, copper–molybdenum and lead–zinc, both the grades and recoveries of the final concentrate being increased. The two circuits can be combined to raise the overall efficiency of the whole flotation circuit.

(4) *Collectorless flotation* of copper sulphide ores has been studied since the early 1970s, but, to date, China alone operated the technique on the industrial scale in the early 1980s at a skarn-type copper mine. Good metallurgical results, comparable with those obtained by normal collector flotation, were achieved.

Pilot-test results on a variety of copper sulphide ores are shown in Table 2. The potential for considerable savings in reagent costs is apparent.

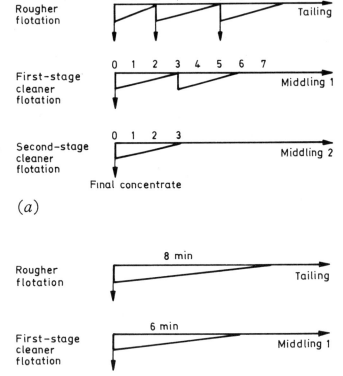

Fig. 6 Froth skimming time of (*a*) differential speed and (*b*) flotation circuits

Table 2   Results of collectorless flotation

Ore type	Feed and product	Collectorless flotation		Collector flotation	
		Grade, wt% Cu	Recovery, %	Grade, wt% Cu	Recovery, %
	Feed	0.53	100.00	0.53	100.00
Porphyry	Rougher concentrate	7.23	90.75	6.66	90.76
Skarn	Feed	1.55	100.00	1.63	100.00
	Rougher concentrate	15.05	93.21	17.25	91.68
Disseminated	Feed	3.45	100.00	3.45	100.00
	Rougher concentrate	22.95	95.30	22.14	95.20

ACKNOWLEDGEMENT

The authors wish to acknowledge generous support by the Beijing Central Research Institute for Mining and Metallurgy, the Central South University of Technology and Beijing General Research Institute for Nonferrous Metals

References

1. Geology of nonferrous metals deposits in China. In *Nonferrous Metals Industry in Modern China* (Beijing: Metallurgy Publishing Co., 1985), Vol. 1.

2. Shi Dashen. Separation of copper, lead, zinc sulphides in China. (Beijing: Metallurgy Publishing Co. 1980).

3. Li Fenglou. New technology for the separation of pyritic lead–zinc sulphide ores by differential bulk flotation. *Chinese J. Nonferrous Metals*, no. 2 1988, 30-5. (Chinese Text)

4. Li Zhimin. China's tin processing plants. (Beijing: Metallurgy Publishing Co., 1985) Vol. 9, 80-241.

5. *An overview of China's nonferrous mineral processing plants* (CNNC Publishing Co. 1989) Vol. 12, 600–700.

6. Qiu Jichun. The rocking–shaking travelling belt vanner. Paper presented to 13th International Mineral Processing. Congress, Warsaw 1980.

7. Du Tiyao. Treatment features of Yunnan's refractory eluvial ores. Paper presented to 13th International Mineral Processing Congress, Warsaw, 1980.

8. Pan Guo Zhu. Development and application of the Yunnan Tin Corporation six-deck slime table. In *Asian Mining '88* (London: IMM, 1988), 117-27.

9. Huang Kaigui. Research on multifeed flotation circuit. Paper presented to the National Nonferrous Flotagents Conference, Beijing, Oct. 1982.

10. Huang Kaiguo. Commercial test of the serial flow multifeed flotation circuit at DeXing Copper Mine. Special Technical report, 1986.

11. Qiu Guanzhou. Coarse Particle Effect on ultrafine Particles Flotation. Paper presented to First National Conference on Mineral Processing of Nonferrous Metals Ores. Beijing, Oct.11, 1986.

12. Huang Kaiguo. Collectorless flotation of copper sulphide ores. *Jl Center Univ. Technol.* vol. 19, no.4, Aug. 1988.

Is flotation the unavoidable way for beneficiating metal sulphide ores?

J. De Cuyper
Ch. Lucion
Laboratoire de Traitement des Minerais, Université Catholique de Louvain, Louvain-La-Neuve,
Belgium   m.

SYNOPSIS

Although it is clear that flotation has become the conventional way for beneficiating metal sulphide ores and will remain so as long as grade is sufficient to bear the cost, other methods have to be taken into consideration. These can be either physical concentration other than flotation or metallurgical processes.

Physical separation methods such as gravity concentration, magnetic separation, electronic sorting are not only used at preconcentration stages, but also with the aim of upgrading sulphide concentrates produced by flotation. Selective flocculation has also been proposed as a possible solution for concentrating ultra-fine sulphide particles, which could not be easily recovered by flotation.

While the metallurgical processes such as roasting, smelting and leaching have been particularly well adapted to the treatment of fine sulphide flotation concentrates, there are some cases where sulphide ores can be advantageously treated directly in roasting or smelting furnaces, e.g. in the segregation process applied to mixed oxide and sulphide ores, which are refractory to physical separations and also in the so-called "pyritic" smelting.

But the main applications of direct metallurgical treatment of sulphide ores are to be found in the direct leaching of low grade sulphide ores or wastes, in combination with bacterial oxidation. While the process has been first developed for copper sulphides, it has been recently extended to the extraction of gold from refractory ores containing pyrites and arsenopyrites.

Finally, this review also includes the special case of massive pyrite ores, which contain up to 90 wt % of pyrite and which are not amenable to selective metal sulphide separation by flotation.

A large number of non-ferrous metals are known to occur as sulphides in ore deposits : geochemists use to classify them as "chalcophile" elements due to their strong affinity to sulphur. From a metallurgical point of view, these elements are considered as "non-reactive" metals, in contrast to other non-ferrous metals which show a strong affinity to oxygen and therefore only occur as oxides within the earth's crust.

The so-called "non-reactive" non-ferrous metals sulphides, to which one should associate selenides, tellurides, arsenides, antimonides and sulpho-salts (sulpharsenides, sulphantimonides and sulphobismuthides) are all characterized by specific gravity values above 4. Their separation from the rest of the mineralized rock should thus be easily obtained by gravimetric concentration methods. Unfortunately, these valuable minerals are in most deposits associated with iron sulphides, which have specific gravities ranging from 4.6 to 5.1, and might constitute in some cases the dominant mineral of the ore.

It rapidly became evident from the limitations observed in the application of gravity methods to the recovery of fine size minerals, that in order to meet the ever-increasing demand for non-ferrous metals, another process had to be developed to treat sulphide ores showing intricate associations of various valuable metals sulphide minerals, eventually very finely disseminated down to liberation sizes of 10 µm or even smaller.

The discovery, about 70 years ago, of the possibility of rendering sulphide minerals hydrophobic after addition of minute amounts of short chain surfactants such as ethyl xanthate has really been the starting point for the development of the flotation process, together with the finding that, by choosing the appropriate chemical environment, it was also made possible to selectively separate the various sulphide minerals from each other. Sulphide minerals flotation still remains today the most extensively used process in providing the world's non-ferrous metals industry with its main sources of copper, lead, zinc, silver, cadmium, molybdenum, platinum and platinum-group elements, germanium, bismuth, selenium, tellurium, mercury, indium, rhenium, arsenic, antimony and with a great deal of its sources of cobalt and nickel.

Nevertheless, this paper will not deal with the remarkable progess which has been made over the years in the understanding of the sulphide minerals flotation phenomena, nor with major successful developments of this process in the mineral industry. On the contrary, it will be attempted to look at alternative processes to flotation in the production line from sulphide ores to metals.

As a matter of fact, despite flotation is remarkably well appropriate to the beneficiation of most metals sulphide ores for the production of selective marketable concentrates, it still has to face some challenges, particularly in relation to the poor response of some of these ores due to their very complex texture and mineralogical composition, the presence of relatively large proportions of iron sulphides and the presence of alteration minerals ...

From a recent world scale survey[1], which covered 22 flotation plants treating complex copper – lead – zinc sulphide ores and representing a total daily capacity of 70,000 tons of ore, it came out that the percentage of the total potential value of the ore actually paid to the mining companies was, for two thirds of the plants, below 40, with a minimum as low as 28. Such low figures clearly reflect the need for improvements.

ALTERNATIVE PHYSICAL SEPARATION METHODS

Gravity concentration

As mentioned above, the valuable constituents of sulphide ores are of a sufficiently high density to allow gravity techniques to be successfully applied, on condition that the concerned minerals are made free at suitable, not too fine grain sizes. Taking account of the complex texture and composition of most sulphide ores, it is thus not surprising that the only application of gravity methods in their treatment is to use such methods as a preconcentration step in order to reject a barren waste product at the coarsest possible size.

To illustrate the advantages of incorporating gravity preconcentration in a complex sulphide ore treatment flowsheet, a comparison is given in Fig. 1 between two flowsheets applied to the same head sample from an Andean deposit. By applying flowsheet II instead of flowsheet I, the weight of ore to be finely ground is reduced, resulting in a 230 % increase in capacity of the grinding section and in a 70 % decrease of energy consumption per ton of raw ore.

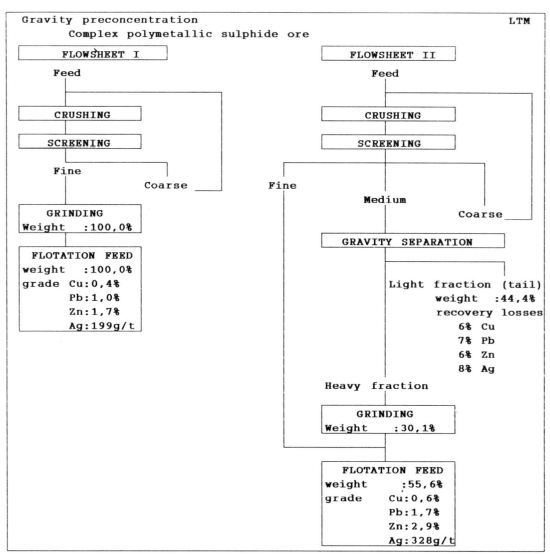

Fig. 1 : Gravity preconcentration of a complex polymetallic sulphide ore.

At present metal prices, the total value lost in the gravimetric concentration tails represents only 6.8 % of the total realizable value in the head sample, while the flotation feed value per ton is upgraded by 65 %. By erecting the gravimetric concentration plant close to the mine, an additional advantage would arise from a substantial reduction of ore transportation costs, since the grinding-flotation plant is located at a distance of about 6 kilometers by steep road.

Most sulphide ores preconcentration plants use heavy medium separation as being the easiest and most efficient gravity separation process to treat at high throughput a very wide size range of material. Of course, this is not a new development in·the treatment of sulphide ores : indeed, it had already been introduced, in 1948, in the famous Sullivan concentrator of Cominco Ltd for the preconcentration of lead and zinc sulphide ores[2]. Later on, it has been included at several other places, such as Vieille Montagne's Lüderich plant and Metallgesellschaft's Meggen plant in the F.R.G., Northgate Exploration's Tynagh plant in Ireland, Zambia Consolidated Copper Mines' Kabwe plant in Zambia, Asarco's Young mill in Tennessee (U.S.A.), the Xikuangshan mill in the Hunan province of P.R. China, the latest and also one of the largest built being the Mount Isa Mines' plant in Australia where cyclones were chosen as separators. This plant was commissioned in 1982 : it achieved target metallurgical performance and reduced concentrator cost/ton of metals in the concentrates[3]. With new progress made in heavy medium separation, particularly for the treatment of finer material, and taking account of the much lower energy it requires compared to fine grinding and flotation, further development of heavy medium separation as preconcentration step in sulphide ore treatment plants is likely.

Of course, other gravity concentration devices, such as jigs, tables and Reichert cones, also find use in sulphide ore treatment plants, ahead of secondary grinding, e.g. in order to recover native metals, such as copper in the Afton concentrator in Canada[4] or gold in the Boliden concentrator in Sweden[5], which would otherwise be hardly recovered by flotation due to their too large grain size.

Magnetic separation
Significant progress has been made in the field of magnetic separation of minerals in the last thirty five years, with the successive developments of wet high-intensity, high-gradient and cryogenic superconducting separators. Such advances were made mostly under the pressure of having to solve the problems encountered in the concentration of large tonnages of low grade weakly magnetic iron ores and in the brightening of kaolin clays. While an important breakthrough already came when Jones devised his steel grooved plates matrix system in the air gap of the magnet in order to generate a high field gradient, at the same time, series of new materials were discovered at an exponential rate, either highly energetic or superconducting, including those which make the critical transition to superconductivity at increasingly high temperatures.

As a result, the range of applicability of magnetic separation can now be extended to many weakly paramagnetic minerals, even of the micrometer size range. This means that magnetic separation might have some potential in future beneficiation of sulphide ores.

About ten years ago, few sulphide minerals were known to be extracted using a high intensity wet magnetic separator, namely : pyrrhotite (Fe_7S_8), renierite ($Cu_6Fe_3GeS_{10}$) and alabandite (MnS)[6]. It is thus not surprising that in some sulphide ore flotation plants, like in the El Monte plant of Cia Fresnillo in Zimapan, Mexico, treating a complex Cu - Pb - Zn - Ag sulphide ore, rich in pyrrhotite, a magnetic separation step found a place in the flowsheet just before flotation in order to discard pyrrhotite which otherwise would contaminate the flotation concentrates. In the copper - zinc sulphides flotation plant of Kipushi in Zaïre, a high grade germanium concentrate has been successfully produced by applying magnetic separation to the copper sulphide flotation concentrate which indeed contained most of the renierite present in the ore[7]. Alabandite which exhibits magnetic properties very like those of renierite might thus similarly be captured by high gradient magnetic separators in order to be removed from the sphalerite flotation concentrates to which it is generally associated ; as a consequence, zinc concentrates of higher market value could be obtained.

Of course, taking account of the latest developments of magnetic separation, new potential applications are to be considered. Indeed, while most of the other sulphide minerals are diamagnetic with negative magnetic susceptibility, some are weakly paramagnetic and therefore should be effectively separated, not only from sulphide minerals belonging to the former group but even between them. Examples of such weakly paramagnetic minerals are given by pyrite (FeS_2), arsenopyrite ($FeAsS$), sphalerite (ZnS) and copper-iron sulphide minerals such as chalcopyrite ($CuFeS_2$), bornite (Cu_5FeS_4) and cubanite ($CuFe_2S_3$).Their magnetic properties however widely vary with their origin and the presence of impurities. This is particularly well illustrated by sphalerite from high-temperature hydrothermal deposits usually rich in iron : such sphalerite exhibits higher paramagnetic susceptibility (up to 5.9 x 10^{-6} m^3 kg^{-1}) while pure zinc sulphide is diamagnetic[8].

Recent attempts[9,10] have been made at laboratory scale at applying magnetic separation to the upgrading of various metal sulphide flotation concentrates, such as chalcopyrite concentrates (to reduce the pyrite content), molybdenite concentrates (to reduce the copper content) and galena concentrates from the copper - lead separation which is commonly applied to bulk Cu - Pb flotation concentrates but which is never complete, even when the minerals are completely liberated. The results obtained were sufficiently encouraging to justify further investigations in order to determine the optimum operating conditions and to evaluate the economic feasibility. Compared to flotation, which is dependent on the surface properties of the particles, magnetic separation is not. Therefore, it should offer better opportunities for getting efficient separations involving sulphide particles contaminated by coatings of another sulphide mineral. However, as a result of the tendency of some flotation concentrates to agglomerate, several mechanical entrapment often occurs which should then be prevented by adding appropriate dispersants or by using vibrating high gradient magnetic separators[11].

Magnetohydrostatic separation
Magnetohydrostatic separation, which is a process using either a solution of a paramagnetic salt or a ferrofluid, is a very attractive method, since it can separate non-magnetic or weakly magnetic particles of different specific gravities without being limited by the maximum cutting density (around 3.5 kg/dm^3) which characterizes the dense media commonly used in gravimetric separation. It should be still more promising to separate heavy metal sulphide minerals from each other, since it can be applied in principle regardless of particle size, shape and size distribution.

Despite these interesting prospects and numerous positive results actually obtained in several laboratories, there have been no commercial applications of the process to sulphide ores beneficiation, except maybe for the refined separation of medicinal cinnabar in P.R. China[12].

The limitations observed are mainly due to the low yield and efficiency of fine particles separation and to the high cost of ferrofluids. However, new advances have been made by Intermagnetics General Corporation[13] that overcome these limitations by introducing the centrifugal force and developing an original design concept of the separator magnetic poles that enables a more efficient utilization of the magnetic field and, as a result, the use of highly diluted and much less expensive magnetic fluids. Testwork using this new separator was performed on a 6.9 % Zn sulphide ore with a size distribution ranging from 0.6 to 600 microns to produce a 50.4 % Zn concentrate with a zinc recovery of 96.2 %[14]. The low capacity of the presently available equipment still limits its commercial applicability.

Sorting
The development of high speed and high capacity ore sorting systems may contribute to greater benefits for the mining industry by avoiding costly selective mining procedures and by increasing ore reserves and mining rates. Since these sorters are used to preconcentrate run-of-mine ore, they also reduce the operating energy consumed for the subsequent treatment and reduce tailings disposal and environmental problems.

Most of the existing systems use sensing devices based on the measurement of a radiation of one kind or another and are operating successfully on different types of ore, under the condition that there is a reliable difference in the physical property concerned between the valuable minerals and waste.

As far as their applicability to sulphide ores is concerned, a gap has been recently filled by Outokumpu's Precon ore preconcentrator based on the measurement of scattered gamma radiation to determine the sum metal content of the ore. This has been used on a commercial scale at the Hammaslahti copper sulphide ore flotation plant in Finland and allowed a lower grade of ore to be mined, with the effect of extending the life of this mine by some six months[15].

Selective flocculation
The processing of very fine ore particles, generated during mining and milling operations, with sizes that might go down to less than 10 microns is one of the most challenging problems in ore processing. Even in the flotation process, which is best adapted to sulphide ores, difficulties arise with fine particles[16]. They are due to several reasons such as reduced collision probability with air bubbles, lower floatability due to higher oxidation, slime coatings, high reagents consumption, mechanical entrainment in the froth. As a result, recoveries become poor and/or concentrates grades are too low.

Various approaches have been proposed to solve these problems. Selective flocculation followed by floc separation by flotation is the one which has received the most attention. Although this process has been successfully used on a commercial scale for the beneficiation of low grade iron oxide ores by Cleveland - Cliffs Inc. in Tilden, Minnesota[17] and for that of potash ore by a Cominco Ltd plant in Saskatchewan[18], its application to sulphide ores remains essentially confined to laboratory scale investigations either with single sulphide minerals or with synthetic mixtures of a sulphide mineral and a gangue mineral[19,20,21]. Recent studies conducted with sulphide minerals pairs have shown that due to the effect of dissolved species released by these sulphides, their respective flocculation characteristics can be completely modified[22]. It is clear that both basic and applied research are still needed before selective flocculation can be successfully applied to the processing of sulphide ores. It is also clear that its combination with the recent successful development of column flotation cells might offer real new potentialities, as already observed in the case of some oxide minerals mixtures[23].

DIRECT METALLURGICAL PROCESSES

Direct roasting
The original development of the roasting process for base metals sulphide ores was justified by the need to remove most of the sulphur from a sulphide ore by oxidation with air in order to prepare the ore as a feed to reduction smelting or distilling. In the case of copper sulphide ores however, only part of the sulphur had to be eliminated, advantage being taken of the strong affinity of copper for sulphur and its weak affinity for oxygen, in comparison with iron. Rich ores and gravimetric concentrates were used and multi-hearth roasters were well adapted to the size range of this feed material. Later on, with the advent of the flotation process, new types of roasting furnaces (such as flash and fluidized bed) were developed to better suit the finer size range of selective flotation concentrates, to reach better kinetics of the process and to get precise control of temperature as required in sulphatizing roast. Parallel to these deve-

lopments, the proportionate importance of pyrites as raw materials for nonfatal sulphuric acid production has been declining at the benefit of elemental sulphur.

As a result, the application of roasting directly to sulphide ores finally became restricted to massive, fine grained, low base-metal sulphide ores containing large amounts of pyrite, from which it is difficult to extract, economically, saleable concentrates by flotation. In such cases, as well as with a few other ores like some mixed sulphide - oxide copper ores, which were considered as refractory to normal and cheaper mineral dressing or leaching processes, roasting may occur in different ways as follows :
- two successive stages, respectively oxidizing and chloridizing. The objectives are to produce sulphuric acid and iron pellets and to recover the non-ferrous metals by hydrometallurgy. Various processes have been developed and are used either as such or in combination : D.K.H., Kowa-Seiko, Montedison[24,25]. In the particular case of copper refractory ores where in fact sulphide copper forms only a minor proportion of total copper, the objective is to take advantage of the segregation of copper metal to produce a high grade metallic copper concentrate. This process, known as the Torco process, has been used during several years at commercial scale in Akjoujt, Mauritania[26];
- two successive stages, respectively a "selective" roasting and a sulphatizing one. The selective roast has to be performed in a multiple hearth furnace so as to ensure the most precise control of temperature and atmosphere in order to avoid any recrystallization of the calcine and any secondary oxidation reaction generating ferrites and spinels. The calcine from this selective roast has a high porosity resulting in a complete selective sulphatization of the non-ferrous metals in the second roasting furnace which is of the fluidized bed type. The final calcine produced has very good non-ferrous metals leaching properties and the leach solutions have excellent settling and filtration properties. Successful results have been obtained at pilot scale by NESA S.A. (Division of Mechim Engineering S.A.) on a copper - zinc - lead pyrite ore[27];
- only one stage, which is sulphatizing and uses a fluidized bed roaster. Sulphuric acid is produced from the gas and the sulphate calcines are leached, zinc being recovered by electrowinning after purification of the leach solution by cementation giving copper and cadmium cements, while the leach residue may be treated by brine and/or cyanide leaching to recover the precious metals and a high grade lead oxide product[28]. On the base of pilot testwork performed on a low grade zinc - lead - copper bulk flotation concentrate, calcines have been produced with levels of ferrite < 10 %. But sulphide ores have also been tested with the same success[29].

Direct smelting
In the early days of copper sulphide smelting before finely divided concentrates were made available on the market, blast furnaces were widely used for copper matte smelting and, in the case of massive pyrite ores, a so-called "pyritic smelting" process had been developed where little or no extraneous fuel was used to provide the necessary heat. It required a massive pyrite ore with little gangue and a flux of almost pure free silica. The process was economical, but not at all flexible : it disappeared in favour of reverberatory smelting, particularly well suited to a wide range of feed materials.

Nowadays, the non-ferrous metals sulphide minerals charged to smelting furnaces are all fine grained materials and there is a trend towards still finer feed sizes in view of the exploitation of the enormous reserves of base metals in complex pyritic ores. One should thus expect direct smelting of ores to be definitely forgotten and smelters designers to be prepared to follow this trend, not only in the design of smelting furnaces but also for materials handling, gas cleaning and sulphur recovery.

Nevertheless, in the case of complex pyritic ores which would require extremely fine grinding and a very complicate differential flotation flowsheet in order to produce saleable concentrates, unfortunately with rather poor marketing value recoveries, the question may arise about future innovations in smelting technology adapted to the direct treatment of such ores. Some pioneering investigations are under way to evaluate the possibilities of microwave smelting of sulphide minerals[30].

From the experience accumulated in both roasting and smelting of pyrite at its Kokkola plant in Finland, Outokumpu has developed a new concept for direct pyrite ore flash smelting, where pyrite would be partially oxidized so as to produce, besides rich SO_2 gas and discardable slag, a small amount of matte recovering most of the precious metals and copper. The applicability of the process has been proven by pilot tests performed on various fine size pyrites[31].

Current investigations into the development of a new direct smelting process to treat complex polymetallic Pb - Zn - Cu sulphide ore feeds without prior beneficiation are also underway at the University of Birmingham in a project sponsored jointly by the Commission of the European Communities and the Mineral Industry Research Organization[32]. As schematically illustrated in Fig. 2, two hearths are sitting side-by-side and linked by an overflow weir at one end and a RH vacuum degassing vessel at the other. Matte is forcibly circulated in a closed loop. As it passes through the oxidation zone of the converting hearth with top blown oxygen, blister copper is produced, while as it passes through the RH vessel, zinc and lead are flashed off and recovered in a condenser.

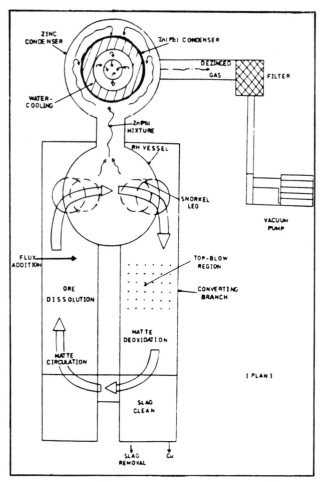

Fig. 2 Schematic layout of new polymetallic smelter proposed by N.A. Warner[33].

These recent developments, together with many others which in the recent years led to several new commercialized smelting processes for non-ferrous metals sulphides concentrates show evidence of the ability of smelting to maintain its competitive position facing the development of leaching processes often considered as more acceptable in environmental terms.

Direct leaching
Direct leaching of non-ferrous metal sulphide minerals requires oxidative conditions. These can be obtained :
- by addition of oxidants such as ferric salts, nitric acid, hypochlorite, chlorine, hydrogen peroxide or even solid oxidants as MnO_2,
- by leaching under high air or oxygen pressures at temperatures ranging from 80 to 250° C,
- by biological leaching using bacteria such as thiobacillus ferrooxidans to enhance the oxidation kinetics of Fe^{++} to Fe^{+++} and that of the sulphide minerals at near ambient conditions in air.

In fact, the first commercial leaching plants were developed to treat easy soluble oxide ores and concentrates. Later on, their use has been extended to the treatment of sulphide concentrates after their previous transformation into oxides or sulphates by roasting. This is still the prevailing method nowadays for concentrates beside direct oxidative leaching processes.

As far as direct leaching of sulphide ores is

concerned, one has to remind the ferric sulphate leaching process developed in 1926 at Inspiration, Arizona, to extract copper from both oxide and sulphide minerals in a single leaching operation. This process has served Inspiration for some 30 years to treat a half sulphide, half oxide copper ore containing about 1 % total Cu. A few years after the process was abandoned, while the first patent for the use of thiobacillus ferrooxidans in the leaching of metals from ores was issued in the U.S.A. Without any longer delay, this was followed by a few early commercial applications of bacterial leaching to dump of low-grade copper sulphide ores. Low capital and operating costs of the copper produced by this process, compared to conventional smelting of sulphide concentrates, led to a world-wide acceptance of this technology by the mining industry, at least in addition to its main traditional mining activities[34].

One has to keep in mind that such rapid success was also related to the development of the solvent extraction process for the treatment of dilute copper solutions before electrowinning. But, on the mining side, the success maybe was too easy because of the low costs which actually masked low metal recoveries. Developed on a strongly empirical base, heap and dump leaching still needs research to be conducted on air circulation, type, concentration and location of bacteria within the dumps, solution flow and the effect of fines and precipitated salts[35]. The need for further research is particularly evident in view of extending the bacterial leaching concept to solution mining of sulphide ore deposits. New mining methods, such as FDLC (Flood Drain Leach Cell), have already been concieved for porphyry copper deposits with the aim to double the in situ recovery of copper compared with black caving and do it at a price competitive with conventional surface mining and processing. A computer model has even been constructed for bacterial leaching of the various types of porphyry ores which can predict the recoveries to be expected for FDLC mines[36].

Other base metals can be extracted from their sulphides by direct ore leaching, but up to now no commercial application is known. While most of the experimental work has been performed on concentrates, some laboratory tests were run with ore samples using bacteria. So, thiobacillus ferrooxidans was shown to accelerate the leaching of silver from a lead - zinc sulphide ore which contained silver mainly as tetrahedrite[37]. This thiobacillus was also shown to have a beneficial effect on the in situ leaching of copper and zinc from a worked-out ore stope at the Prieska copper - zinc mine in South Africa but the full - scale bacterial leach project was abandoned because of the channeling problems owing to the particular shape and dip of this orebody[38].

During the last decade, a considerable interest has grown in the heap and dump leaching of lower grade gold/silver oxidized ores, with the result that much more careful investigations were initiated on this apparently simple process in order to optimize metal extraction. The wider experience that will be gained on gold ores will undoubtedly be profitable to any future development in the whole field of heap leaching. Also, more attention is now drawn on the importance of coordinated geotechnical, geochemical, hydrogeological and metallurgical evaluation from the onset of any new project[39].

But the attractive possibilities offered by the biological oxidation of sulphide minerals as well as by pressure oxidation opened the way to the development of alternate processes to roasting in the treatment of refractory gold ores of the pyritic and arsenopyritic types[40]. Several commercial scale plants are in operation for pretreating refractory gold ores, either by pressure oxidation or by bioleaching[41]. In the latter case, leaching occurs in agitated vessels[42] : indeed, biological oxidation on heaps is still in a development stage.

CONCLUSIONS

As surprising as it may be, it is a negative answer that has to be given to the question whether flotation is a necessary step in the treatment of sulphide ores or not. Indeed, it has been shown that some sulphide ores were not subjected to any commercial flotation process, nor even to any other physical beneficiation method at all. These ores however are not high grade ores, but rather marginal ones or massive pyrites. The former are treated by bacterial leaching on heaps or dumps, while the latter are treated by direct roasting. All the other cases reported concerned either experimental testwork or abandoned commercial processes.

Among the physical concentration methods other than flotation, several are actually used on a commercial scale for beneficiating sulphide ores, but always in addition to flotation, either as preconcentration method or for upgrading flotation concentrates. It is thus clear that the flotation process, although not unavoidable, is used in nearly every case and still knows substantial developments, not only in equipment design, but also in reagents chemistry and process control.

Nevertheless, numerous problems still remain to be solved in the treatment of complex sulphide ores, such as the polymetallic sulphide ores rich in pyrite (or pyrrhotite), characterized by very complex textures and the presence of alteration minerals as well as minor minerals rich both in precious metals and in elements which are highly penalized on the market. The relative importance of such problems widely varies with the orebodies and may result in a strong diminution of the total "marketing value recovery" of the flotation concentrates, to such a low level as corresponding to a limited percentage of the total potential value of the valuable metals of the ore. This percentage might become so low as to eliminate any profit for the miner.

To face such perspectives made more acute by the ever decreasing grades and higher complexity of the ores, the mining industry will have to reduce its mining costs by using simpler and thus less selective methods, which will result in the necessity of introducing preconcentration stages at the coarsest possible grain size. In front of decreasing marketing value recoveries of the flotation concentrates, several answers might be looked for to :
- improve the flotation process itself or develop new beneficiation processes able to give better results ;
- limit the concentration process to a preconcentration stage followed by metallurgical treatment (e.g. roasting) of the preconcentrates ;
- apply a direct metallurgical treatment to the ore (case of massive pyrites).

For any new complex polymetallic sulphide ore deposit to develop, it is thus of ever increasing importance that a constructive dialogue be established between all the actors concerned (both on the mining and metallurgical sides) in order to find fair solutions to ensure optimum entire beneficiation of the deposit, without neglecting any aspect. These aspects are multiple and of course far from being limited to technical ones. However, even from a strictly technical point of view, the development of new feasible processes to treat materials, which will in the future less and less respond to the feed characteristics of presently known conventional metallurgical processes, will require pluridisciplinary teams of specialists such as they may be found in large research institutions and universities. These should have a greater role to play within joint programs cofinanced by the concerned private industries and by national and supranational organizations. Sufficient funds should be granted to such programs, as it is a question of maximum utilization of mineral resources, which are an important source not only of major base metals but also of several rare and precious metals.

References
1. Gorgemans N. Etude de la valorisation des gisements de sulfures polymétalliques du point de vue commercial. Mémoire fin d'études, Univ. Cath. Louvain, Belgium, 1988. p. 108-111.
2. Doyle E.N. The sink-float process in lead - zinc concentration. In : AIME World Symposium on Mining and Metallurgy of Lead and Zinc, vol. 1, New York : AIME, 1970. p. 814-851.
3. Fiedler K.J., Munro P.D. and Pease J.D. Commissioning and operation of the 800 tph heavy medium cyclone plant at Mount Isa Mines Limited. In : Austral. Inst. Min. Met. Annual Conference, Darwin, 1984, p. 259-271.
4. Lovering L. and Allen M. The Afton concentrator. Canadian Mining Journal, vol.100, n° 3, March 1979, p. 37-38.
5. White L. Boliden improves processing economics. Engineering and Mining Journal, vol. 186, n° 7, July 1985, p. 32-36.
6. Israelson A.F. Magnetic separation of minerals. Mining Magazine, vol. 139, n° 9, Sept. 1978, p. 211-219.
7. Bouchat M.A., Detiege A. and Robert D. Magnetic recovery of germanium sulphide with the Frantz ferrofilter. Paper presented at AIME annual meeting, New York, 1960.
8. Magnetic methods of the treatment of minerals. J. Svoboda. Amsterdam : Elsevier, 1987, 692 pp.
9. El Tawil M.M. and Morales M.M. Application of wet high intensity magnetic separation to sulfide mineral beneficiation. In : Complex sulfides, Processing of ores, concentrates and by-products

symposium. Sponsored by the Metallurgical Society of AIME and the Canadian Institute of Mining and Metallurgy. San Diego : AIME, 1985. p. 507-524.

10. Kim Y.S., Fujita T., Hashimoto S. and Shimoiizaka J. The removal of Cu sulphide minerals from the Pb flotation concentrate of black ore by high gradient magnetic separation. In : XVth International Mineral Processing Congress. Sponsored by Société de l'Industrie Minérale and others. Cannes : Gedim, 1985, vol. 1. p. 381-390.

11. Xuemin Y., Pingbo H. and Zhongyuan S. Studies on vibrating high gradient magnetic separation of mixed sulphide flotation concentrates. In : Proceedings of the XVIth International Mineral Processing Congress. Stockholm : Elsevier, 1988, part A. p. 1065-1074.

12. Quingxing S. Personal communication. April 1989.

13. Andres U.T., Devernoe A.L. and Walker M.S. (Assigned to Mag - Sep Corporation). Apparatus and method employing magnetic fluids for separating particles. U.S. Patent 4,594,149. June 10, 1986.

14. Walker M.S., Devernoe A.L., Stuart R.W. and Urbanski W.S. Application of a new mineral separation process : the MC process. In : Proceedings 17th Annual Meeting of the Canadian Mineral Processors. Sponsored by the Canadian Institute of Mining and Metallurgy. Ottawa : The Canadian Mineral Processors, 1985. p. 238-255.

15. Kennedy A. Mineral processing developments at Hammaslahti, Finland. Mining Magazine, vol. 152, n° 2, Febr. 1985, p. 122-129.

16. Fuerstenau D.W. Fine particle flotation. In : Fine particles processing : Proceedings International Symposium. Sponsored by Society of Mining Engineers of AIME. New York, 1980. p. 669-719.

17. Colombo A.F. Selective flocculation and flotation of iron-bearing materials. In : Fine particles processing : Proceedings International Symposium. Sponsored by Society of Mining Engineers of AIME. New York, 1980. p. 1034-1056.

18. Banks A.F. Selective flocculation - flotation of slimes from sylvinite ores. In : Fine particles processing : Proceedings International Symposium. Sponsored by Society of Mining Engineers of AIME. New York, 1980. p. 1104 -1111.

19. Somasundaran P. Selective flocculation of fines. In : The Physical Chemistry of Mineral-reagent interactions in sulfide flotation : proceedings of symposium. Sponsored by the U.S. Bureau of Mines. College Park, Md. : U.S. Bureau of Mines Information Circular n° 8818, 1980. p. 150-167.

20. Termes S.C., Wilfong R.L. and Richardson P.E. Flocculation of sulfide mineral fines by insoluble cross - linked starch xanthate. U.S. Bureau of Mines Report of investigations n° 8819, 1983.

21. Ocepek D. Selective flocculation and flotation. Rudarsko - Metalurski Zbornik, vol. 31, 1984, p. 363-375.

22. Acar S. and Somasundaran P. Effect of dissolved mineral species on flocculation of sulfides. Minerals and Metallurgical Processing, vol. 2, n° 4, Nov. 1985, p. 231-235.

23. Soto H. and Barbery G. Separation of fine particles by floc flotation. In : Production and Processing of fine particles. New York : Pergamon, 1988, p. 297-308.

24. Strauss G.K. and Gray K.G. Complex pyritic ores of the Iberian Peninsula and their beneficiation, with special reference to Tharsis Company mines, Spain. In : Complex Sulphide Ores. Conference sponsored by the Institution of Mining and Metallurgy and others. London : I.M.M., 1980. p. 79-87.

25. Motta Guedes R.M. and Severino Rodriguez A. Integrated treatment of pyrite cinders at Quimigal - Barreiro. In : Complex sulfides, Processing of ores, concentrates and by-products symposium. Sponsored by the Metallurgical Society of AIME and the Canadian Institute of Mining and Metallurgy. San Diego : AIME, 1985. p. 457-470.

26. Rey M. The copper segregation process. In : SME Mineral processing handbook. Sponsored by Society of Mining Engineers of AIME. New York : AIME, 1985. 14F p. 1-7.

27. Cauwe Ph., Minet Ph. and Sheridan R. Selective roasting of complex sulfide material. In : Advances in sulfide smelting : International Sulfide Smelting Symposium. Sponsored by the Metallurgical Society of AIME and others. New York : AIME, 1983. p. 427-449.

28. Salter R.S., Synnott J., Gilders R., Doucet G. and Boorman R.S. Design, construction and commissioning of the demonstration plant for the RPC sulphation roast process. In : Complex sulfides, Processing of ores, concentrates and by-products symposium. Sponsored by the Metallurgical Society of AIME and the Canadian Institute of Mining and Metallurgy. San Diego : AIME, 1985. p. 609-622.

29. Wilkomirsky I., Boorman R.S. and Chalkley M.E. Process and reactor design for the RPC sulphation roast process. In : The Reinhardt Schuhmann international symposium on innovative technology and reactor design in extractive metallurgy. Sponsored by TMS - AIME and others. Warrendale : The Metallurgical Society, 1986. p. 937-950.

30. Worner H.K. Microwave energy as an aid to smelting. In : Non-ferrous smelting symposium : 100 years of lead smelting and refining in Port Pirie. Sponsored by the Australasian Institute of Mining and Metallurgy. Parkville, Victoria : AusIMM, 1989. p. 17-20.

31. Asteljoki J.A. and Hanniala T.P.T. Flash smelting of pyrite and sulphur recovery. In : Productivity and technology in the metallurgical industries : Proceedings International Symposium. Sponsored by the Minerals, Metals and Materials Society. Warrendale, Pennsylvania : T.M.S., 1989. p. 341-356.

32. Warner N.A. Towards polymetallic sulphide smelting. In : Complex sulfides, Processing of ores, concentrates and by-products symposium. Sponsored by the Metallurgical Society of AIME and the Canadian Institute of Mining and Metallurgy. San Diego : AIME, 1985. p. 847-865.

33. Hanna R.K. and Warner N.A. Process requirements for the direct condemnation of both zinc and lead as metals in the polymetallic smelting of Zn-Pb-Cu sulphides. In : Non-ferrous smelting symposium : 100 years of lead smelting and refining in Port Pirie. Sponsored by the Australasian Institute of Mining and Metallurgy. Parkville, Victoria : AusIMM, 1989. p. 227-236.

34. Murr L.E. Theory and practice of copper sulphide leaching in dumps and in situ. Minerals Science and Engineering, vol. 12, n° 3, July 1980, p. 121-189.

35. Fuerstenau M.C. and Han K.N. Challenges in hydrometallurgy. In : Challenges in mineral processing : symposium proceedings. Sponsored by So-

ciety of Mining Engineers. Littleton, Colorado :
Society of Mining Engineers, 1989. p. 749-765.
36. Paul B.C., Sohn H.Y. and McCarter M.K. Model
for bacterial leaching of copper ores containing
a variety of sulfides. In : Metallurgical processes for the year 2000 and beyond. Proceedings
of International Symposium. Sponsored by the Minerals, Metals and Materials Society and others.
Warrendale : Minerals, Metals and Materials Society, 1988. p. 451-464.
37. Ehrlich H.L. Bacterial leaching of silver
from a silver - containing mixed sulfide ore by
a continuous process. In : Fundamental and applied biohydrometallurgy. Proceedings of VIth
International Symposium on Biohydrometallurgy.
Sponsored by British Colombia Research and o-
thers. Amsterdam : Elsevier, 1986, p. 77-88.
38. Miller P.C. Large - scale bacterial leaching
of a copper - zinc ore in situ. In : Fundamental
and applied biohydrometallurgy. Proceedings of
VIth International Symposium on Biohydrometallurgy. Sponsored by British Colombia Research
and others. Amsterdam : Elsevier, 1986, p. 215-
239.
39. Lopes R.F. and Leloux P.L. A review of problems in optimizing extraction in gold heap leaching. CIM Bulletin, vol. 81, n° 915, July 1988,
p. 86-89.
40. Bhappu R.B. Hydrometallurgical advances in
precious metals extraction. In : Advances in mineral processing. Arbiter Symposium. Sponsored
by the Society of Mining Engineers. Littleton,
Colorado : Society of Mining Engineers, 1986.
p. 463-480.
41. Von Michaelis H. Present and future of gold
and silver metallurgy. In : Challenges in mineral processing : symposium proceedings. Sponsored by Society of Mining Engineers. Littleton,
Colorado : Society of Mining Engineers, 1989.
p. 605-637.
42. Sutill K.R. Bio-oxidation for refractory
gold. Engineering and Mining Journal, vol. 190,
n° 9, Sept. 1989, p. 31-32.

Concentrate processing and tailings disposal

Improved model for design of industrial column flotation circuits in sulphide applications

R.A. Alford

Julius Kruttschnitt Mineral Research Centre, University of Queensland, Indooroopilly, Queensland, Australia

Synopsis

The dispersed plug flow model has gained acceptance for predicting mineral recovery from flotation columns, following the work of Dobby and Finch[1]. The model combines a steady state residence time distribution with first order flotation kinetics. It has been recognised that the first order rate parameter is related to a number of operating variables.

The present work has focused on developing a better understanding of the effect of the operating variables, to facilitate the use of the model in practical situations for column circuit simulation, scale-up, and design.

The influence of a range of operating variables was systematically studied at several sulphide concentrators throughout Australasia. The columns surveyed ranged from pilot to full scale, and were used to recover a variety of minerals, including galena, sphalerite, chalcopyrite, and iron sulphides.

The basic column model structure has been modified to include the observed effects of the operating variables. In this refined model the behaviour of the operating variables is clearly separated from the mineral specific terms. Consequently the model is a useful foundation for circuit simulation, as changes in performance between stages are not clouded by changes in column operation. In circuit modelling to date only discrete distributions have been used to describe the full range of mineral behaviour (ie. the individual species were divided into fast and slow fractions).

Introduction

The scope of flotation column utilisation has increased markedly in recent years. Early applications were restricted to replacing or adding final cleaning stages in existing circuits. Recently designed and constructed circuits include multiple stages of columns in a variety of applications from roughing to cleaning. A general column model for design, scale-up, and simulation of such multi-stage column circuits is required. Column studies at the Julius Kruttschnitt Mineral Research Centre (JKMRC) have been directed towards this objective.

A multi-stage column model should use a single set of parameters, applicable throughout a given circuit. That is, the parameters should be independent of the location in the circuit (or stage). To achieve this criterion, the effects of the operating variables (eg. air flow rate, froth characteristics, etc.) must be clearly separated from the measurement of mineral floatabilities.

A distribution of mineral floatabilities (for each mineral) is required to model multi-stage column circuits; commonly, fast and slow fractions are arbitrarily specified for each specie. If the behaviour of the operating variables is not clearly defined in the model, the calculated floatability distribution would likely be erroneous.

Testwork discussed in this paper was designed to develop a better understanding of the action of the column operating variables. The general model form has been modified to incorporate all the significant components. This refined model has been used as the basis of column circuit simulations.

Modelling Background

Dobby and Finch[1] proposed that the dispersed plug flow model be used for simulating flotation columns. This model essentially combines a steady state residence time distribution (RTD) with first order flotation kinetics, as follows:

$$Rec_i = f (k_i , \tau_p , D) \qquad \text{eqn (1)}$$

where Rec_i is the recovery of specie i, k_i is the first order rate parameter for specie i (min^{-1}), τ_p is the mean particle residence time (min), and D is a measure of the axial dispersion in the collection zone. In this general form the behaviour of the collection (or pulp) and froth zones are combined. Ideally a two stage model, separating the zones, should be used.

A derivation of the dispersed plug flow model can be found in Levenspiel[2]. In general terms the change in mineral concentration in the pulp with time is a function of three factors:

1. Transport due to eddy dispersion.
2. Transport due to bulk or plug flow.
3. Disappearance due to collection by the air phase (assumed to be first order).

These factors have been combined in the following differential equation:

$$D \frac{d^2 M}{d x^2} + u_p \frac{d M}{d x} - k M = 0. \qquad \text{eqn (2)}$$

where M is the mineral concentration, D is the dispersion parameter, x is the vertical distance from the pulp/froth interface, and u_p is the interstitial particle velocity.

The three solutions to the above differential equation most commonly used are:

Plug flow model:

$$Rec_i (\%) = 100. (1. - \exp (- k_i \tau_p)) \qquad \text{eqn (3)}$$

Perfect mixer model:

$$Rec_i (\%) = \frac{100. k \tau_p}{1. + k \tau_p} \qquad \text{eqn (4)}$$

Dispersed plug flow model:

$$Rec_i (\%) =$$
$$100. \left(1 - \frac{4 a \exp(1/2/Np)}{(1+a)^2 \exp(a/2/Np) - (1-a)^2 \exp(-a/2/Np)} \right) \qquad \text{eqn (5)}$$

$$a = \sqrt{1 + 4 k_i \tau_p Np}$$

where Np is the dispersion number (D/u_pL) and the L is the column height less the froth depth. The plug and perfect mixer models are special cases of the dispersed plug flow model.

The dispersion parameter is related to the column diameter, air rate, and the pulp viscosity. Laplante, Yianatos, and Finch[3] proposed the following correlation:

$$D = 2.98 D_c^{1.31} V_g^{0.33} \exp(-0.025 S) \qquad \text{eqn (6)}$$

where D is in cm^2/sec, D_c is the column diameter (cm), V_g is the superficial air rate (cm/sec), and S is the percent solids in the feed. Equation 6 can be used in cases where D has not been measured.

The general model form (equation 1) can not be applied uniformly to all columns. In some cases the collection of particles (in the pulp) may be limited by the carrying capacity of the bubbles. Espinosa, Finch, Yianatos, and Dobby[4] developed a correlation for bubble carrying capacity based on the work of Szatkowski and Freyberger[5]. The general relationship for the column carrying capacity is:

$$C_a = \frac{60 K_1 \pi d_p \rho_p V_g}{d_b} \qquad \text{eqn (7)}$$

where C_a is the carrying capacity ($g/min/cm^2$), K_1 is the proportion of the bubble covered with a monolayer of particles, d_p is the particle diameter (cm), ρ_p is the particle density (g/cm^3), and d_b is the mean bubble size (cm). Based on industrial measurements the general relationship was simplified to:

$$C_a = 0.0682 d_{80} \rho_p \qquad \text{eqn(8)}$$

where d_{80} is the 80% passing size (mm).

Jameson, Nam, and Moo-Young[6] suggested that the first order rate parameter was related to the operating variables and mineral specific terms, as follows:

$$k_i \propto (V_g / d_b) E_c E_{ai} \qquad \text{eqn (9)}$$

where E_c is the collision efficiency, and E_{ai} is the attachment efficiency (which is mineral specific). Dobby and Finch[7] elaborated on the descriptions of the collision and attachment efficiencies. For small particles (low inertia) the collision efficiency is approximately proportional to the ratio of particle size to bubble size. The attachment efficiency is more difficult to define, but basically it is proportional to the time the particle and the bubble are in contact (the sliding time), and inversely proportional to the time for the particle to rupture the bubble film (the induction time) - greater details are given by Dobby and Finch[7]. The induction time is assumed to be proportional to the film viscosity. Simply, the attachment efficiency is proportional to the hydrophobicity of the particles and the bubble size, and inversely proportional to the particle size and the film viscosity. Luttrell and Yoon[8] have also developed models for E_c and E_{ai}. The models are different to those proposed by Dobby and Finch[7], but the basic trends to the key variables are similar.

In two stage column models the froth and pulp zones can be linked, as follows:

$$Rec_i = \frac{100\, Rec_{pi}\, Rec_{fi}}{1 - Rec_{pi} + Rec_{pi}\, Rec_{fi}} \qquad \text{eqn (10)}$$

where Rec_{pi} is the recovery in the pulp zone and Rec_{fi} is the recovery in the froth zone. Strictly speaking the general model form (equation 1) only applies to the pulp zone. However, using standard metallurgical testing procedures it is not possible to independently determine separate recoveries for each zone. Alternative measurement procedures have been proposed by Yianatos, Laplante, and Finch[9] and Falutsu and Dobby[10], which enable a froth recovery to be calculated.

Yianatos et al.[9] collected samples at a variety of depths in the column froth phases and these results were used to determine the parameters in a froth model. In the proposed froth model, selective mineral attachment was assumed to be proportional to the residence time of the bubble in the froth phase. A plug flow model form was used:

$$F_i(x) = F_i(0)\, \exp\left(-k_{di} \frac{\int_0^x E_g\, dx}{V_g} \right) \qquad \text{eqn (11)}$$

where x is the distance from the interface, $F_i(x)$ is the concentrate mass rate of specie i at x, k_{di} is detachment constant for specie i, Eg is the holdup (or the fraction of air) in the froth at x, and Vg is the air rate. This approach does not give a direct measurement of the froth phase recovery.

The method used by Falutus and Dobby[10] gives a more direct measurement of the froth phase recovery. A two stage pilot scale column was constructed to physically separate the two zones. Falutus and Dobby[10] measured the effect of a number of operating variables on froth 'drop-back' or recycle. The most significant factor was the solids flux to the froth zone. It was concluded that the froth stability (and consequently recycle) was dependent on bubble loading. Bubble loading is defined as the ratio of concentrate mass flow to the bubble carrying capacity.

The results of previous researchers provided a comprehensive list of operating variables which were considered in this study:

1. Air flow rate.
2. Residence time (mean and distribution).
3. Bubble size.
4. Pulp viscosity or percent solids
5. Froth depth.
6. Wash water flow rate or bias flow rate.

Standard metallurgical results were used to modify the general model form to include the significant operating terms.

Experimental

Surveys of flotation columns were conducted at a number of industrial sites processing sulphides, in Australia and Papua New Guinea. A summary of the database resulting from these surveys is presented in Table No.1; not included in this database are a number of isolated tests at each of the named sites. As well, testwork was conducted on the copper and zinc circuits at Mount Isa Mines (MIM), and at the AMDEL pilot plant facilities in Adelaide where a variety of sulphide ores were treated. Although the results of these additional tests were not directly included in the model development programme, the background information provided was valuable. The columns studied ranged from pilot scale (50mm to 200mm in diameter) to full scale (1.2m to 3.25m in diameter). The survey procedures were consistent between sites and the main objective of these tests was to measure systematically the metallurgical significance of all the operating variables. In the pilot scale surveys a continuously conditioned sample was taken from the full scale plants for processing in the parallel pilot circuit.

Early in the project it was observed that air flow rate and residence time had the greatest influence on the metallurgical performance of columns. Consequently a base level of column performance (over a wide range of recoveries) was determined, at each site, by varying only these two variables. The effects of the other variables (froth depth, etc.) were evaluated by comparison with the base set of results.

Each test was conducted at steady state for a duration of at least one residence time. A sample was cut from each stream a number of times and combined for assay and percent solids determination. The mass flow of the concentrate stream and occasionally the tailing stream was measured during the tests. This measured information was mass balanced using software developed at the JKMRC. The software is based on a modified Levenberg-Marquardt procedure, subject to the constraint

that INPUT = OUTPUT. From the mass balanced
results mineral recoveries for each specie were calculated.

The following measurements were completed
during each test series, in order to determine the average
particle residence and extent of axial dispersion in the
pulp:

1. Air holdup (or volume fraction of air) in
   the collection zone versus air rate data. An
   example from testwork at BCL is
   presented in Figure No.1.
2. Typical tailing residence time
   distributions; see Figure No.2 for an
   example from investigations at MIM.
3. Wash water flow rates.
4. Mean concentrate particle size.
5. Froth depth.

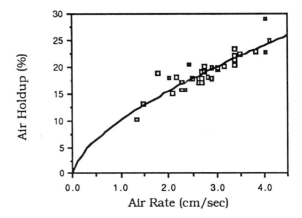

Figure No.1: Air holdup measurements - BCL stage #1

Table No.1: Summary of column modelling database,
for development of a single column model.

---

1. <u>Bougainville Copper Limited (BCL)</u>. Investigations were conducted on the trial column cleaning circuit - this circuit
was of full scale dimensions but only treated a portion of the total combir.ed rougher and scavenger concentrate. Two
stages, of unequal dimensions, were used in series.
Dimensions: 1.8m x 12m (stage #1) and 1.2m x 12m (stage #2).
No. of tests: 20 (stage #1) and 18 (stage #2)

2. <u>Mount Isa Mines (MIM)</u>. The low grade lead/zinc middlings column circuit was surveyed. This is a cleaning
application comprising three equal size columns in series. The concentrates from each column are combined. For
further details see Espinosa, Johnson, and Finch[11].
Dimensions: 2.5m x13m.
No. of Tests: 16 (stage #1) and 19 (stage #3).

3. <u>Paddington Gold Mining Areas</u>. Two columns in parallel are used in a roughing application to produce a bulk
sulphide concentrate. A pilot scale column was operated in parallel to the full scale circuit during this investigation;
further details are given in Newell, Gray and Alford[12]. Only the pilot scale results have been used to evaluate column
performance.
Dimensions: 3.25m x 12m and 150mm x 14m.
No. of tests: 16

4. <u>Pasminco Broken Hill (Zinc Mine)</u>. Pilot scale surveys were conducted in parallel to the conventional zinc recleaner
cells.
Dimensions: 200mm x 6m.
No. of tests: 10

5. <u>Windarra Nickel Mine</u>. Columns are used for cleaning and roughing duties in a circuit producing a gold bearing bulk
sulphide concentrate. A pilot scale column was operated in parallel to the rougher column; only tests on the pilot scale
unit have been included in the modelling database.
Dimensions: 2.5m x 12m (rougher column) and 150mm x 10.5m.
No. of tests: 14

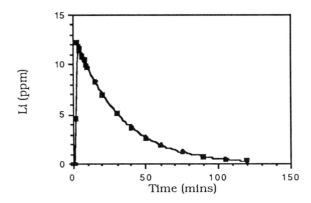

Figure No.2: MIM LGM column stage #3 Tailing RTD
(Tanks in series model)

The columns examined were not limited by the carrying capacity of the bubbles. Using equation 8 a carrying capacity for each site (in Table No.1) was calculated; the maximum concentrate mass rates measured during tests at these sites were significantly lower than the calculated maximums (see Table No.2). Consequently the general column model form (equation 1) should be valid at each site.

Table No.2: Carrying capacities and maximum concentrate mass rates.

	d_{80}	r_p	C_a	max. C_r
BCL	65	4.1	18.2	5.24
MIM	41	4.4	12.3	4.07
Paddington	90	4.0	24.6	1.53
Pasminco	90	4.1	25.2	3.82
Windarra	50	3.6	12.3	1.75

Note: C_r is the concentrate mass rate (Q_c) divided by the column cross-sectional area $(g/min/cm^2)$

A Single Column Model

The database (see Table No.1) was used to evaluate several modelling alternatives, including two stage models. The best results were achieved with a single stage model (see equation 1) and a phenomenological rate expression. The rate expression includes all the significant operating variables and was developed as a result of studies within this project. The general form of the rate expression is:

$$k_i = (C_i\ V_g^*) / (\mu\ d_b)\qquad\text{eqn (12)}$$

where C_i is a mineral-specific proportionality constant (dimensionless), V_g^* is the air rate corrected by a factor

related to the froth stability (cm/sec), μ is the pulp viscosity (Poise), and d_b is the mean bubble size (cm).

Selective upgrading due to the action of the froth is not directly included in this model. Only the results of tests conducted at 'standard' froth depths and bias rates were included in the database (see Table No.1). Froth depths ranged from 70 to 120 cm, and bias rates were greater than 0. cm/sec.

Although overall mineral selectivity was not significantly dependent on the froth phase characteristics, a term related to froth stability was required. In the rate expression (see equation 12) froth stability is included by modifying the air rate. In the tests conducted mineral recovery could be easily correlated to air rate, but the trends did not pass through the origin. A minimum air rate was required to recover any minerals, and in fact in earlier publications of this work[13] the air rate correction is referred to as the minimum air rate. The minimum air rate was found to be inversely dependent on the bubble loading at the column lip. Consequently, unselective recycle from the froth phase decreases with increased bubble loading. This conclusion has been confirmed by pilot scale studies conducted by Falutsu and Dobby[10]. Based on this finding, the corrected air rate is calculated as follows:

$$V_g^* = V_g - (b\ A_c\ V_g^{0.75})/ (Q_c)\qquad\text{eqn (13)}$$

where Q_c is the total concentrate mass flow rate at the column lip (tph), A_c is the column cross-sectional area (m^2) and b is a fitted constant. The air rate correction is independent of the mineral specie.

The mean bubble size for a given column and sparger type was found to be most significantly dependent on the air rate:

$$d_b = c\ V_g^{0.25}\qquad\text{eqn (14)}$$

Mean bubble sizes were determined from air holdup readings using the drift-flux method[14]. The coefficient (0.25), determined by regression, is quantitatively consistent with the findings of Xu[14]. Rubber and filter cloth spargers were used in the testwork reported in this paper. In practice the mean bubble size in the collection zone is also dependent, to a minor extent, on the pulp viscosity and the pulp velocity. Inclusion of such terms in the rate expression does not improve the model fits (see the subsequent section on comparison with alternative model forms).

Based on viscosity measurements of samples from MIM and Pasminco the pulp viscosity (in Poise) has been correlated to the tailing percent solids in the following general form:

$$\mu = \mu_w + a\ (Tsol)^b\qquad\text{eqn (15)}$$

where μ_w is the viscosity of the fluid (water) and the Tsol is the tailing percent solids. For the modelling results reported a value of 1.388×10^{-5} was used for the coefficient 'a' and 2.08 for 'b' in equation 15. Using rotational type viscometers a variety of shear rates were considered for each sample. A Fann viscometer was used to measure the viscosity of the MIM samples, while a prototype derivative of the Debex viscometer[15,16] was used for the Pasminco samples. Over the entire range of shear rates the pulps exhibited non-Newtonian behaviour (see Figure No.3 for example), but at the low value of shear stress applicable for mineral particles the behaviour was best approximated by the Newtonian relationship. Shear stress is dependent on particle size[17]. For the minerals tested the applicable shear rates would be lower than the lowest reading possible with the viscometers.

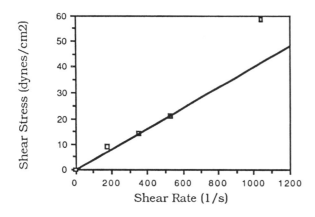

Figure No.3: Shear rate versus shear stress for MIM
Tailing sample
(Percent solids = 40.0)

Equations 12,13 and 14 were combined to give the following simplified rate expression:

$$k_i = \frac{C_i \left(V_g - \frac{b \, A_c \, V_g^{0.75}}{Q_c} \right)^{0.75}}{\mu} \qquad \text{eqn (16)}$$

The mineral specific constant in the equation 16 is insensitive to changes in the column operating variables, as all the significant operating variables are included in the model. Studies to date have shown that, the generalised form can be readily applied to all column stages within a given column circuit, and the model can be applied using standard industrial data.

The fitted parameters for each column in the database are presented in Table No.3. The model

Table No.3: Fitted Model Parameters

Site/ Stage/ Specie		$C_i \times 10^4$ (eqn 12)	b (eqn 13)
BCL			
Stage #1	Chalcopyrite	13.2	0.88
	Non sulphides	0.15	0.88
Stage #2	Chalcopyrite	30.8	0.87
	Non sulphides	0.75	0.87
MIM			
Stage #1	Galena	4.6	0.55
	Sphalerite	9.9	0.55
	Iron sulphide	1.0	0.55
	Gangue	0.65	0.55
Stage #2	Galena	4.0	0.30
	Sphalerite	6.1	0.30
	Iron sulphide	1.1	0.30
	Gangue	0.85	0.30
Paddington			
Arsenopyrite		7.3	0.13
Pasminco Broken Hill			
Galena		1.69	1.0
Sphalerite		3.26	1.0
Iron sulphide		0.775	1.0
Gangue		0.499	1.0
Windarra			
Iron sulphide		5.61	0.38
Arsenopyrite		1.41	0.38
Gangue		0.217	0.38

parameters (C_i's and b) were determined using non linear fitting software developed at the JKMRC. The software uses the Levenberg-Marquardt method to minimise the chi-squared statistic (further details are given by Kojovic[18]). All the model parameters can be determined simultaneously by this method. In this case chi-squared is given by:

$$\chi^2 = \sum_i \sum_j \frac{Recm_{ij} - Recp_{ij}}{Recm_{ij} \, RSD_i} \qquad \text{eqn (17)}$$

where $Recm_{ij}$ is the measured recovery for species i and test j, $Recp_{ij}$ is the predicted recovery, and RSD_i is relative standard deviation $(= \sigma_i / Recm_{ij})$. A chi-squared value can be determined for each specie, as well as the total value given by equation 17. The relative standard deviations for each mineral specie were estimated by conducting a number of 'duplicate' tests. In industrial environments it is difficult to duplicate tests due to changes in the feed conditions. The standard deviations are therefore representative of the errors from a number of sources, including sampling errors, assaying errors, and error due to changes in ore type. The results of the tests used to estimate the relative standard deviations for the MIM data are given in Appendix No.1.

The calculated chi-squared values were used to evaluate the model fits in each case; in simple terms, the smaller the chi-squared value the better the model fit. The chi-squared values can be used to calculate directly the probability that the deviations between the model predictions and the measured data are due to random errors rather than modelling inaccuracies[19]. Press, Flannery, Teukolsky, and Vetterling[19] suggest that model fits are satisfactory when the probability is greater than 10^{-3} and for truly wrong models the probability will be of the order of 10^{-18}. For a good fit the chi-squared value will be approximately equal to the number of degrees of freedom. As well as evaluating the model fits, t-tests were used to test the significance of each of the fitted parameters.

The results from MIM have been used to demonstrate the model evaluation procedures (see Table No.4). The model fits were satisfactory for all mineral species, except the iron sulphide in stage #3. The overall probability for stage #3 was low, chiefly due to the contribution from the iron sulphide, but not small enough to warrant rejecting the model. It is considered that the main reason for the poor model fit was that the estimate of the relative standard deviation for iron sulphide was too low. Consequently, the relative standard deviations have been modified to obtain satisfactory model fits. The modified values are given in Table No.4. These new values are well within the range

Table No.4: Model evaluation. (a) Model Fits

	υ	Estimated std. devs			Modified std. devs		
		RSD_i	χ^2	q	RSD_i	χ^2	q
MIM Stage #1							
Galena	12	0.10	20.60	5.7×10^{-2}	0.10	20.64	5.6×10^{-2}
Sphalerite	12	0.075	21.46	4.4×10^{-2}	0.08	18.84	9.2×10^{-2}
Iron Sulphide	12	0.113	22.44	3.3×10^{-2}	0.15	12.77	0.39
Gangue	12	0.213	9.04	0.70	0.21	9.22	0.68
Overall	51	-	73.54	2.1×10^{-2}	-	61.47	0.15
MIM Stage #3							
Galena	12	0.10	18.80	9.3×10^{-2}	0.10	18.35	0.11
Sphalerite	12	0.075	31.19	1.8×10^{-3}	0.08	27.43	6.7×10^{-3}
Iron Sulphide	12	0.113	40.01	7.2×10^{-5}	0.15	22.97	2.8×10^{-2}
Gangue	12	0.213	19.36	8.0×10^{-2}	0.21	20.00	6.7×10^{-2}
Overall	51		109.36	3.9×10^{-6}	-	88.75	$1. \times 10^{-3}$

Note: υ is the number of degrees of freedom, and q is the probability.

Table No.4: Model evaluation. (b) t-tests

Parameter	Value	Std. Dev.	T	υ	Signif.
MIM Stage #1					
Mineral-specific constants					
Galena	4.6	0.6	7.7	9	>99.%
Sphalerite	9.9	1.3	7.6	9	>99.%
Iron Sulphide	1.0	0.14	7.6	9	>99.%
Gangue	0.65	0.092	7.1	9	>99.%
b	0.55	0.16	3.4	9	>99.%
MIM Stage #3					
Mineral-specific constants					
Galena	4.0	0.28	16.1	9	>99.%
Sphalerite	6.1	0.38	16.4	9	>99.%
Iron Sulphide	1.06	0.075	14.2	9	>99.%
Gangue	0.85	0.077	11.0	9	>99.%
b	0.30	0.014	21.4	9	>99.%

of results calculated from the duplicate tests. Based on the t-tests the fitted parameters were all significant, at the 99% confidence level or better.

An alternative to evaluating the model fits using the chi-squared statistic is to examine the randomness of the residual errors. This can be completed by regressing the errors against the model variables and any new factors. To date, this procedure has not lead to any improvements in the model form, but continued research is required. The measured versus predicted recoveries for each specie are given in Figures No.4 (a) to (h). These plots give an indication of the random nature of the modelling errors.

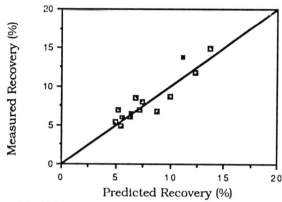

Figure No.4(c): Measured versus predicted iron sulphide recovery (%)
(MIM LGM column stage #1)

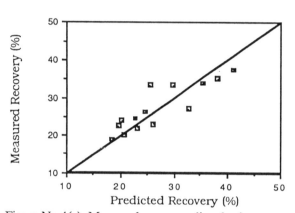

Figure No.4(a): Measured versus predicted galena recovery (%)
(MIM LGM column stage #1)

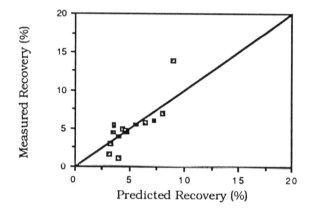

Figure No.4(d): Measured versus predicted gangue recovery (%)
(MIM LGM column stage #1)

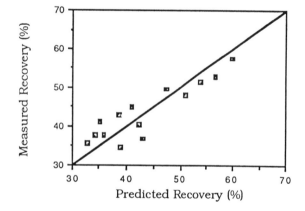

Figure No.4(b): Measured versus predicted sphalerite recovery (%)
(MIM LGM column stage #1)

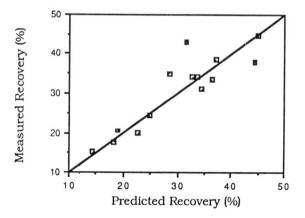

Figure No.4(e): Measured versus predicted galena recovery (%)
(MIM LGM column stage #3)

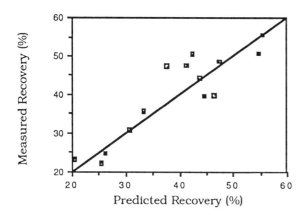

Figure No.4(f): Measured versus predicted sphalerite
recovery (%)
(MIM LGM column stage #3)

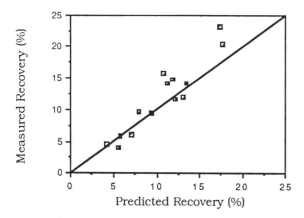

Figure No.4(g): Measured versus predicted iron sulphide
recovery (%)
(MIM LGM column stage #3)

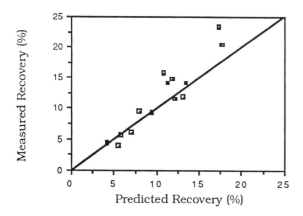

Figure No.4(h): Measured versus predicted gangue
recovery (%)
(MIM LGM column stage #3)

In the new model the metallurgical performance of a flotation column is assumed to be independent of variations in froth depth and bias flow rate (within accepted ranges). This assumption can be validated in a qualitative manner by comparing concentrate grades (at equivalent recoveries). Ynchausti, McKay, and Foot[20] observed that the froth depth had an effect on concentrate grade, but the results are coupled with changes in overall recovery, that is a direct comparison at equivalent recoveries was not made.

The model residuals were used to quantitatively examine the influence of froth-based characteristics on column performance. Using the refined model form, measured values for the mineral-specific constants for each test at MIM were calculated. The ratio of the sphalerite constant to each of the slower floating species was regressed against the froth depth, the bias flow rate, and the bubble residence time in the froth. The ratio of the mineral-specific constants is representative of the selectivity between the species, as was used in perference in the actual residuals in this case. The results of additional tests conducted using froth depths and bias rates outside the accepted ranges were included in this analysis.

The only froth based operating variable that had a consistent effect on mineral selectivity was the bubble residence time in the froth. As an example, the regressions for the non-sulphide gangue in stage #3 are given in Figures 5 to 7. Both the relationships against bubble residence time and froth depth were significant at the 95% level[21] in this case, but the relationship against froth residence time had a higher correlation coefficient. The relationship against the bias rate was not significant. For the regressions against froth depth and froth residence time the 95% significance level was r=0.455, and for the bias the level was r=0.532.

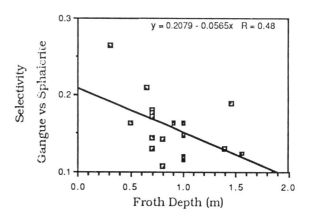

Figure No.5: The effect of froth depth on mineral
selectivity
(MIM LGM column stage #3)

197

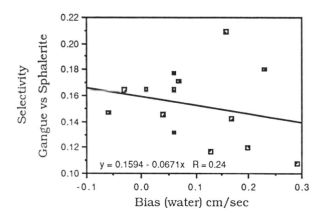

Figure No.6: The effect of bias rate on mineral
selectivity
(MIM LGM column stage #3)

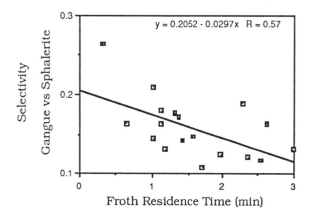

Figure No.7: The effect of the bubble residence time in
the froth on mineral selectivity
(MIM LGM column stage #3)

The bubble residence time in the froth was
included in the residual analysis following the work of
Yianatos *et al.*[9], who proposed a plug flow model for
selective mineral detachment in the froth (see equation
11). Modifying the present rate expression to include the
effects of the froth residence time did not significantly
improve the fits to the single column data (see the
subsequent section on comparison with alternative
model forms). Consequently it has been concluded
that, although selective upgrading in the froth does
occur, the effect on overall column performance is
minor.

In the present model mineral selectivity is
assumed to be independent of bubble size, although total
mineral recovery is inversely proportional to bubble
size. Preliminary investigations at a pilot scale at MIM
support this assumption. The bubble size was controlled
by changing the sparger. Two spargers were tested - a
porous stainless steel sparger and a perforated rubber

sparger. The grade/recovery curve produced using these
spargers did not differ significantly; see Figure No.8.

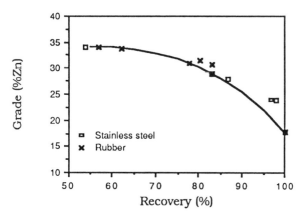

Figure No.8: The effect of bubble size on mineral
selectivity
(MIM LGM feed - pilot scale results)

The average bubble sizes using the stainless
steel and rubber spargers were 0.86mm and 1.98mm,
respectively. Bubble sizes were calculated from air
holdup measurements using the drift-flux method[14].
This difference in bubble size did alter the maximum
bubble carrying rate.

The observed results for the tests using different
bubble sizes are consistent with predictions made by
Dobby and Finch[7]. Using a particle/bubble collision and
attachment model Dobby proposed that bubble size
would not greatly affect the selectivity between minerals
of differing hydrophobicities.

The form of the air rate correction term
(equation 13) is largely empirical, and further refinement
may be required. Total solids recovery from the froth
phase has been calculated for the MIM data and plotted
against the calculated bubble loading at the froth/pulp
interface; see Figure No.9. The froth recoveries were
calculated from the measured overall recoveries and the
calculated collection zone recoveries. The collection zone
recoveries for each species were determined using
uncorrected air rates.

The calculated froth zone recoveries are
significantly larger than measurements made by Falutus
and Dobby[10], who used a two stage pilot scale column,
designed to separate the behaviour of the collection and
froth zones, to measure the froth recovery. The froth
recoveries measured by this procedure ranged from 0 to
60% depending on the solids flux at the interface. Froth
recovery is likely to be dependent on a number of
variables, including frother dosage, air rate, machine
type, and mineral system.

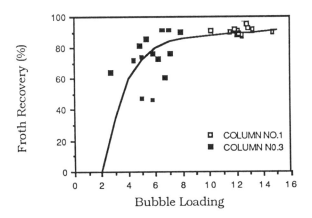

Figure No.9: Calculated total solids froth recoveries (MIM LGM column circuit)

The relationship between froth recovery and bubble loading is clear from the MIM data. A model separating the froth and pulp phases would be more rigorous than the current model, but the additional complexity does not improve the model fits. In the developed model the actions of the froth and the collection zone are 'lumped' together.

In the developed model form (equations 1 and 16) the operating variables are clearly separated from the mineral specific terms. As a result, this improved model is a suitable building block or basis for circuit modelling and simulation.

Comparison with Alternative Model Forms

The refined column model (equations 1 and 16) has been compared with a variety of other model forms. The alternative models evaluated in this paper do not represent a complete list of models available, but have been chosen specifically to illustrate a number of important points associated with the development of the new model.

The model evaluation procedures described earlier were used as the basis of the comparison. Only the data from the two column stages tested at MIM has been used for comparing the various models. The following relative standard deviations were used for both stages: 0.10, 0.08, 0.15, and 0.21 for galena, sphalerite, iron sulphide, and gangue respectively (see Table No.4).

Alternative Models:

No.1 - "no viscosity or air rate correction term in the rate expression".

This model combines equation 1 and the following rate expression:

$$k_i = C_i V_g / d_b \qquad \text{eqn (18)}$$

This is probably the simplest rate expression that can be used in column flotation modelling. It assumes that the collision and attachment efficiencies are constant and independent of the operating variables. As well, all froth phase based effects are excluded.

No.2 - "no viscosity term"

For this model form viscosity is omitted from the rate expression, as follows:

$$k_i = C_i V_g^* / d_b \qquad \text{eqn (19)}$$

This model was used to specifically show the advantages of including viscosity in the rate expression. A single stage model form was used (equation 1).

No.3 - "general rate expression"

In equation 16 it has been assumed that the attachment efficiency (and consequently the rate parameter) is inversely proportional to the pulp viscosity. As well the bubble size is a function of viscosity and the pulp velocity, but these relationships are assumed to be insignificant in equation 16. The validity of these assumptions has been checked by using a more general form for the rate expression and equation 1, as follows:

$$k_i = (C_i V_g^* u_l^a) / (\mu^c d_b) \qquad \text{eqn (20)}$$

where u_l is the superficial pulp velocity, and a and b are fitted constants.

No.4 - "fast and slow floating sphalerite"

A significant difference between the fitted mineral specific constants for sphalerite in the two MIM stages tested was observed. As a result it is reasonable to assume that this mineral can be divided into a variety of fractions depending on floatability. In this case sphalerite was divided into fast and slow fractions. The mineral specific constant for the slow fraction was constrained to a value close to that of the slow floating gangue minerals (greater details on this parameterisation technique are given in the section on column circuit fitting). Little difference in the mineral specific constants between stages was observed for the other species, and consequently single mineral specific constants are adequate for these species. Equations 1 and 16 were used for this model. In this model form an additional parameter is fitted, the fraction of fast floating sphalerite, as there are effectively five mineral species instead of four.

No.5 - "inclusion of a froth residence time term"

The standard rate expression (equation 16) was modified to include a simplified version of Yianatos' froth model[9] (equation 11). In this model selective upgrading in the froth is assumed to be proportional to the residence time of the bubbles in the froth. The froth residence time is approximated by the ratio of froth depth to air rate. In the evaluation of the model residuals it was observed that this term was slightly significant, and consequently required further investigation. The modified rate expression is:

$$k_i = \frac{C_i \, (\exp\,(\,-k_{di}\,FD/V_g)\, V_g{}^*}{\mu \, d_b} \qquad \text{eqn (21)}$$

A single stage model (equation 1) was used in this case.

No.6 - "two stage model"

In this model form the actions of the froth and pulp phases have been separated. The two phases are linked using equation 10. For the recovery from the pulp zone equations 1 and 16 were used, with the exception that in the rate expression the air rate was not corrected. Recycle from the froth phase was assumed to be unselective and proportional to the loading, as follows:

$$Rec_f = b \; Q_{cp} / (A_c \, V_g{}^{0.75}) \qquad \text{eqn(22)}$$

where Q_{cp} is the total mass flow from the pulp to the froth.

An evaluation of the alternative model fits is presented in Table No.5. The chi-squared and probabilities for the standard model are also shown. Models 1, 2, 5, and 6 can be rejected on the basis of the model fits, and further evaluation of these models is not warranted. The probabilities associated with the chi-squared values for these models was at least an order of magnitude worse than the values using the standard model. The fits for models 3 and 4 were in fact slightly better than the standard.

The significance of the fitted parameters for models 3 and 4, must be considered to complete the comparison with the standard model. The fitted

parameters for these models, as well the significance tests, are given in Table No.6.

Model 3 can be rejected on the basis of the significance tests. The pulp velocity coefficient was not fitted in the results shown. The standard deviation for this parameter, when fitted, was much larger than the fitted value.

The significance test for the fitted parameters using model 4 was satisfactory, although the standard deviations for the fast floating sphalerite were relatively large. A noticeably better result was obtained using the standard model. The parameterisation of the sphalerite was inconsistent between stages, and it considered that the standard model gives a more reliable result for single column stages. It can be concluded that a distribution of floatabilities should be used for sphalerite, but the parameterisation used in this case is inadequate.

The model 4 results have interesting ramifications for scale-up applications. Using a simple discrete parameterisation of the mineral floatabilities a single stage of flotation is inadequate for fully characterising the feed material. If more than one stage of column flotation is planned for the full scale application, it is advisable to test at least two stages at the pilot scale level.

For the model form developed in this work a compromise has been reached between the complexity of the model and the accuracy of the data. Ideally more elaborate descriptions of particle attachment and detachment in both the pulp and froth phases should be used, but the parameters in such models can not be determined with any reliability. The model comparisons made in this section ably demonstrate this point.

Table No.5: Evaluation of Alternative Models

Model No.	MIM Stage #1			MIM Stage #3		
	υ	χ^2	q	υ	χ^2	q
1	52	141.1	$4.\times10^{-10}$	52	232.9	$1.\times10^{-24}$
2	51	112.7	$1.\times10^{-6}$	51	101.1	$4.\times10^{-5}$
3	50	50.5	0.45	50	88.6	$6.\times10^{-4}$
4	50	56.1	0.26	50	88.2	$7.\times10^{-4}$
5	49	93.3	$1.\times10^{-4}$	49	96.1	7×10^{-5}
6	51	53.4	0.38	51	112.9	$1.\times10^{-6}$
Standard	51	61.5	0.15	51	88.8	$1.\times10^{-3}$

Table No.6: Fitted Parameters for Models 3 and 4.

Parameter		Value	Std. Dev.	T	υ	Signif.
Model No.3						
Stage #1						
Mineral-specific constants						
Galena		8.2	4.3	1.91	8	95
Sphalerite		17.5	9.2	1.91	8	95
Iron Sulphide		1.9	1.0	1.90	8	95
Gangue		1.2	0.6	1.90	8	95
b (air rate correction coeff.)		0.59	0.14	3.98	8	>99
a (coeff. to pulp velocity)		0.	-	-	-	-
c (coeff. to viscosity)		0.86	0.13	6.82	8	>99
Stage #3						
Mineral-specific constants						
Galena		7.9	10.3	0.77	8	<90
Sphalerite		12.1	15.6	0.77	8	<90
Iron Sulphide		2.1	2.7	0.77	8	<90
Gangue		1.7	2.2	0.77	8	<90
b (air rate correction coeff.)		0.30	0.016	18.8	8	>99
a (coeff. to pulp velocity)		0.	-	-	-	-
c (coeff. to viscosity)		0.84	0.30	2.8	8	99
Model No.4						
Stage #1						
Mineral-specific constants						
Galena		4.9	0.7	7.68	8	>99
Sphalerite	fast	22.	8.7	2.54	8	97.5
	slow	1.0	-	..	-	-
	frac. fast	0.66	0.10	6.35	8	>99
Iron Sulphide		1.1	0.15	7.66	8	>99
Gangue		0.69	0.096	7.18	8	>99
b		0.61	0.15	4.09	8	>99
Stage #3						
Mineral-specific constants						
Galena		4.1	0.3	15.4	8	>99
Sphalerite	fast	7.6	3.0	2.53	8	97.5
	slow	1.0	-	-	-	-
	frac. fast	0.86	0.23	3.65	8	>99
Iron Sulphide		1.1	0.08	13.7	8	>99
Gangue		0.86	0.069	10.7	8	>99
b		0.31	0.016	19.6	8	>99

Parameterisation of Column Circuits

The results from tests on the low grade lead/zinc middlings column circuit at MIM have been used to evaluate column circuit simulation. A schematic diagram of the MIM LGM column circuit[11] is given in Figure No.10. It comprises three columns in series and all the concentrates are combined.

A single set of model parameters (applicable to the whole circuit) was fitted to seven surveys of the circuit.

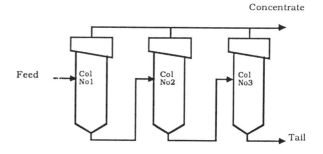

Figure No.10: Simplified MIM LGM Circuit.

The fitted parameters and an evaluation of the model fit are given in Table No.7. The mineral specific constant for sphalerite was divided into fast and slow fractions. The values for these fractions were fitted using non linear techniques rather than by the graphical method proposed by Kelsall[22]. This type of method has been used by a number researchers to model conventional circuits (eg. Lauder[23]). The mineral specific constant for the slower floating portion of the sphalerite was constrained to a value close to that of the gangue minerals - iron sulphide and non-sulphide gangue. This restriction is likely to be conservative as it implies that selective flotation of the slow floating sphalerite and the gangue minerals is not possible. Without a constraint it is unlikely that a unique solution can be derived[24]. A distribution of mineral-specific terms was not required for the slower floating minerals (galena, etc) as little variation was observed between stages.

Table No.7(a): Model parameters for the MIM LGM circuit.

Parameter	Value	S.D.	Signif
Mineral specific constants, C_i			
Galena	4.5	0.22	>99%
Sphalerite Fast floating	12.6	1.3	>99%
Slow floating	1.0	-	
Fraction fast fl.	0.85	0.05	>99%
Iron sulphide	1.00	0.05	>99%
Non sulphide gangue	0.680	0.04	>99%
Air rate correction coefficient, b	0.43	0.03	>99%

Table No.7(b): Evaluation of Circuit Model Fitting

	υ	RSD_i	χ^2	q
Galena	19	0.12	28.1	8×10^{-2}
Sphalerite	18	0.10	28.6	5×10^{-2}
Iron sulphide	19	0.16	32.6	3×10^{-2}
Gangue	19	0.21	38.6	5×10^{-3}
Overall	78	-	127.9	3×10^{-4}

The measured and predicted sphalerite grades and recoveries for the circuit fitting are given in Table No.8. The model fit to circuit data was evaluated using the same procedure as that used for single column fitting. The adjusted relative standard deviations are within the range of values recorded from the duplicate tests. These adjusted values give a satisfactory measure of the confidence levels for the model predictions.

A simple discrete parameterisation of the mineral specific terms is adequate for the MIM circuit, but more complicated circuits may require greater detail.

Table No.8: Circuit fitting results - MIM LGM column circuit.

Survey	Sphalerite Grade		Sphalerite Rec	
	meas	pred	meas	pred
1	64.1	65.6	66.9	71.5
2	60.1	65.7	77.9	74.3
3	61.5	64.3	85.9	85.8
4	64.6	65.6	85.5	85.2
5	60.9	62.2	73.9	70.8
6	62.7	65.7	71.6	66.6
7	64.8	65.5	68.7	67.4

Ideally continuous distributions should be used for each specie (for example see Tikhonov and Kavetsky[25]).

Column Circuit Simulations

The column flotation model developed in this work can be used for optimising and simulating column circuits. The fitted rate parameters from the MIM LGM column circuit have been used to demonstrate the model's application. Two types of circuits were examined - series circuits with combined concentrates, and circuits including recycled streams.

In addition to the mineral recovery model, a relationship for the recovery of water to the concentrate is required. An empirical approach was used to determine a water recovery relationship, and consequently the developed water model can only be used with confidence for simulations of the MIM LGM circuit.

The measured concentrate water flow rates from the MIM LGM column stages 1 and 3 were regressed against concentrate mass flows, air rates, and the pulp viscosity, using multiple linear regression techniques. Only concentrate mass flow was significant for both stages analysed. The data sets for the two stages were combined to develop the following general water recovery relationship:

$$Q_w = 2.41 \, Q_c^{1.23} \qquad \text{eqn (23)}$$

where Q_w is the water flow rate (m^3/hr). This general relationship can be used for each stage in the circuit. The measured results and equation 23 are shown in Figure 11.

Table No.9: Circuit Simulation Results.

No. (%)	Description	Air Rate (cm/sec)				Sphal. Rec.
		Stage 1	Stage 2	Stage 3	Stage 4	
	Series Circ.					
1	2 stages	1.38	1.16	-	-	81.50
2	3 stages	0.86	0.88	1.04	-	83.83
3	4 stages	0.64	0.66	0.87	1.24	84.88
	Recycle inc.					
4	3rd to 2nd	1.19	1.43	3.00	-	85.90
5	3rd to 1st	1.34	1.35	3.00	-	85.94

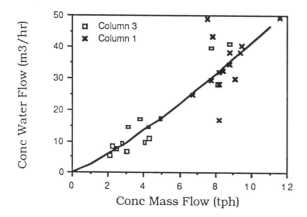

Figure No.11: Concentrate water flow versus mass flow for MIM LGM column stages #1 and #3

The inputs to the simulation program are:

1. The rate model parameters (for each mineral specie).
2. Feed solids and water flow rates.
3. Feed mineral grades.
4. Initial estimates for the air rates to each stage.
5. Froth depths for each stage.
6. Wash water flow rates to each stage.
7. Circuit configuration.
8. Target overall concentrate grade for the valuable mineral specie (in this case sphalerite).

For each simulation the wash water flow rates and the froth depths were kept constant, and the air rates were determined to maximise overall sphalerite recovery (at the target concentrate grade). A Powell optimisation routine[19] was used to determine the best air rates. A true comparison between circuits is only valid when the metallurgical performance of the circuits has been optimised. For this example, a target grade of 63%

sphalerite (or 37.8% zinc) and the average feed flows from the material surveys were used. The results for the five circuits examined are summarised in Table No. 9.

For the optimisation procedure, the air rate to one column stage was constrained to meet the grade requirement. For example, for two stages in series with the concentrates combined (circuit No.1), the air rate to the first stage was optimised while the air rate to the second stage was constrained. For the circuits where the concentrate from the third stage was recycled (circuits 4 and 5), maximum overall recovery was obtained when the air rate to the third stage was maximised. In this example a maximum air rate of 3 cm/sec was arbitrarily chosen; in practice higher air rates may be possible.

The air rates determined by the optimisation procedure were not unique, except for the two stage circuit. That is, the maximum overall recovery could be obtained using a variety of air rates. The air rates were dependent on the initial estimates, and the rates given in Table No. 9 are only examples of the possible solutions.

For the circuits examined, the highest recovery was achieved when the concentrate from the third stage was recycled to the feed to the first column (circuit 5) - 2.1% higher than the recovery using the current circuit (No.2).

The circuits considered in this section are only examples and were used to demonstrate the use of the column flotation model.

Conclusions

An improved model for column flotation has been developed based on the model proposed by Dobby and Finch[1]. In this model, mineral recovery is dependent on a first order rate parameter, the average particle residence time, and the distribution of the residence times about the mean. The improvement is achieved by using a fuller description of the first order rate parameter (see equation 16) in terms of the operating variables.

All the significant operating variables are now included in the model. Froth depth and bias rate are not included, as these variables have been shown to have little effect on mineral selectivity in flotation columns.

The model provides a satisfactory basis for circuit simulations, as the behaviour of the operating variables is separated from the mineral dependent terms. In simulations to date only discrete distributions for the mineral specific terms have used; that is, individual mineral species have been arbitrarily divided into fast and slow floating fractions.

Acknowledgements

The author would like to thank the Australian Mineral Industries Research Association (AMIRA) for providing the funding for this work. Metallurgical staff at Bougainville Copper Limited, Mount Isa Mines Ltd, Paddington Gold Mining Areas, Pasminco Broken Hill, and Windarra Nickel Operations are thanked for providing access to their flotation columns. In particular, Enzo Artone (formly of BCL), Rodolfo Espinosa-Gomez (MIM), and Alister Grimes (Pasminco) are thanked for their many useful discussions relating to this work. Juan Yianatos deserves thanks for his assistance and discussions in the early stages of this project.

References

1. Dobby G. S. and Finch J. A., 1986. Flotation column scale-up and modelling, CIM Bulletin, May, 89-96.

2. Levenspiel O., 1972. Chemical reaction engineering, second ed., Wiley, 253 -314.

3. Laplante A.R., Yianatos J.B., and Finch J.A. 1988. On the mixing characteristics of the collection zone in flotation columns, Proceedings - Column Flotation '88, AIME.

4. Espinosa-Gomez R., Finch J.A., Yianatos J.B., and Dobby G.S., 1988. Flotation column carrying capacity: particle size and density effects, Minerals Engineering, Vol 1, 77-79.

5. Szatkowski M., and Freyberger W.L., 1985. Kinetics of flotation with fine bubbles, Trans. Inst. Min. Met., 94: C61-C70.

6. Jameson G.J., Nam S., Moo-Young M., 1977. Physical factors affecting the recovery rates in flotation, Minerals Sci. Eng., 9(3), 103-118.

7. Dobby G. S. and Finch J. A., 1987. Particle size dependence in flotation derived from a fundamental model of the capture process, Int. J. Min. Proc., 21, 241-260.

8. Luttrell G.H., Adel G.T., and Yoon R.H., 1988. Hydrodynamics and mathematical modelling of fine coal flotation. XVI International Mineral Processing Congress.

9. Yianatos J.B., Finch J.A., and Laplante A.R., *in press*. Selectivity in column flotation froths, Int. J. Min. Proc.

10. Falutsu M. and Dobby G.S., 1989. Direct measurement of froth drop back and collection zone recovery in a laboratory flotation column, Minerals Engineering, Vol 2, No.3, 377-386.

11. Espinosa-Gomez R., Johnson N.W., and Finch J.A., 1989. Evaluation of flotation column scale-up at Mount Isa Mines Limited, Minerals Engineering, Vol 2, No.3, 369-375.

12. Newell A., Gray D., and Alford R.A., 1989. The application of flotation columns to gold recovery at Paddington gold mine, W.A. IPMI conference, Montreal, July.

13. Alford R.A., 1989. The application and scale-up of flotation columns at Paddington, Column Flotation Workshop, Adelaide, July.

14. Xu M., 1987. Sparger study in flotation columns, Masters thesis, McGill University, Montreal.

15. Reeves T.J., 1985. On-line viscometer for mineral slurries, Trans. Inst. Min. Met., 94: C201-C208.

16. Napier-Munn T.J., Reeves T.J., and Hansen J.Y., 1989. The monitoring of medium rheology in dense medium cyclone plants, Aus. IMM Bulletin, Vol 294 No.3, 85-93.

17. Batchelor G.K., 1967. An introduction to fluid dynamics, Cambridge University Press.

18. Kojovic, T., 1988. Automated model building, Ph. D thesis, University of Queensland.

19. Press W.H., Flannery B.P., Teukolsky S.A., Vetterling W.T., 1986. Numerical recipes, Cambridge University Press.

20. Ynchausti R.A., McKay J.D., and Foot Jr. D.G., 1988. Column flotation parameters - their effects, Proceedings - Column Flotation '88, AIME.

21. Davies O. L. (ed.), 1961. Statistical methods in research and production, Oliver and Boyd (ICI).

22. Kelsall D.F., 1961. Application of probability in the assessment of flotation systems, Transactions of the institute of Min. and Met., Vol 70.

23. Lauder D.W., 1988. Flotation circuit performance theory, Ph.D thesis, University of Queensland.

24. Lanczos C., 1964. Applied analysis, Pitman and Sons, London.

25. Tikhonov O. and Kavetsky A., *in press..* Flotometric analysis and predictive calculations for industrial flotation flowsheets, Aus IMM Bulletin.

Nomenclature

a,b,c	fitted constants
A_c	cross-sectional area of column, m^2
C_i	mineral specific constant for specie i, dimensionless
d_b	bubble diameter, cm
d_p	mean particle diameter cm
d_{80}	80% passing size µm
D	dispersion parameter cm^2/sec
D_c	column diameter cm
E_{ai}	particle attachment efficiency for specie i, dimensionless
E_c	particle/bubble collision efficiency, dimensionless
E_g	air holdup, fraction
F_i	concentrate mass flow of mineral specie i, tph
FD	froth depth, m
k_i	first order rate parameter for specie i, min^{-1}
k_{di}	froth detachment constant for specie i, sec^{-1}
L	column height, m
M	mineral concentration, mass/volume
N_p	dispersion number ($D/u_p L$)
q	probability for model fitting
Q_c	concentrate mass flow rate, tph
Q_{cp}	mass flow from pulp to froth, tph
Q_w	concentrate water flow rate, m^3/hr
Rec_i	recovery of specie i
RDS_i	relative standard deviation for specie i
S	feed percent solids
T	student t value (parameter/standard deviation)
Tsol	tailing percent solids
u_l	superficial pulp velocity, cm/sec
u_p	interstitial particle velocity, cm/sec
V_g	superficial air rate, cm/sec
x	distance, m
µ	pulp viscosity, P
τ_p	particle residence time, min
σ_i	standard deviation
χ^2	chi-squared
υ	number degrees of freedom
ρ_p	particle density g/cm^3

Appendix No.1 - Duplicate Tests at MIM

(a) Operating Conditions

		Feed Grade (%Zn)	τ_p	V_g	Tailing % Solids
Set (a) - Stage #1	1	20.24	32.54	0.57	31.10
	2	22.25	33.34	0.57	31.80
Set (b) - Stage #1	1	24.21	45.22	0.71	25.70
	2	24.59	51.18	0.71	24.50
Set (c) - Stage #3	1	11.65	17.57	0.78	15.40
	2	9.88	17.79	0.71	13.10
Set (d) - Stage #3	1	11.53	23.95	1.05	9.99
	2	10.20	23.14	1.06	11.31

(b) Recoveries and calculations

		Galena	Sphalerite	Iron Sul.	Gangue
Set (a)	1	22.58	37.70	6.98	3.07
	2	20.13	37.67	6.00	5.34
	mean	21.36	37.69	6.49	4.21
	s.d.	1.73	0.021	0.69	1.61
coefficient of variance(%)		8.10	0.06	10.63	38.24
Set (b)	1	34.05	51.42	13.72	6.02
	2	35.33	52.85	11.66	7.02
	mean	34.69	52.14	12.69	6.52
	s.d.	0.91	1.01	1.46	0.71
coefficient of variance(%)		2.62	1.94	11.72	10.89
Set (c)	1	17.66	22.27	3.93	3.00
	2	15.20	23.30	4.41	3.52
	mean	16.43	22.78	4.17	3.26
	s.d.	1.74	0.73	0.34	0.37
coefficient of variance(%)		10.59	3.20	8.15	11.35
Set (d)	1	31.15	39.54	11.58	10.55
	2	38.43	48.53	14.11	12.20
	mean	34.79	44.04	12.84	11.38
	s.d.	5.15	6.36	1.79	1.17
coefficient of variance(%)		14.80	14.44	13.94	10.28
RMS coefficient of variation		10.0	7.5	11.3	21.3

Comparison of methods of gold and silver extraction from Hellyer pyrite and lead-zinc flotation middlings

D.W. Bilston
W.J. Bruckard
D.A. McCallum
G.J. Sparrow
J.T. Woodcock
CSIRO Division of Mineral Products, Port Melbourne, Victoria, Australia

ABSTRACT

The Hellyer massive sulphide deposit in north-west Tasmania, Australia, which is a complex fine-grained Cu-Pb-Zn-Au-Ag-pyrite-arsenopyrite orebody, has recently been brought into production. Various base metal flotation concentrates are produced for sale. However, a substantial proportion of the gold and silver report in flotation middlings or tailings and is currently not recovered. Research into a number of methods for extraction of this gold and silver was conducted on various plant products, which had different mineralogy, and which assayed 1.5-3.3 g/t Au and 27-144 g/t Ag, as well as 0.7-8.0 % Zn and 1.1-11.4 % Pb. Product sizings were 90 % minus 38 μm.

Direct cyanidation extracted 12-19 % of the gold in 24 h and 30-50 % of the silver in 8 h. Ultrafine grinding did not increase the extraction, and additives had little effect. Increasing the pH from 10 to 12 slightly increased silver extraction, but also markedly increased lime consumption (e.g. from 20 to 40 kg/t CaO on a flotation middling). Cyanide consumptions were in the range 3-7 kg/t NaCN.

Roasting/sulphuric acid leaching/cyanidation, roasting/acid brine leaching/cyanidation, and roasting/sulphuric acid leaching/thiourea leaching were all investigated in varying detail, with roasting/sulphuric acid leaching/ cyanidation appearing to be the preferred option. The optimum roasting temperature for gold extraction was 600-700°C, but for silver was 500°C. Leaching of the calcine with dilute sulphuric acid for 4 h, or less, was effective in removing cyanide-consuming base metals. Cyanidation of the leached calcine for 24 h extracted about 80 % of the gold and silver with moderate reagent consumptions (<3 kg/t NaCN and <5 kg/t CaO).

Acid pressure oxidation at 200-220°C, under conditions designed to produce either a hematite/lead jarosite product or a basic iron sulphate/lead sulphate product, was investigated, and XRD techniques were used to confirm the nature of the products. Each type of product was treated with lime to raise the pH to 10 and to 12 (this required CaO additions of 28-52 kg/t for hematite/lead jarosite and 300-480 kg/t for basic iron sulphate/lead sulphate). The lime-treated product was then cyanided. Cyanidation of the hematite/lead jarosite product extracted 82-94 % of the gold but only 25-44 % of the silver. Cyanidation of the basic iron sulphate/lead sulphate product extracted 91-95 % of the gold and 88-93 % of the silver. Cyanide consumptions were moderate (<5 kg/t NaCN).

Acid pressure oxidation followed by thiourea leaching was also investigated. Thiourea leaching of the hematite/lead jarosite product extracted only 46-88 % of the gold and 1-7 % of the silver with thiourea consumptions of 8-70 kg/t thiourea. Thiourea leaching of the basic iron sulphate/lead sulphate product extracted 83-92 % of the gold and 39-96 % of the silver, with thiourea consumptions of 33-132 kg/t.

Assessment of the results of this research indicate that direct cyanidation may be a marginally economic method for currently produced products. While good extractions were achieved after roasting, it would be necessary to dispose in an environmentally acceptable way of the SO_2 and the As_2O_3 produced during roasting. Acid pressure oxidation followed by cyanidation gave the best extractions of gold and silver and warrants further examination to attempt to reduce the high lime additions currently required for this route. Thiourea leaching after pressure oxidation can also give good extractions but thiourea consumptions are high.

INTRODUCTION

Treatment of complex sulphide ores containing copper, lead, zinc, iron, arsenic, gold, and silver is an important problem at many mines, and many routes have been investigated for recovery of the metals present. The most common route is to first fine grind the ore and then produce, by flotation, various concentrates for further treatment. A typical suite of products comprises copper-lead and zinc concentrates (which can be treated in various ways), a lead-zinc-iron middling (which may be economically treatable), and a pyrite-arsenopyrite-gangue tailing (which is usually discarded). Gold and silver present in the ore are distributed in various proportions in all the flotation products and recovery of that present in the middlings and tailings can present special problems.

This paper is concerned with a comparison of selected methods for the extraction of gold and silver from flotation middlings and tailings from the Hellyer orebody in Tasmania. The products investigated were a bulk lead-zinc scavenger tailing assaying 2.2-3.3 g/t gold and 82-144 g/t silver, a zinc middling assaying 2.3-2.6 g/t gold and 53-92 g/t silver, and arsenopyrite-pyrite-gangue tailings assaying 1.5-2.8 g/t gold and 27-45 g/t silver. These products contained some 70 % of the gold in the ore and about 30 % of the silver, so a satisfactory treatment method for recovery of the gold and silver, and possibly other metals, was needed.

The treatment routes investigated, as outlined in Fig. 1, were

A. direct cyanidation,
B. roasting, sulphuric acid leaching, cyanidation,
C. roasting, acid brine leaching, cyanidation,
D. roasting, sulphuric acid leaching, thiourea leaching,
E. pressure oxidation and acid leaching, lime treatment, and cyanidation, and
F. pressure oxidation and acid leaching, and thiourea leaching.

Each of these routes was investigated in varying detail, with various sub-routes being considered in some cases, especially for pressure oxidation. The results for each route are presented in unpublished reports[2-5,10,16,17] and by Bruckard et al.[6-8] Illustrative results, particularly for the bulk lead-zinc scavenger tailing are given here, and the 'best' result from each route is used for an overall comparison.

EXPERIMENTAL

Experimental work was concerned with preparation and characterisation of the samples, roasting and leaching, pressure oxidation and leaching, cyanidation of the original materials, and cyanidation and thiourea leaching of the test products. The simplified flowsheets shown in Fig. 1 indicate the major operations conducted. The experimental procedures are summarised here and details are given in unpublished reports[2-5,10,16,17] and by Bruckard et al.[6-8]

Table 1 Pressure oxidation conditions

Temperature (°C)	200 or 220
Oxygen overpressure (kPa)	700
Initial free acidity (g/L H_2SO_4)	0 or 36.8
Dry feed weight (g)	120, 250, or 450
Total solution volume (mL)	2250
Initial pulp density (% solids)	5.1, 10.0, or 16.7
Reaction time (h)	4
Calcium lignosulphonate conc. (g/L)	0.2
Quebracho ATO conc. (g/L)	0.1
Heat-up atmosphere (kPa N_2)	1400
Stirring speed (rev/min)	1200

Sample preparation

Samples were received either wet or dry in amounts ranging from 2 kg to 30 kg. The wet samples were air dried before use. Each dry sample was brushed through a 1.651 mm aperture screen to break up lumps, thoroughly mixed, and then riffled into sub-samples ranging in weight from 100 g to 500 g for different purposes. A 100 g lot from each material was pulverised for chemical analysis.

Roasting

Roasting tests were conducted in an electrically-heated muffle furnace with free access of air. A thermocouple, which was used for automatic temperature control of the furnace, was located near the top of the muffle about 15 cm from the charge. The temperature shown by this thermocouple was not exactly that of the charge but was the value quoted in this work. The charge, usually 200 g of material, was placed in fused silica dishes which, in turn, were placed in either a cold (i.e. a cold start) or a hot (i.e. a hot start) furnace. In each test roasting was conducted at the selected temperature (500, 600, 700, 850, and 1000°C) for 100 min with frequent rabbling of the charge. Each roasted product was cooled in the open air and then weighed. Those samples which were relatively high in lead and zinc tended to form crusts on the charge and cake badly so that a lumpy product was obtained. These lumps were gently broken with a mortar and pestle before leaching.

Pressure oxidation

Pressure oxidation leaches to give either a hematite/lead jarosite product or a basic iron sulphate/lead sulphate product were conducted in an Autoclave Engineers Inc. 4 L titanium autoclave. A summary of the general operating conditions is given in Table 1. For formation of a hematite/lead jarosite product, an initial pulp density of 5.1 % solids, with no added acid, was used. For formation of a basic iron sulphate/lead sulphate product, an initial pulp density of 16.7 % solids, with an initial free acidity of 36.8 g/L H_2SO_4, was used. Between two and eight replicate runs under each set of conditions were conducted to generate sufficient solid product for subsequent thiourea leaching or lime treatment and cyanidation. After 4 h reaction time each charge was cooled and filtered. The filtrate was filtered again through an 0.5 μm aperture Millipore membrane before analysis. The filter cake was washed three times with 500 mL lots of water. The washed residues were air dried, weighed, and prepared for further treatment. Other details are given by Bruckard, Sparrow, and Woodcock.[7]

Sulphuric acid leaching

Sulphuric acid leaching occurred during pressure oxidation, as noted above, and was also conducted as a separate

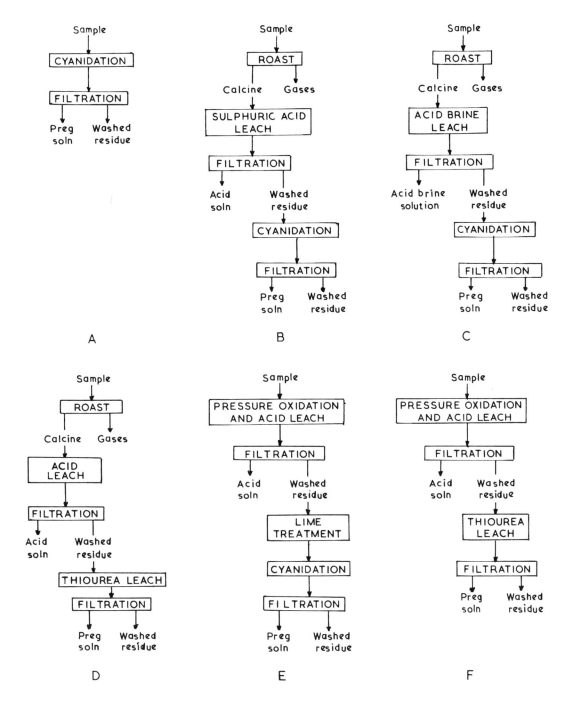

Fig. 1 - Experimental flowsheets.

A: Direct cyanidation.
B: Roasting, acid leaching, cyanidation.
C: Roasting, acid brine leaching, cyanidation.
D: Roasting, acid leaching, thiourea leaching.
E: Pressure oxidation, lime treatment, cyanidation.
F: Pressure oxidation, thiourea leaching.

operation on calcines from roasting. For calcines, most acid leaches were conducted at ambient temperature in an open stirred beaker for 24 h at an initial liquid:solid (L:S) ratio of 5:1 using 10 % v/v sulphuric acid. One test was conducted at 90°C for 8 h. Solution samples were taken at appropriate intervals for analysis. After leaching, the pulp was filtered and the filtrate retained for assay. The filter cake was washed with de-ionised water and then sent to lime treatment and cyanidation while still wet.

Acid brine leaching

Acid brine leaching of roasted calcine was conducted at a L:S ratio of 5:1 using a solution containing 5 % v/v HCl and 300 g/L NaCl for 24 h. After leaching the pulp was filtered to recover pregnant solution. The residue was first washed with acid brine to remove dissolved metals while avoiding hydrolysis of silver and lead, and then with water to remove acid and brine.

Lime treatment

Before commencing cyanidation, the pH of the pulp of the original middling or the product from acid leaching of a calcine or pressure oxidation was raised to the desired value (e.g. 10, 11, or 12) by adding small amounts (1-10 kg/t) of solid powdered lime. After each lime addition the pH would rapidly rise to a peak value and then fall over a period of minutes or hours to a stable level. As discussed in more detail later, the time required for pH adjustment before direct cyanidation on the original middling samples was only a few minutes, whereas on acid leached calcines it was several hours, and on some pressure oxidation products it was up to 2-3 days.

Cyanidation

Cyanidation tests were conducted on samples weighing between 100 g and 500 g at L:S ratios between 3:1 and 5:1. Cyanidation times used were 4, 6, 8, 18, 24, 48, and 72 h, with the most common time being 24 h. Tests were conducted in a rolling bottle or in stirred beakers. A standard initial cyanide strength of 0.08 % NaCN was used which was obtained by adding the required amount of stock cyanide solution (25 g/L NaCN). pH levels ranged from 9 to 12. Periodic checks were made of cyanide, lime, and pH levels and appropriate additions of cyanide and lime were made to maintain the desired conditions. In some tests, an addition of 0.5 kg/t litharge (PbO) or ammonium sulphate was made.

At the end of the desired cyanidation period the pulp was weighed to check the final L:S ratio on which the calculations were based, and then filtered to recover the pregnant cyanide solution. The solutions were assayed for gold and silver and other metals as desired. The filter cakes were washed by repulping with tap water and refiltering, and then dried, weighed, and assayed for gold and silver. In a few tests the washed residue was repulped with cyanide solution and cyanided for a further period or reground to a finer size and then re-cyanided.

Thiourea leaching

Thiourea leaches were conducted in 2 L stirred beakers on 100 g charges at L:S ratios of 10:1 for 4 h. Control of pH was achieved by manual or automatic addition through a pH-stat of either 50 % v/v sulphuric acid or 10 % w/v sodium hydroxide. In some tests the pH was not controlled but was allowed to assume its natural level of 2.3-3.4. Control of redox potential (Eh) was achieved by manual or automatic addition through an Eh-stat of 10 % w/v sodium sulphite (reductant), 40 % w/v $Fe_2(SO_4)_3.9H_2O$ in 1M H_2SO_4 (oxidant), or 10 % v/v H_2O_2 (oxidant). In some tests the Eh was allowed to assume its natural level of 200-580 mV. In this paper, all redox potentials are given as the potential of a platinum electrode versus a saturated calomel electrode (SCE). In some tests, sufficient ferrous ammonium sulphate dissolved in acidified water was added to give an initial concentration of 20 000 ppm Fe^{2+} in solution.

Thiourea addition rates used were 1, 2, and 4 g/L/h with an addition at 0, 1, 2, and 3 h, or 2.0 g/L/h with an addition every 15 min (i.e. 0.5 g/L/0.25 h). The thiourea was added as a 10 % w/v slurry in water. Solution samples were taken at appropriate intervals during each leach for determination of thiourea and silver concentrations.

After leaching, the pulp was weighed to check the final L:S ratio on which the calculations were based, and then filtered to recover the pregnant solution for assay. The filter cake was washed once with 300 mL of acidified water (pH 3) and then twice with 300 mL water. It was then dried, weighed, and assayed.

Chemical analysis

Analyses conducted included those for selected metals (Ag, Au, Cu, Pb, Zn, Fe) in solutions and solids, free acidity, thiourea, cyanide, and free lime in solution, and various constituents in the head samples.

Leach liquors were diluted as appropriate and analysed for Ag, Cu, Pb, Zn, and Fe by atomic absorption spectrometry (AAS). For gold analysis, a 20-50 mL aliquot was treated by aqua regia digestion, followed by baking to remove nitric acid, re-acidification with 6 M hydrochloric acid, extraction of gold into methyl isobutyl ketone (MIBK), and then determination of the gold in the extract by AAS.

Free acidity in pressure oxidation liquors was determined by a potentiometric titration with standard NaOH solution using MgEDTA as a complexing agent for ferric ion.[20] Thiourea concentrations were determined by titration of a 3.0 mL aliquot with potassium iodate using VITEX indicator (a soluble starch) after dilution and acidification with 1 M orthophosphoric acid. Cyanide concentrations were determined by titration with silver nitrate using potassium iodide as an indicator. In solutions of high lead content, especially on some roasted products, addition of potassium iodide gave a whitish precipitate of lead iodide which had to be removed by filtration so that it would not interfere with the normal end point. Free lime in solution was determined after cyanide determination, using the same aliquot, by titration with standard sulphuric acid using phenolphthalein as an indicator.

Chemical analyses on the head and residue samples were conducted using standard analytical methods. Gold and silver were determined using variations of the aqua regia digestion method noted above for solutions.

Sizing analysis

Most sizing analyses were done using a standard wet and dry screening technique based on a 38 μm aperture screen (400 mesh Tyler). However some Cyclosizer sizings were conducted and the results were reported as nominal pyrite size after correction for the operating conditions.

X-ray diffraction

X-ray diffraction patterns of the head samples, pressure oxidation leach residues, selected lime-treated products, and selected leach residues were recorded on a Philips PW 1050 goniometer with a PW 1710 diffraction controller using Cu Kα radiation. Materials present in the samples were identified by comparison of peaks in the diffraction patterns with data published by the Joint Committee on Powder Diffraction Standards (JCPDS).

CHARACTERISATION OF SAMPLES

The samples used in this investigation were derived from the Hellyer, Tasmania, orebody of Aberfoyle Resources Limited. They were either bulk samples produced in Aberfoyle's Cleveland mill, as outlined by Richmond and Lai,[19] or laboratory samples produced either in Aberfoyle's research laboratory in Burnie, Tasmania, or by Amdel Limited in Adelaide, South Australia.

In general, the samples represented a scavenger tailing from bulk re-flotation of lead and zinc concentrates, a zinc-rich middling product produced from zinc scavenger tailing, and a pyrite-rich middling product produced from zinc scavenger tailing. They represented various plant products, and together contained about 70 % of the gold and 30 % of the silver in the original ore. The samples were characterised in this work in terms of their chemical composition, mineralogical content and associations, and their sizings.

Chemical composition and sizing

A chemical analysis of each of the samples used in this work is given in Table 2. The samples contained 1.5-3.3 g/t Au and 27-144 g/t Ag as well as different amounts of lead, zinc, and other elements. The A, B, and C samples are simply similar samples of a particular type received at different times.

All the samples contained 60-70 % pyrite and a major difference was the lead plus zinc content which ranged from 12-19 % in the bulk scavenger tail, through 5-7 % in the zinc middling, to 2-3 % in the pyrite middling.

Sizing analyses of the samples are given in Table 3. It can be seen that each sample was very fine (at least 88 % finer than 38 μm), as expected for middling products from an ore which is known to require fine grinding for mineral liberation.[19]

Mineralogy

As noted by Richmond and Lai,[19] the ore has a high specific gravity (4.5) and contains fine intergrowths of the major minerals pyrite, sphalerite, and galena. Other sulphide minerals include chalcopyrite and arsenopyrite. The main non-sulphide minerals are barite, quartz, sericite, chlorite, and calcite.

Silver is present in silver minerals (tetrahedrite and others), and in a sub-microscopic form in galena, sphalerite, and pyrite. Gold is largely associated with pyrite and arsenopyrite, either as very fine particles or as sub-microscopic gold in the minerals. Some of the gold is thought to be in the sphalerite.

XRD patterns of the samples used in this work, e.g. those reported by Bruckard, Sparrow, and Woodcock,[7] confirmed that the major phases present were pyrite and sphalerite with lesser amounts of galena, chalcopyrite, and quartz. No silver minerals or arsenopyrite were specifically identified. The sulphate sulphur content is mainly present as barite ($BaSO_4$), but some sulphate was formed by oxidation of sulphide minerals during storage and drying of the samples. Anglesite ($PbSO_4$), formed by oxidation of galena, was identified in most samples.

RESULTS OF DIRECT CYANIDATION

Direct cyanidation is potentially the most attractive method for extracting gold and silver from the various middling

Table 2 Analytical data for test head samples

Constituent	Bulk scav. tail			Zinc middling		Pyrite middling	
	A	B	C	A	B	A	B
Au (g/t)	2.27	2.6	3.3	2.62	2.3	1.50	2.8
Ag (g/t)	82.0	144	105	53.4	92	27.2	45
Pb (%)	4.36	11.4	3.8	2.28	4.4	1.68	1.1
Zn (%)	7.85	7.7	8.0	3.35	2.6	0.70	1.8
Cu (%)	1.01	0.17	0.24	0.81	0.14	0.27	0.12
Fe (%)	28.0	36.6	-	31.5	44.4	31.4	41.5
Na (%)	0.09	-	-	0.03	-	0.04	-
Ca (%)	0.81	-	-	0.60	-	2.04	-
Mg (%)	0.19	-	-	0.09	-	0.22	-
Ba (%)	0.18	-	0.18	0.10	-	0.86	-
As (%)	0.50	2.3	1.6	0.31	2.1	0.24	-
Sb (%)	0.11	0.12	0.2	0.03	0.11	0.01	-
SiO_2 (%)	1.66	-	6.6	0.72	-	4.40	-
S total (%)	37.4	41.7	-	39.8	45.9	41.1	-
S as SO_4 (%)	6.1	-	-	7.4	-	3.2	-

211

products because of its apparent simplicity and well known technology. Results of direct cyanidation using the flowsheet in Fig. 1A, are summarised in Figs. 2 and 3. Most of the results given here were obtained on the A samples in Table 2 but similar results were obtained with other samples.[2-5]

Effect of agitation time

Fig. 2 shows that at pH 10.5, between 30 and 45 % of the silver was extracted from the samples by direct cyanidation for 24 h. Most of this silver was extracted in the first 4-6 h agitation. Increasing the cyanidation time for the pyrite middling to 48 h gave only 2 % additional silver extraction.

Gold extractions increased slowly over the 24 h cyanidation period, but the final extractions were only in the range 10-20 %. Reagent consumptions were greater for the bulk scavenger tailing and increased slightly with cyanidation time to about 7 kg/t NaCN and 20 kg/t CaO after 24 h cyanidation.

Effect of pH

Fig. 3 shows that when using a 6 h cyanidation period, the silver extraction, in general, increased appreciably as the pH 'was raised from about 9 to about 12, at which pH the extractions achieved were the greatest obtained by direct

Table 3 Screen sizing analysis of the three A head samples

Screen aperture (μm)	Bulk scav. tail (% wt.)	Zinc middling (% wt.)	Pyrite middling (% wt.)
+104	Trace	Trace	Trace
-104+74	Trace	0.1	0.5
-74+53	0.1	1.4	2.4
-53+38	2.5	5.8	9.2
-38	97.4	92.7	87.9
Total	100.0	100.0	100.0

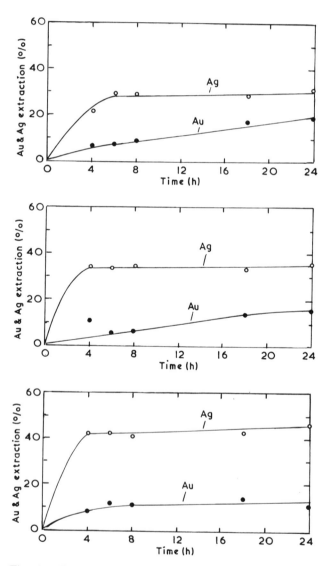

Fig. 2 - Effect of cyanidation time at pH 10.5 on gold and silver extractions from the three A samples. Top: Bulk scavenger tailing. Centre: Zinc middling. Bottom: Pyrite middling.

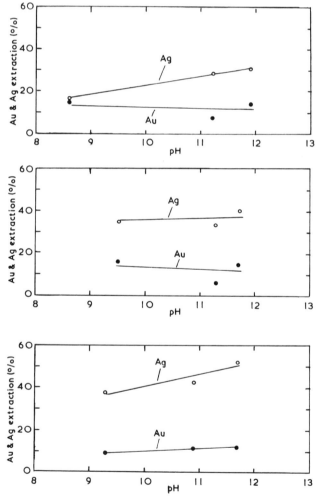

Fig. 3 - Effect of pH on gold and silver extractions for 6 h cyanidation of the three A samples. Top: Bulk scavenger tailing. Centre: Zinc middling. Bottom: Pyrite middling.

cyanidation. These values are therefore used in the overall comparison discussed later. Cyanidation at pH 12 for 24 h may have given even better results.

Gold extractions were not conclusively affected by pH and any trends may have been obscured by sampling and assay errors. The higher cyanidation pH resulted in lower cyanide consumptions (reduced by about 1 kg/t NaCN) but markedly higher lime consumptions (up to 20 kg/t CaO more).

Effect of regrinding and additives

The above results were in general accord with those obtained with other samples in earlier work. However, some of the earlier work had also been concerned with the effect of ultrafine grinding[4] and the effect of additives[3] on different samples.

The effect of ultrafine grinding on cyanidation extraction was investigated[4] on a sample of pyrite middling (sample B) containing 29 % minus 7 μm material. Direct cyanidation for 24 h extracted 12 % of the gold and 19 % of the silver. Cyanidation after regrinding the cyanided pyrite middling to 86 % minus 7 μm extracted only a further 6 % of the gold but 26 % of the silver. This confirmed that over 80 % of the gold present probably occurred in a submicroscopic form in the pyrite and could not be exposed by grinding. Silver extractions, on the other hand, were improved by fine grinding. This result was consistent with the premise that at least some of the silver was present as one or more specific minerals which are moderately refractory to cyanidation.

An addition of 0.5 kg/t litharge made little difference to the extractions, and while the addition of 0.5 kg/t ammonium sulphate gave a small apparent increase in gold and silver extractions,[3] this was little more than experimental error.

Pre-oxidation of a sample with 500 L/t hydrogen peroxide (100 volume) at an Eh of 100 mV at pH 9 for 24 h resulted in a small improvement in extractions but could not be seen to be economic.

RESULTS OF ROASTING, ACID LEACHING, AND CYANIDATION

An alternative to direct cyanidation, to obtain increased extraction, is to follow the flowsheet outlined in Fig. 1B. In this scheme the material was first roasted to break down sulphide mineral lattices and liberate the contained gold and silver. The calcine was then leached with dilute sulphuric acid to dissolve cyanide-consuming base metals, and finally the acid-leached calcine was cyanided for long periods to extract the gold and silver. Results obtained from this procedure, using pyrite middling sample B, are summarised in Figs. 4 and 5, and were similar to those obtained on other samples.[10,16,17]

Roasting resulted in a weight loss of about 30 % of the charge for the roasting conditions used.[16,17] The calcines were various colours including red-brown, purplish-grey, and dark brown depending on the feed composition and the

roasting conditions. The lumps which formed during roasting, and which were partly broken before leaching, disintegrated satisfactorily during acid leaching and cyanidation. Most of the detailed roasting work was conducted on pyrite middling sample B (Table 2) and was confirmed with other samples. There appeared to be no significant difference in the results between hot start and cold start roasting.

Effect of roasting temperature

The effect of roasting temperature on gold and silver extraction by 48 h cyanidation of the acid-leached calcines is shown in Fig. 4. It can be seen that the maximum gold extraction (77 %) was obtained after roasting at 600-700°C whereas the maximum silver extraction (77 %) was obtained after roasting at 500°C.

Base metal extractions during the acid leaching stage, which was conducted to minimise cyanide consumption during subsequent cyanidation, were also affected by the roasting temperature. For example, over 80 % of the copper was extracted after roasting at 500°C but the extraction dropped to under 20 % after roasting at 1000°C. Much of the acid-insoluble copper was cyanide-soluble, leading to high cyanide consumptions. Zinc extractions followed a very similar pattern. Iron extractions were different, however, and there was a minimum iron extraction after roasting at 700°C. However, very little iron dissolved during cyanidation even in the absence of a prior acid leach.

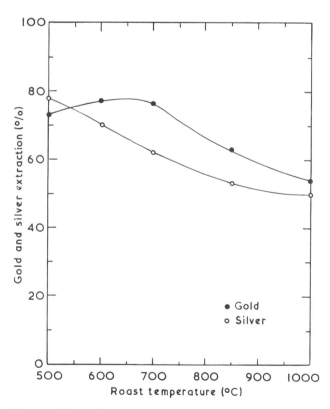

Fig. 4 - Effect of roasting temperature on gold and silver extraction for 48 h cyanidation of acid leached calcines from pyrite middling (sample B).

213

A hot acid leach (90°C) for 8 h extracted well over 90 % of the copper and zinc. About 40 % of the iron was extracted, giving a substantial weight reduction in the feed to cyanidation but little reduction in cyanide consumption.

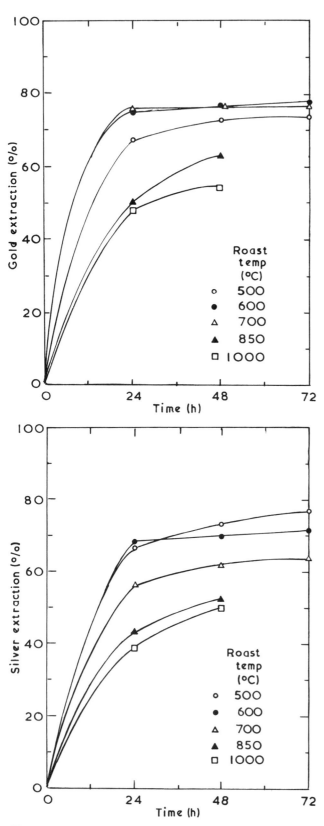

Fig. 5 - Effect of cyanidation time on gold and silver extractions from acid leached calcines from pyrite middling (sample B) roasted at different temperatures. Top: Gold extractions. Bottom: Silver extractions.

About 11 % of the silver, but effectively no gold, was dissolved in the hot acid leach. Subsequent cyanidation extracted 85 % of the gold. This was the highest value obtained in this work on calcines and was thought to be due to dissolution of iron compounds enclosing the gold. Silver extractions by cyanidation were similar after hot and cold acid leaching but when the 11 % of the silver dissolved in the hot acid solution was taken into account, the overall extraction was greater with hot, rather than cold, acid leaching.

Effect of cyanidation time

Fig. 5 shows the effect of cyanidation time on gold and silver extractions for each roasting temperature. Although most of the gold and silver was extracted in the first 24 h cyanidation period, additional extraction was obtained by extending the cyanidation time to 72 h. The extractions at 72 h for roasting at 500°C and 700°C were used in the general comparison discussed later.

Studies of the effect of time of leaching during the acid leaching stage[16] showed that most of the dissolution of the base metals took place during the first two hours of leaching. However there were small increases with time up to 24 h, especially for iron. These differences were not correlated in detail with subsequent gold and silver extractions or reagent consumptions. Variations in sulphuric acid strength over the range 5-15 % did not seem to be important.

Effect of sample composition

Several different concentrates were treated by the flowsheet shown in Fig. 1B[10,16,17] but no definite relationship between feed composition and gold and silver extraction was established. This was probably because of uncontrolled variations in composition, as illustrated by the assays in Table 2. Even for the same zinc level, for example, the lead, arsenic, gold, and silver levels were different.

However, there were indications, when using specially prepared flotation products from the same zinc cleaner tailing,[17] that the extractions could be lower from products higher in zinc and lead. Gold extractions dropped from 74 % using zinc middling B to 70 % using scavenger tailing B, while silver extractions dropped from 79 % to 72 %. This suggests that the silver content of the galena may be the most important factor. Lead is known to interfere with the cyanide extraction of gold from roasted pyrite.[15,23]

RESULTS OF ROASTING, BRINE LEACHING, AND CYANIDATION

As noted above, lead may interfere with gold and silver extraction from roasted products.[15,23] One way to avoid this interference is to use the flowsheet shown in Fig. 1C, in which the sulphuric acid leaching step is replaced by an acid brine leach. This leach dissolves both lead and silver before cyanidation.

Results of the work conducted on bulk scavenger tailing sample C are summarised in Table 4. It can be seen that after roasting at 550°C the brine leach route resulted in the extraction of nearly 83 % of the gold by cyanidation compared with 76 % by cyanidation after sulphuric acid leaching. This was, apparently, a reasonable improvement although the silver extraction was slightly lower. The brine leach dissolved 64.5 % of the silver and 80 % of the lead. The subsequent cyanidation step dissolved an additional 7.6 % of the silver giving a total silver extraction of 72 %, which was slightly lower than the 76 % achieved by cyanidation after sulphuric acid leaching.

These results indicated that the brine leaching route did not offer sufficient improvement in extraction over normal sulphuric acid leaching to warrant further investigation at this stage in view of the extra complexities involved with brine leaching.

RESULTS OF ROASTING AND THIOUREA LEACHING

An alternative to cyanidation for gold and silver extraction from calcines is thiourea leaching. This has the advantage that the large lime additions required for cyanidation would not be needed. In the work conducted here it was thought to be necessary to retain the acid leaching step, as in Fig. 1D, since it has been shown[18] that soluble copper interferes with thiourea leaching of gold, and it was thought that soluble zinc would also interfere.

Results of the work conducted[10] are summarised in Table 5. The thiourea leach was conducted for 24 h at pH 1.5, an Eh of 150 mV, and a thiourea concentration of 12 g/L. These conditions may need modification as discussed elsewhere.[18] However it is clear that the gold and silver extractions obtained by thiourea leaching (about 77 %) were very similar to those obtained by cyanidation. Further investigation of the thiourea route may be worthwhile in an attempt to reduce thiourea consumption.

RESULTS OF PRESSURE OXIDATION, LIME TREATMENT, AND CYANIDATION

In a separate approach to breaking down the sulphide minerals the treatment route outlined in Fig. 1E, which involves pressure oxidation and acid leaching followed by lime treatment and cyanidation, was investigated in some detail by Bruckard, Sparrow, and Woodcock[7] using the A samples listed in Table 2. It was known that aqueous pressure oxidation of pyrite and arsenopyrite could liberate the contained gold for subsequent cyanidation[1] but, in spite of encouraging results reported by Thompson,[21] it was less obvious whether silver in galena, or gold and silver in sphalerite, could be liberated and then extracted by cyanidation.

Pressure oxidation

Each sample was pressure oxidised under various conditions of pulp density, initial acid concentration, temperature, oxygen pressure, and time, to give various mixtures of hematite, basic iron sulphate, lead jarosite, and lead sulphate. Dissolution of the base metals and a little silver, but no gold, took place during pressure oxidation. Conditions and the main results of pressure oxidation are summarised in Table 6.

Two products were prepared for subsequent treatment. These were either a hematite plus lead jarosite or a basic iron sulphate plus lead sulphate product. Preparation of the hematite/lead jarosite product, which was made at low acidity, gave high extractions of zinc and copper in the pressure oxidation solutions and relatively moderate extractions of iron and silver. Preparation of the basic iron sulphate/lead sulphate product, which was made at high acidity, also gave high extractions of copper but slightly lower extractions of zinc. Iron extractions were appreciably higher (28-35 %) and silver extractions were low.

Lime treatment

The washed pressure oxidation residues were treated with lime at ambient temperatures to give specified pH levels before cyanidation. This was a slow process, particularly when a basic iron sulphate/lead sulphate product was treated, and total lime additions required were as much as 480 kg/t CaO. Reaction rates were slow and, in some tests, the lime treatment took 2-3 days. Typical neutralisation curves showing the change in pH with lime addition for calcined bulk scavenger tail and pressure oxidation products from the zinc middling are shown in Fig. 6. It is interesting to note

Table 4 Comparison of results of roasting/sulphuric acid leaching/cyanidation with those of roasting/brine leaching/cyanidation on bulk scavenger tailing sample C.

	H₂SO₄ leach/cyanidation					Brine leach/cyanidation				
	Au	Ag	Cu	Zn	Pb	Au	Ag	Cu	Zn	Pb
Extr. in acid (%)	0.6	0.3	34.9	43.3	0.0	0.0	64.5	29.6	53.8	80.0
Extr. in cyan (%)	75.4	75.5	0.0	0.0	0.0	82.6	7.6	0.9	0.3	0.0
Total extr. (%)	76.0	75.8	34.9	43.3	0.0	82.6	72.1	30.5	54.1	80.0

that the curve for lime treatment of the roasted calcine (hematite) is very similar to that for the hematite/lead jarosite product from pressure oxidation.

XRD examination of the feed and products from lime treatment showed[7] that the hematite remained unchanged throughout lime treatment, and that the lead jarosite remained unchanged below pH 9 but was slowly converted to amorphous lead and iron hydroxides at higher pH levels. The basic iron sulphate was converted to amorphous ferric hydroxide at and above pH 9, and the lead sulphate to lead hydroxide at pH 11. Gypsum (calcium sulphate) was formed in all cases. Lime additions required were 28-52 kg/t CaO for hematite/lead jarosite products and 300-480 kg/t CaO for basic iron sulphate/lead sulphate products (Table 7).

However, much of the lime additions could be replaced with limestone, at lower cost, particularly for neutralisation up to about pH 4 as is done at Sao Bento.[11]

Cyanidation

The lime-treated hematite/lead jarosite products were cyanided for 24 h at pH 10.5-11.0, and the lime-treated basic iron sulphate/lead sulphate products were cyanided at both pH 10.6 and 12.0. Results for the three A samples listed in Table 2 are summarised in Table 7. Detailed results for all three types of head sample were given by Bruckard, Sparrow, and Woodcock.[7]

Silver extractions from the hematite/lead jarosite material were poor, being only 25-73 %. However, from the basic iron sulphate/lead sulphate material, after conversion to amorphous hydroxides and gypsum at pH 12, silver extractions were better and were in the range 87-90 %.

Gold extractions were better than silver extractions, in spite of the lower feed assay. In most tests they were also greater from the basic iron sulphate/lead sulphate material than from the hematite/lead jarosite product, and were slightly greater at pH 12 than at pH 10.6. The best gold extractions were in the range 91-95 %, and these figures have been used in the general comparison.

RESULTS OF PRESSURE OXIDATION AND THIOUREA LEACHING

As an alternative to pressure oxidation, lime treatment, and cyanidation for extraction of gold and silver from Hellyer

Table 5 Comparison of results of roasting/acid leaching/cyanidation with those of roasting/acid leaching/thiourea leaching on bulk scavenger tailing sample C.

Method	Extractions		Reagent consumptions
	Au (%)	Ag (%)	(kg/t)
Thiourea leach	77.4	77.5	Acid,* 112 Thiourea, 67 Hydrogen peroxide, 49
Cyanid-ation	76.0	75.8	Acid,* 112 Sodium cyanide, 4.9 CaO, 6.0

* Leaching with 10 % H_2SO_4.

Table 6 Conditions and results of pressure oxidation and acid leaching of the three A samples using the conditions given in Table 1

Sample	Temp. (°C)	Initial pulp density (% solids)	Free acidity (g/L H_2SO_4) Init.	Free acidity (g/L H_2SO_4) Final	Residue weight (% of feed)	Major phases in pressure oxidation product	Metal extractions Zn (%)	Cu (%)	Fe (%)	Ag (%)
Bulk scav. tail	200	5.1	0	48.5	60.1	Hematite and lead jarosite	82.6	89.1	6.1	5.3
	220	16.7	36.8	108	80.5	Basic iron sulphate and lead sulphate	76.1	92.7	28.2	1.9
Zinc middling	200	5.1	0	54.3	52.3	Hematite and lead jarosite	95.2	90.3	17.3	8.4
	200	16.7	36.8	125	78.7	Basic iron sulphate and lead sulphate	67.2	88.3	32.3	1.8
Pyrite middling	200	5.1	0	90	61.5	Hematite and lead jarosite	70.8	97.2	23.3	7.6
	200	16.7	36.8	119	88.6	Basic iron sulphate and lead sulphate	35.7	92.6	35.1	5.7

products it was considered worthwhile to investigate the application of pressure oxidation and thiourea leaching (Fig. 1F) since the acid pressure oxidised product would be directly suitable for thiourea leaching and this would eliminate the large lime addition required before cyanidation. Detailed results of the work conducted have been presented by Bruckard, Sparrow, and Woodcock.[8]

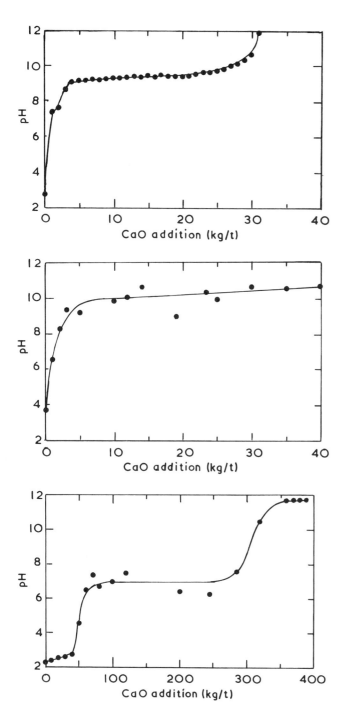

Fig. 6 - Change in pH with lime addition for acid leached roasted calcines and pressure oxidation products. Top: Calcine from bulk scavenger tail (sample B). Centre: Hematite/lead jarosite product from pressure oxidation of zinc middling (sample A). Bottom: Basic iron sulphate/lead sulphate product from pressure oxidation of zinc middling (sample A).

Pressure oxidation

Each sample was pressure oxidised under similar conditions to those noted previously to give a hematite/lead jarosite product and a basic iron sulphate/lead sulphate product. Results of the pressure oxidation/leaching stage were effectively the same as those shown in Table 6.

Thiourea leaching

A variety of conditions were used during thiourea leaching of the pressure oxidised products. After finding the optimum conditions for the basic iron sulphate/lead sulphate product from the zinc middling, these conditions were applied to the other middling samples.

Fig. 7 shows the effect of Eh/pH conditions on the rate of silver extraction from the basic iron sulphate/lead sulphate product obtained from zinc middling (sample A). It was concluded that natural Eh (200 mV) and a natural pH (2.3-3.4) gave the best results in spite of the evidence of reprecipitation of silver after 3 h leaching. The final value was equivalent to a silver extraction of 92.6 %.

Fig. 8 shows the effect of thiourea addition rate between 0 and 4 g/L/h on silver extraction from the basic iron sulphate/lead sulphate product from the zinc middling. It can

Table 7 Conditions and results of cyanidation of the pressure oxidation products from the three A samples

Sample	Major phase in cyan. feed	Cyan. pH	Consumption		Extraction	
			NaCN (kg/t)	CaO* (kg/t)	Au (%)	Ag (%)
Bulk scav. tail	Hematite and lead jarosite.	10.8	4.9	52	93.6	44.2
	Am. hydroxides, lead sulphate, and gypsum.	10.6	4.4	350	90.5	90.4
	Am. hydroxides, and gypsum.	12.0	1.4	480	91.5	90.0
Zinc mid.	Hematite and lead jarosite.	11.0	3.8	42	73.6	72.7
	Am. hydroxides, lead sulphate, and gypsum.	10.6	4.4	360	91.8	82.6
	Am. hydroxides, and gypsum.	12.0	2.9	390	94.5	90.1
Pyrite mid.	Hematite and lead jarosite.	10.5	1.3	28	89.3	25.1
	Am. hydroxides, lead sulphate, and gypsum.	10.6	2.3	300	89.8	69.5
	Am. hydroxides, and gypsum.	12.0	1.3	350	92.8	87.5

*Includes lime used in pre-cyanidation treatment.
Am: Amorphous.

be seen that the silver extraction rate increased as the thiourea addition rate increased, when adding thiourea every hour. Moreover it was found that the rate increased further, as shown by the point at 2 g/L/h, when one-quarter of the total addition per hour was made every 0.25 h. This more

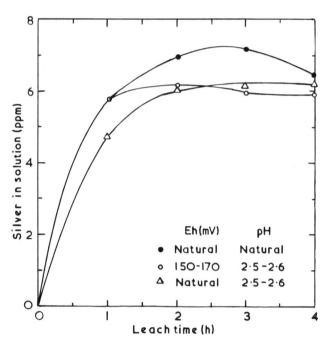

Fig. 7 - Effect of Eh and pH on silver extraction as a function of time for thiourea leaching of basic iron sulphate/lead sulphate product from pressure oxidation of zinc middling A for a thiourea addition rate of 4 g/L/h.

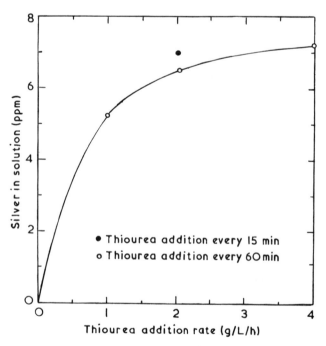

Fig. 8 - Effect of thiourea addition rate on silver extraction by thiourea leaching for 3 h at natural Eh and pH from a basic iron sulphate/lead sulphate product from pressure oxidised zinc middling (sample A).

frequent addition of small amounts of thiourea resulted in a slightly lower consumption of thiourea (e.g. a decrease from 67.5 kg/t to 56.6 kg/t).

Application of the best conditions found to each of the three A samples listed in Table 2 gave the results shown in Table 8. High gold and silver extractions were obtained from each sample, except for silver extraction from the pyrite, indicating that pressure oxidation to give basic iron sulphate/lead sulphate had effectively liberated the gold and silver from the original minerals.

However, application of the best conditions for thiourea leaching of the basic iron sulphate/lead sulphate product to the hematite/lead jarosite products gave much poorer results. For example, on the pyrite middling, a gold extraction of only 47.4 % was obtained and the silver extraction was only 0.8 %.[8]

In an attempt to increase these extractions the Eh was raised to 250 mV with additions of ferric sulphate and the pH adjusted to 1.5 with sulphuric acid. In another test, ferrous ammonium sulphate was added to increase the iron concentration in solution from 2-3 ppm to an initial value of 20 000 ppm. The final iron concentrations were 22 500, 19 300, and 8 800 ppm for the bulk scavenger tailing, the zinc middling, and the pyrite middling respectively. These changes raised the extractions slightly to those shown in Table 8 but the results were relatively poor, especially for silver.

It is thought that oxidation of the original material to hematite/lead jarosite locks up silver in either hematite or lead jarosite or both, as suggested by others.[9,12-14] Relatively little gold is locked in these compounds. XRD patterns before and after thiourea leaching of the hematite/lead jarosite product showed no detectable changes in the compounds present. However, XRD patterns of the basic iron sulphate/lead sulphate products before and after thiourea leaching showed that some lead jarosite had been formed during the thiourea leach. This could have locked up some of the silver during the leach and contributed to the low silver extractions for the pyrite middling.

Table 8 Results of thiourea leaching of pressure oxidation products from the three A middling samples.

Sample	Basic iron sulphate/ lead sulphate product*			Hematite/ lead jarosite product[†]		
	Au extr. (%)	Ag extr. (%)	Tu used (kg/t)	Au extr. (%)	Ag extr. (%)	Tu used (kg/t)
Bulk tailing	82.6	96.0	34.4	88.3	1.8	16.1
Zinc middling	92.1	92.6	56.6	86.6	6.9	64.5
Pyrite middling	91.5	38.7	73.4	45.9	5.7	70.0

* Leach conditions: Natural Eh and pH, thiourea addition rate 0.5 g/L/0.25 h, leach time 4 h.
[†] Leach conditions: pH 1.5, Eh 225-250 mV (using H_2O_2), addition of Fe^{2+} to give 20 000 ppm Fe in solution.

COMPARISON OF TREATMENT METHODS

A comparison of the 'best' results, i.e. greatest extraction, for selected conditions from each treatment method is given in Table 9 for each type of sample used. It is not possible to specify precisely what is meant by the 'best' result because, for example, of the variation in assay value and the gold:silver ratios in the samples. Consequently the monetary value of a 10 % increase in silver extraction may more than offset the value of a corresponding decrease in gold extraction as well as additional treatment costs. Furthermore, because different samples of each middling type were used at different times, the results may not be comparable in very fine detail. However, similar results were obtained on similar samples so reasonable confidence can be placed on the validity of the results.

Table 9 shows three groups of gold and silver extraction values of different magnitude — those obtained by direct cyanidation, those by cyanidation after roasting, and those by cyanidation or thiourea leaching after pressure oxidation. Extractions by direct cyanidation were in the range 10-20 % of the gold and 30-50 % of the silver. Long cyanidation times at high pH were required for the highest extractions but this brought a consistent increase in reagent consumption. The extractions were not increased by fine grinding. In general terms it appears that direct cyanidation is perhaps

marginally economic given that the material would be already available in an alkaline pulp that would normally be discharged to a tailing dam. In other words, only the costs of cyanidation and gold and silver recovery need to be charged against the metals produced.

Extractions after roasting to liberate gold and silver and then acid leaching to remove base metals were in the range 70-80 % for gold and silver whether cyanide or thiourea was used as the leaching agent. Roasting at 500°C gave the best silver extraction but a slightly lower gold extraction, whereas roasting at 700°C gave the best gold extraction but an appreciably lower silver extraction. There is considerable scope to optimise the various stages, particularly for the roast/sulphuric acid leach/cyanidation sequence which, in general terms, would seem to be the preferred option in this group. This route would be preferable to cyanidation after brine leaching, in view of the extra complexity of such a circuit, and preferable to thiourea leaching in view of the high thiourea consumption. A possible benefit of this circuit is that dissolved zinc and copper would be available for recovery. However, the roasting step is probably environmentally unacceptable if sulphur dioxide gas is released to the atmosphere, particularly in Tasmania, which is a relatively high rainfall area. Nevertheless, if sulphuric acid could be made and sold, the economic attractiveness of this route would be increased.

Table 9 Comparison of gold and silver extractions from bulk tailing, zinc middling, and pyrite middling by different treatment methods.

Treatment method	Bulk tailing		Zinc middling		Pyrite middling	
	Au (%)	Ag (%)	Au (%)	Ag (%)	Au (%)	Ag (%)
Direct cyanidation at pH 10.5 for 24 h	19.3	31.3	16.3	36.1	11.5	46.9
Direct cyanidation at pH 12 for 6 h	14.3	30.8	14.5	40.8	12.4	52.8
Roast at 500°C, acid leach, cyanidation for 72 h	n.d.	n.d.	n.d.	n.d.	74.7	82.9
Roast at 700°C, acid leach, cyanidation for 72 h	69.9*	71.5*	74.0*	79.1*	77.1	63.7
Roast at 550°C, brine leach, cyanidation for 24 h	82.6	72.1†	n.d.	n.d.	n.d.	n.d.
Roast at 550°C, acid leach, thiourea leach for 24 h	77.4	77.5	n.d.	n.d.	n.d.	n.d.
Pressure oxidise (+leach) to form hematite/lead jarosite, lime treatment, cyanidation at pH 11	93.6	44.2	73.6	72.7	89.3	25.1
Pressure oxidise (+leach) to form basic iron sulphate/ lead sulphate, lime treatment, cyanidation at pH 12	91.5	90.0	94.5	90.1	92.8	87.5
Pressure oxidise (+leach) to form hematite/lead jarosite, thiourea leach	88.3	1.8	86.6	6.9	45.9	5.7
Pressure oxidise (+leach) to form basic iron sulphate/ lead sulphate, thiourea leach	82.6	96.0	92.1	92.6	91.5	38.7

* Roast at 600°C, and cyanidation for 48 h.
† Includes extraction in acid brine leach.
n.d.: Not determined.

After pressure oxidation to form a basic iron sulphate/lead sulphate product (with simultaneous extraction of copper and zinc into solution) and lime treatment to raise the pH to 12, cyanidation extracted 88-95 % of the gold and silver. These values represented the best extractions obtained with any of the flowsheets used, and in view of experience at the McLaughlin mine,[22] at Sao Bento,[11] and at other places[1] where a similar process is used economically, it warrants further examination to attempt to reduce the high lime additions currently required for this route. Experience at Sao Bento[11] is particularly relevant since reagent consumption included 395 kg/t limestone and 159 kg/t lime.

Pressure oxidation to form hematite/lead jarosite products followed by cyanidation or thiourea leaching gave inferior gold and silver extractions, and probably need not be considered further.

CONCLUSION

This work has shown that any of the three main routes investigated for extraction of gold and silver (i.e. direct cyanidation, roasting/acid leaching/cyanidation, and pressure oxidation/lime treatment/cyanidation) are technically viable. Each route has its advantages and disadvantages. Direct cyanidation is potentially the most attractive method because of its apparent simplicity and well known technology. Environmental considerations probably preclude the use of the roasting routes without sulphuric acid production. However, since the route involving pressure oxidation/lime treatment/cyanidation can give the greatest gold and silver extraction from any of the samples investigated, it is worth investigating in more detail.

ACKNOWLEDGEMENTS

The authors would like to thank Aberfoyle Resources Limited for financial support of part of this work, for providing all of the samples used, and for approval to present data on Hellyer samples from unpublished reports to the company. They would also like to thank Miss C.M. McInnes of the CSIRO Division of Mineral Products, Melbourne, for providing analytical and typing assistance.

REFERENCES

1. Berezowsky, R.M.G.S., and Weir, D.R., 1989. Refractory gold: the role of pressure oxidation, in *Gold Forum on Technology and Practices - 'World Gold '89'* (Ed. R.B. Bhappu and R.J. Harden), pp. 295-304 (Society for Mining, Metallurgy, and Exploration, Inc.: Littleton).

2. Bilston, D.W., McCallum, D.A., and Woodcock, J.T., 1985. Preliminary tests on gold extraction from Hellyer pyrite, CSIRO Division of Mineral Chemistry, Mineral Chemistry Communication, MCC 679, 21 pp.

3. Bilston, D.W., and Woodcock, J.T., 1986. Treatment of barite-gold ore from Hellyer, CSIRO Division of Mineral Chemistry, Mineral Chemistry Communication, MCC 686, 16 pp.

4. Bilston, D.W., and Woodcock, J.T., 1986. Effect of grinding on gold and silver extraction by cyanidation of Hellyer pyrite, CSIRO Division of Mineral Chemistry, Mineral Chemistry Communication, MCC 711, 12 pp.

5. Bilston, D.W., and Woodcock, J.T., 1989. Gold and silver extraction by cyanidation from flotation concentrates produced in the Hellyer milling trials at Cleveland, CSIRO Division of Mineral Products, Mineral Products Communication, MPC/M-076, 19 pp.

6. Bruckard, W.J., Canterford, J.H., Dyson, N.F., and Sparrow, G.J., 1989. Oxygen pressure leaching of a bulk flotation concentrate from a complex Cu-Pb-Zn sulphide ore, in *Non-Ferrous Smelting Symposium*, pp. 111-117 (The Australasian Institute of Mining and Metallurgy: Melbourne).

7. Bruckard, W.J., Sparrow, G.J., and Woodcock, J.T., 1990. Gold and silver extraction from Hellyer lead-zinc flotation middlings using pressure oxidation, lime treatment, and cyanidation, in *Proc. 14th CMMI Congress (Minerals, Materials and Industry)*, Edinburgh, July 1990, (The Institution of Mining and Metallurgy: London) (in press).

8. Bruckard, W.J., Sparrow, G.J., and Woodcock, J.T., undated. Gold and silver extraction from Hellyer lead-zinc flotation middlings using pressure oxidation and thiourea leaching, paper submitted to *Hydrometallurgy* (1990).

9. Bull, W.R., Spottiswood, D.J., and Tourre, J.M., 1985. Oxidative acid pressure leaching of sulphidic ores and concentrates - the control of silver losses, in *Complex Sulphides* (Ed. A.D. Zunkel, R.S. Boorman, A.E. Morris, and R.J. Wesely), p. 141 (The Metallurgical Society Inc.: Warrendale).

10. Canterford, J.H., Dyson, N.F., Graham, W.E., McCallum, D.A., and Woodcock, J.T., 1986. Treatment of high-gold pyrite from Hellyer, CSIRO Division of Mineral Chemistry, Mineral Chemistry Communication, MCC 696, 13 pp.

11. da Silva, E.J., Haines, A.K., Carvalho, T.M., de Melo, M.P., and Doyle, B.N., 1989. Process selection, design, commissioning and operation of the Sao Bento Mineracao refractory gold ore treatment complex, in *Gold Forum on Technology and Practices - 'World Gold '89'* (Ed. R.B. Bhappu and R.J. Harden), pp. 322-332 (Society for Mining, Metallurgy, and Exploration, Inc.: Littleton).

12. Dutrizac, J.E., and Chen, T.T., 1984. A mineralogical study of the jarosite phase formed during the autoclave leaching of zinc concentrate, *Canadian Metallurgical Quarterly*, 23: 147-157.

13. Dutrizac, J.E., Dinardo, O., and Kaiman, S., 1980. Factors affecting lead jarosite formation, *Hydrometallurgy*, 5: 305-324.

14. Dutrizac, J.E., and Jambor, J.L., 1987. Behaviour of silver during jarosite precipitation, *Transactions Institution of Mining and Metallurgy (Section C: Mineral Processing and Extractive Metallurgy)*, 96: C206-C218.

15. Leaver, E.S., Woolf, J.A., and Jackson, T.A., 1933. Cyanidation of calcined gold ores made refractory by the presence of lead minerals, American Institute of Mining and Metallurgical Engineers, Contribution 5, 15 pp.

16. McCallum, D.A., and Woodcock, J.T., 1986. Treatment of Hellyer pyrite by roasting/acid leaching/cyanidation, CSIRO Division of Mineral Chemistry, Mineral Chemistry Communication, MCC 732, 23 pp.

17. McCallum, D.A., and Woodcock, J.T., 1988. Treatment of Hellyer zinc flotation tailing by roasting/acid leaching/cyanidation, CSIRO Division of Mineral Products, Mineral Products Communication, MPC/M-048, 17 pp.

18. McInnes, C.M., Sparrow, G.J., and Woodcock, J.T., 1989. Thiourea leaching of gold from an oxidised gold-copper ore, in *Gold Forum on Technology and Practices - 'World Gold '89'* (Ed. R.B. Bhappu and R.J. Harden), pp. 305-314 (Society for Mining, Metallurgy, and Exploration, Inc.: Littleton).

19. Richmond, G.D., and Lai, K.F., 1988. Metallurgical development of the Hellyer ore, in *Third Mill Operators Conference, Cobar*, pp. 9-14 (The Australasian Institute of Mining and Metallurgy: Melbourne).

20. Rolia, E., and Dutrizac, J.E., 1984. The determination of free acid in zinc processing solutions, *Canadian Metallurgical Quarterly*, 23: 159-167.

21. Thompson, P., 1986. Acid pressure oxidation of sulphide flotation concentrates, TMS Technical Paper No. A86-8, 11 pp. (The Metallurgical Society of AIME: Warrendale).

22. Turney, J.R., Smith, R.J., and Janhunen, W.J., Jr, 1989. The application of acid pressure oxidation to the McLaughlin refractory ore, in *Precious Metals '89* (Proceedings of an International Symposium, Las Vegas) (Ed. M.C. Jha, S.D. Hill, and M. El Guindy), pp. 25-45 (The Minerals, Metals & Materials Society: Warrendale).

23. Woodcock, J.T., 1953. Treatment of Victorian auriferous sulphide concentrates, *Chemical Engineering and Mining Review*, 45: 431-434.

Variables in the shear flocculation of galena

T.V. Subrahmanyam
Division of Mineral Processing, Luleå University of Technology, Luleå, Sweden
Z. Sun
Division of Inorganic Chemistry, Luleå University of Technology, Luleå, Sweden
K.S.E. Forssberg
Division of Mineral Processing, Luleå University of Technology, Luleå, Sweden
W. Forsling
Division of Inorganic Chemistry, Luleå University of Technology, Luleå, Sweden

SYNOPSIS

Like froth flotation the shear flocculation is governed by the physical, chemical and geometrical variables. Among the important factors which influence the particle aggregation are the particle size, hydrophobicity, charge and the intensity of agitation. The present paper investigates the effect of different variables on the shear flocculation of (1). galena and (2). synthetic PbS. Since flocculation is influenced both by the charge and the hydrophobicity of the particle the zeta potentials were measured for different experimental conditions. The reaction mechanisms of PbS surfaces in aqueous solutions are examined based on the results of electrokinetic- and potentiometric titration- studies. In comparison to conventional flotation, higher recoveries were obtained with shear flocculated aggregates. The results are discussed.

INTRODUCTION

Sulphide ores form the most important group and unquestionably, froth flotation is the method adopted for obtaining high grade concentrates. As per the US Bureau of Mines statistics approximately 225 million metric tons of sulphide ores were treated by flotation in 1980.

In mineral processing operations the fine particles (<10 μm) generated during comminution cause problems in flotation circuits and about 10-20% of fine mineral values are lost in tailings. Shear flocculation of fines with coarse particles is one of the methods to minimize the losses in tailings. The aggregates formed by shear flocculation can directly be subjected to flotation.

Aggregation of fines with coarse particles of either the same mineral species as that of fines (autogenous carrier flotation) or a different coarse mineral (carrier flotation) is a known method adopted to improve the flotation recovery of fines. Depending on the mechanism involved in effecting aggregation, earlier workers named the methods as ultra- or piggyback flotation[1,2]; floc flotation[3,4]; autogenous carrier- or ramification carrier flotation[5,6] and shear flocculation[7,8]. In shear flocculation, higher stirring speeds are applied to overcome the energy barrier between the similarly charged hydrophobic particles. The requirements that are common to shear flocculation and carrier- or autogenous carrier flotation are: (1). the presence of coarse particles, (2). higher agitations and (3). the hydrophobicity of both the carrier (coarse) and the carried (fine) particles. A recent review on shear flocculation and carrier flotation deals with related aspects in detail[9].

Among the important factors that govern the aggregation by flocculation mechanism are: particle size and its hydrophobicity, charge and the speed of stirring. Particle hydrophobicity is a critical factor since the aggregates are considered to form due to the overlapping of hydrocarbon chains, once the particles collide with each other. For non sulphide minerals viz., scheelite, quartz, the aggregates were observed[7,10] only in the presence of a collector, without which no aggregates formed even under turbulent stirring conditions. The behaviour of sulphide minerals which exhibit hydrophobicity in the absence of collectors with respect to flocculation mechanism has not been studied.

The hydrophobic species responsible for the floatability of some sulphides in the absence of collectors is still a debatable issue[11-15]. However, for the hydrophobic nature of minerals like galena, it was proposed that both the concentration of surface oxides as well as the degree of sulphur enrichment to be the likely entities[16].

The present work investigates the effect of several variables viz., time and intensity of agitation, size and proportion of coarse particles, pH, concentration of the collector and the cell geometry i.e. tests in cells with and without baffles, on the shear flocculation of (1). coarse and fine galena and (2). coarse galena and synthetic PbS. Synthetic PbS particles were used in some tests in the place of -5 μm galena to evaluate the effect of particle size and also to assess the influence of surface oxidation on shear flocculation. The galena and synthetic PbS particles can be considered to be heterogeneous in several aspects— particle size of galena fines, degree of oxidation of galena, surface impurities due to grinding etc., (see experimental - material preparation). The electrokinetics of galena and synthetic PbS was studied from the view point of the aggregation mechanism.

EXPERIMENTAL

Galena

Several coarse size fractions -20+5 μm, -38+20 μm, -53+38 μm and +53 μm, were prepared from a pure galena sample obtained from Ward's Natural Sciences Establishment Inc., USA, by crushing, followed by grinding in a steel rod mill. The -5 μm fraction was prepared by microsieving (the median size measured by Cilas Granulomètre-715 was

2.4 μm). The samples were dried and packed in plastic bags and no special precautions were taken to prevent oxidation.

Synthetic PbS

Lead sulphide was precipitated by titrating a nitrate solution with stoichiometric amounts of freshly calibrated sodium sulphide solution. The suspension was rinsed several times in 0.1 M NaNO$_3$ medium until a constant value was obtained with a sulphide ion selective electrode, which indicates that the solution has the same concentration of sulphide ions corresponding to the solubility of PbS in solution. Pure nitrogen gas was used to keep an inert atmosphere during the titrations. Lead sulphide thus prepared was stored in 100 ml volumetric flasks and a desired volume was taken before the experiment. The precipitate was black in colour and in the form of flocs with a high settling velocity. But when subjected to ultrasonic dispersion the solution turned turbid with ultrafine particles ($\leqslant$ 5 μm) in suspension, thus indicating the breakage and dispersion of flocs.

Reagents

All solutions were prepared in deionized water before the experiment. Xanthate series collectors were obtained from Hoechst Co., West Germany. Unless or otherwise stated the collector used was sodium ethyl xanthate and methylisobutyl carbinol (Merck, 97% pure) as frother for flotation tests. The pH was adjusted with NaOH or dil HNO$_3$.

METHODS

Aggregation tests

To a 100 ml xanthate solution, desired quantities of -5 μm galena or synthetic PbS and coarse galena particles were added and subjected to ultrasonic dispersion for 15 min. To this suspension 150 ml xanthate solution were added (total volume 250 ml) and the pH re-adjusted. The suspension was stirred in a plexiglas container (80-120 mm high; 63 mm in dia.) fitted with six baffle plates (each 50 mm x 7 mm x 1.5 mm) with a single bladed paddle stirrer (14 mm x 25 mm x 3 mm). Stirrer speeds upto 1500 rpm were obtained without air entrainment or vortex formation. At a high speed of 2000 rpm air entrainment was considerably high.

The aggregates of fine-coarse or fine-fine particles formed as a result of stirring decrease the turbidity of the solution— i.e. the concentration of -5 μm particles in suspension, which was measured by a Hach Ratio XR Turbidimeter.

Electrokinetic measurements

The zeta-potentials were measured with a Laser Zee Meter Model 501. A known amount of -5μm galena or synthetic PbS was added to deionized water or xanthate solution of desired concentration and dispersed in ultrasonic for 15 min. after adjusting the pH. The zeta potential measurement was completed as quickly as possible after injecting the sample into the cell in order to avoid the side effects of temperature.

Flotation tests

Flotation tests were carried out in a Labor Flotation Machine (West Germany). The maximum number of rotations reached in this cell was 300 rpm. After the shear flocculation test the pulp was transferred to the flotation cell. A conditioning time of 2 min. was allowed after the addition of frother and air was let in to continue flotation for 2 min. The flotation recoveries thus obtained were compared with the conventional flotation recoveries. In the conventional test the pulp was not subjected to shear flocculation. A conditioning time of 15 min. after the addition of xanthate and a flotation time of 2 min. after the addition of frother, were allowed. This procedure was adopted to evaluate the difference in recoveries between the two methods i.e. conventional flotation and shear flocculation followed by flotation. Any improvement in flotation recovery with the pulp subjected to shear flocculation could only be due to particle aggregation.

Scanning electron micrographs

Selected samples were evaporated on to carbon stubs from acetone solution containing the particles while under magnetic stirring. Scanning electron micrographs were taken on the aggregates after shear flocculation tests and also in some cases on the flotation concentrates.

RESULTS AND DISCUSSION

In the present investigation the evidence in support of the formation of aggregates comes from the scanning electron micrographs, besides the turbidity results. It must be pointed out that in the case of minerals like galena (density 7) settling of even individual fine particles during turbidity measurement may be mistaken for flocs. Therefore, a constant time interval was maintained in all the tests i.e. collection of the solution from the cell after shear flocculation test till the turbidity measurement.

The variables studied are grouped into physical, chemical and geometrical, for discussion.

VARIABLES
Physical

Three types of aggregates are formed when a mixture of coarse and fine particles is subjected to stirring in aqueous collector solutions viz. coarse-coarse, fine-coarse and fine-fine, the last two decrease the turbidity of the solution. Fig. 1 shows the apparent concentration of -5 μm particles in suspension for different agitation speeds. With -5 μm galena particles alone the turbidity decrease is gradual with time for a stirring speed of 1500 rpm. Whereas in the presence of coarse galena (-38+20 μm) the rate of decrease of fines is higher during the first 15 min. and later, is gradual. The general trend of the curves suggests increased aggregation of fine and coarse particles with stirring speed and figs. 2 and 3 show the scanning electron micro-

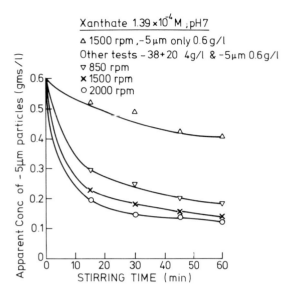

Fig. 1 Concentration of -5 μm particles in suspension with time of stirring (both coarse and fine particles: Galena).

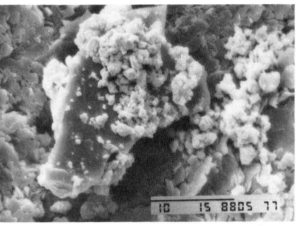

xanthate conc.: 1.39 x 10^{-4} M

Fig. 2 Aggregates formed after stirring -38+20 μm and -5 μm galena particles at 2000 rpm; time of stirring: 60 min.; pH: 7.

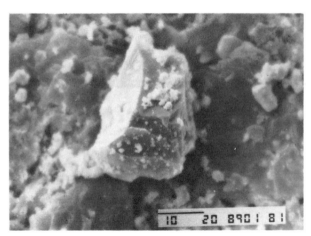

Fig. 3 Effect of stirring speed (850 rpm) on aggregate formation (other conditions same as Fig. 2).

graphs of the aggregates for two different stirring speeds. The slime coatings effect as seen in fig. 3 was also observed[7] with ultrafine (1μm) and coarse (20 μm) scheelite at 850 rpm. However, higher stirring (1000 rpm) was found to be necessary to shear flocculate the ultrafines alone. The number of collisions between the coarse and the fine particles decrease if the difference in particle sizes is large i.e. fines may begin to flow past the coarse instead of colliding with each other. As the size of the particle decreases the number of particle collisions also decrease and hence higher energy input. The adhesion rate of ultrafines to coarse was estimated[17] to be 10^3-10^4 times than the rate of cohesion between fines alone.

In order to evaluate the particle size effect of fines on shear flocculation the synthetic PbS sample was used in place of -5 μm galena along with coarse galena (-38+20 μm). The results are shown in fig. 4. For comparison the results of flocculation between fine and coarse galena are also included in the same figure. The turbidity values were found to increase and decrease for synthetic PbS even for prolonged periods of stirring and such fluctuation is due both to the formation and breakage of aggregates. Though the weights of -5 μm galena and synthetic PbS used in

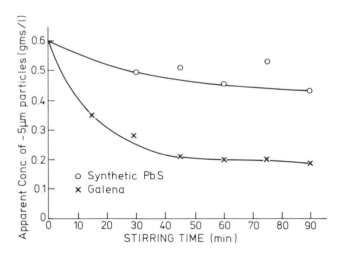

Fig. 4 Effect of stirring time on the concentration of fines in suspension in presence of coarse galena (-38+20 μm).

the tests were the same (0.6 g/l), the number of particles in the case of synthetic PbS would be more due to much smaller size (5 μm)--i.e. a variation in the proportion of fines to coarse. Either an increase in the proportion of fines to coarse or vice versa, decrease the adhesion of fines to coarse. In the case of the former the coarse particle surface available to fines is reduced; whereas an increase in the proportion of coarse to fines may result in higher collision rates between coarse particles alone thus resulting in detachment of already adhered fines.

Chemical

Influence of particle charge and hydrophobicity

The double layer repulsion of two particles depends on the stern potential which is a function of the electrolyte concentration and zeta potential is a good measure of the repulsive interactions. Since both hydrophobicity and charge are critical factors, the zeta potentials were measured to evaluate the flocculation behaviour.

Fig. 5 shows the effect of collector concentration on the fines in suspension as a function of stirring time. While there is a general decreasing trend in turbidities of solutions for different collector concentrations, two situations draw attention: (a). aggregation in the absence of xanthate and (b). absence of flocculation at a xanthate concentration of 6.94×10^{-4} M (100 mg/l).

Aggregation in the absence of xanthate

A decrease in solution turbidity in the absence of xanthate leads to the conclusion that the aggregation could be due to the hydrophobic species formed as a result of the surface oxidation of the mineral. Fig. 6 shows the aggregates formed in the absence of xanthate.

The surface potential of galena at pH 7 in the absence of xanthate corresponds to -20 mV (fig.7). In aqueous solutions the variation of charge i.e. less or more negative, among other factors is influenced by the pH,

$$\begin{matrix} \equiv SH \\ \equiv SH \end{matrix} + Pb^{2+} \rightleftharpoons \begin{matrix} \equiv S \\ \equiv S \end{matrix} Pb + 2H^+ \quad .. \; pH < 7.2 \; ..(1)$$

$$\begin{matrix} \equiv S \\ \equiv S \end{matrix} Pb + H_2O \rightleftharpoons \begin{matrix} \equiv S \\ \equiv S \end{matrix} PbOH + H^+ \quad .. \; pH > 7.2 \; ..(2)$$

For pH 7 the ion exchange mechanism between Pb^{2+} and the proton (H^+) adsorption is proposed. The surface becomes less negative as equilibrium is attained between the number of protons adsorbed

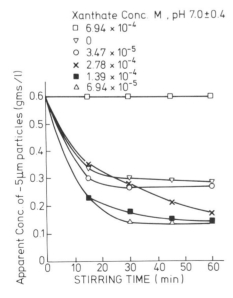

Xanthate Conc. M , pH 7.0±0.4
□ 6.94 × 10⁻⁴
▽ 0
○ 3.47 × 10⁻⁵
✕ 2.78 × 10⁻⁴
■ 1.39 × 10⁻⁴
△ 6.94 × 10⁻⁵

Fig. 5 Effect of collector concentration on the shear flocculation of galena. Coarse galena: -38+20 μm 4 g/l and -5 μm galena 0.6 g/l.

Fig. 6 Aggregates formed in deionized water after stirring -38+20 μm and -5 μm galena mixture at 1500 rpm; pH 7; time of stirring: 60 min.

227

and the release of Pb^{2+} into the bulk solution. Whereas in the alkaline region the surface lead sites are hydrolysed, thereby leading to a more negative potential.

The adsorption of protons at low pH's 2-4, however, is not quite stoichiometric in relation to the bulk solution and this explains the behaviour observed in figs. 8 and 9, which are presented for a comparison. In the region of pH 2-4, as the pH increases the surface becomes less negative and a further increase in pH (beyond 4) leads to more negative charges.

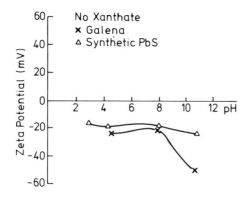

Fig. 7 Zeta potentials in the absence of xanthate.

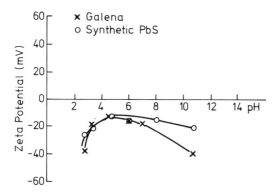

Fig. 8 Zeta potentials as a function of pH in the absence of xanthate (ionic strength 0.01 M NaClO₄).

Absence of flocculation at a high xanthate concentration

Figs. 10 and 11 show the electrokinetic behaviour of galena and synthetic PbS. The zeta potential of galena corresponding to a xanthate concentration of 6.94×10^{-4} M (100 mg/l) is -65 mV. At this potential the repulsive forces between the

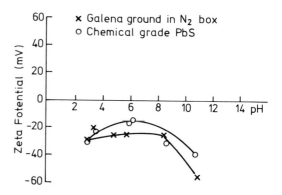

Fig. 9 Zeta potentials as a function of pH in the absence of xanthate (ionic strength not controlled).

particles seem to be higher than can be overcome by the forces of collision for the stirring speed applied (1500 rpm). The turbidity values remain unchanged with time (fig. 5) and the absence of aggregates is also observed from the scanning

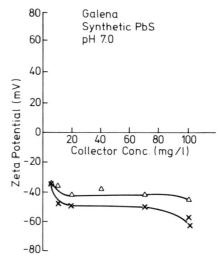

Fig. 10 Zeta potentials as a function of xanthate concentration.

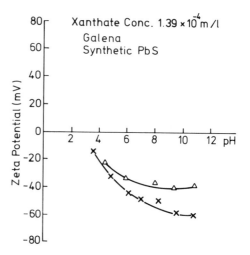

Fig. 11 Zeta potentials as a function of pH.

228

electron micrograph (fig. 12). From fig. 5 it is seen that the maximum flocculation occurs for collector concentrations of $6.94 \times 10^{-5} - 1.39 \times 10^{-4}$ M (10-20 mg/l) and the zeta potential vlues vary between -46 to -50 mV. The lowest xanthate concentration used in shear flocculation test was 3.47×10^{-5} M (5 mg/l) and the zeta potential as oberved from fig. 10 is -32 mV. If surface charge alone is the factor then maximum flocculation should be expected at the lowest xanthate concentration. But it is seen from fig. 5 that the turbidity decrease at 3.47×10^{-5} is less than that for concentrations of $6.94 \times 10^{-5} - 1.39 \times 10^{-4}$ M and is slightly more than that in the absence of xanthate. But, at a high xanthate concentration even though the particle is hydrophobic enough, repulsive forces due to high negative potentials disfavour the flocculation. From this it becomes evident that both surface charge and surface coverage (hydrophobicity) are important for shear-flocculation.

xanthate conc.: 6.94×10^{-4} M

Fig. 12 Effect of high collector concentration on aggregate formation. -38+20 μm and -5 μm galena mixture at 1500 rpm; pH 7; time of stirring 60 min.

The surface reactions in xanthate solutions are,

$$\equiv SH_2 + Pb^{2+} + 2X^- \longrightarrow \equiv SH_2 + PbX_2 \text{ (s)} \quad pH<7..(3)$$
and
$$\equiv SPb + X^- \longrightarrow \equiv SPbX^- \quad pH>7..(4)$$

i.e. lead xanthate is precipitated in acidic pH

while in the alkaline region xanthate will be adsorbed on the surface sites by chemisorption, thus leading to a more negative potential.

Shear flocculation of coarse galena and synthetic PbS

In another series of tests, both coarse galena and synthetic PbS particles were treated with sodium sulphide before subjecting the pulp to shear flocculation in aqueous xanthate solution. The flocs observed were found to be stable and is attributed to uniform xanthate adsorption upon sulphidization. These results are discussed elsewhere.[18] The electrokinetic behaviour of the synthetic PbS is shown in fig. 13 for two different conditions. With excess of S^{2-} the surfaces attain more negative potentials,

$$\equiv SH_2 + Pb^{2+} + S^{2-} \rightleftharpoons \equiv SH_2 + PbS_{(s)} \quad pH<7.. (5)$$
and
$$\equiv SPb(H_2O) + HS^- \rightleftharpoons \equiv SPbSH^- + H_2O \quad pH>7.. (6)$$

Whereas in the presence of excess of lead the

$$\begin{matrix} \equiv SH \\ \quad + 2Pb^{2+} \\ \equiv SH \end{matrix} \rightleftharpoons \begin{matrix} \equiv SPb^+ \\ \quad + 2H^+ \\ \equiv SPb^+ \end{matrix} \quad ..(7)$$

Pb^{2+} ions start to adsorb on the mineral surface leading to a less negative charge and a further increase in pH i.e. in alkaline region the particles become more negative due to hydrolysis. Equations 1-7 are proposed based on the evidence obtained from the electrokinetic studies and the results of potentiometric titrations.[19]

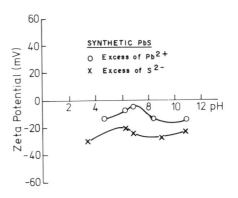

Fig. 13 Zeta potentials as a function of pH.

Geometrical

For a successful flotation operation the bubble and particle collisions are important. Similarly, the shear flocculation mechanism is governed by the collisions between the fine and coarse particles. The conditions maintained to flocculate the particles by shear force are not much different from those of a flotation operation. However, certain differences like the agitation speed, pulp density, cell geometry, do exist. At this stage an important question arises i.e. whether particle aggregation occurs in a conventional flotation operation ? The experiments carried out in the cell with baffles (described in the experimental section) and in a cell without baffles (adapted to Agitair-LA 500 flotation machine) reveal that particle collisions are effective in baffled cell, though sparse aggregates were also observed in the cell without baffles.

FLOTATION

Particles < 10 μm in size have poor floatabilities due to low collision efficiencies with gas bubbles in flotation circuits. The objective in aggregating the fines with coarse particles is to improve the true floatabilities of fines. Among the variables which influence the adhesion mechanism are the size of the coarse and the proportion of coarse to fine. Table I shows the effect of particle size on the adhesion and flotation recovery of fines.

Table I The effect of coarse particle size on the recovery of -5 μm particles.

Particle size of galena (μm)	flotation recovery of -5 μm galena, %
-5	58.00
-20 + 5	62.66
-38 + 20	76.86
-53 + 38	67.33

The flotation recovery of -5 μm galena increases with particle size up to -38 + 20 μm. A further increase in coarse size to -53 + 38 μm reduces the recovery of -5 μm and this is attributed to effects like attrition i.e. detachment of fines when slime covered coarse particles collide with each other either by a direct hit or abrasion. A similar observation was made by earlier workers.[5-8]

Fig. 14 shows the effect of pH on the overall flotation recoveries of galena (1). by conventional flotation and (2). subjected to stirring at 1500 rpm fpr 60 min. followed by flotation. In both cases, since the number of revolutions used in the flotation cell was 300 rpm, any improvement in the flotation recovery with the pulp subjected to shear flocculation is clearly due to the aggregates.

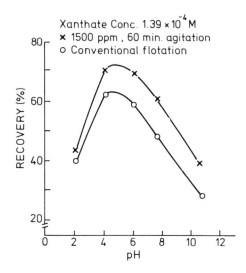

Fig. 14 Flotation recoveries of galena

SUMMARY AND CONCLUSIONS

The electrokinetics of galena and synthetic PbS are examined to explain the mechanism since both particle hydrophobicity and charge influence the shear flocculation. At high collector concentrations even though the particles are hydrophobic enough, repulsive interactions between the particles predominate due to high negative potentials. Similarly at high pH's the particles are highly negatively charged. A decrease in turbidity in the absence of xanthate is attributed to flocculation caused by hydrophobic surface oxidation products. The recoveries of fines are improved by shear flocculation/flotation in comparison to conventional flotation.

References

1. Greene E. W. and Duke J. B. Selective froth flotation of ultrafine minerals or slimes. Mining Engineering, 14, 1962, p. 51-55.

2. Deryagin B. V. Samygin V. D. and Livshits A. K. Study of the flocculation of mineral particles under turbulent conditions 1. Flocculation Mechanism. Colloid Journal, 26, 1964, p. 149-154.

3. Gaudin A. M. Schuhmann R. Jr. and Schlechten A. W. Flotation kinetics II. The effect of size on the behaviour of galena particles. Journal of Physical Chemistry, 46, 1942, p. 902-910.

4. Clement M. Harms H. and Trondle H. M. In: Proceedings IX International Mineral Processing Congress, 1, 1970, p. 179-187.

5. Hu W. Wang D. and Qu G. Principle and application of carrier flotation. Journal of Central South Institution of Mining and Metallurgy, 4, 1987, p. 408-414.

6. Autogenous Carrier Flotation. W. Hu and others. In: Proceedings of XVI International Mineral Processing Congress, Part A, 1988, p. 445-452.

7. Warren L. J. Slime coating and shear flocculation in the scheelite-sodium oleate system. Transactions of the Institution of Mining and Metallurgy (Section C: Mineral Processing & Extractive Metallurgy) 84, 1975, C99-C104.

8. Warren L. J. Shear flocculation of ultrafine scheelite in sodium oleate solutions. Journal of Colloid Interface Science, 50, 2, 1975, p. 307 - 318.

9. Subrahmanyam T. V. and Forssberg, K. E. S. Fine Particles Processing: Shear flocculation and carrier flotation—A review (submitted to International Journal of Mineral Processing, Elsevier, The Netherlands, and accepted).

10. Bhaskar Raju G. Subrahmanyam T. V. Sun Z. and Forsling W. Shear flocculation of quartz (submitted to International Journal of Mineral Processing, Elsevier, The Netherlands).

11. Finkelstein N. P. Allison S. A. Lovell V.M. and Stewart B. V. Natural and induced hydrophobicity in sulphide mineral systems. In: American Institute of Chemical Engineers Symposium Series, 150, 71, 1975, p. 165-175.

12. Heyes G. W. and Trahar W. J. The natural floatability of chalcopyrite. International Journal of Mineral Processing, 4, 1977, p. 317-344.

13. Heyes G. W. and Trahar W. J. Oxidation- reduction effects in the flotation of chalcocite and cuprite. International Journal of Mineral Processing, 6, 1979, p. 229-252.

14. Gardner J. R. and Woods R. An electrochemical investigation of the natural floatability of chalcopyrite. International Journal of Mineral Processing, 6, 1979, p. 1-16.

15. Gardner J. R. and Woods R. A study of the surface oxidation of galena using cyclic voltammetry. Journal of Electroanalytical Chemistry Interfacial Electrochemistry, 100, 1979, p. 447-459.

16. Sulfur enrichment at sulfide mineral surfaces. A. N. Buckley and Walker G. W. In: Proceedings of the XVI International Mineral Processing Congress, Part A, 1988, p. 589-599.

17. Samygin V. D. Barskii A. A. and Angelova S.M. Mechanism of mutual flocculation of particles differing in size. Colloid Journal, 30, 1968, p. 435-439.

18. Subrahmanyam T. V. Sun Z. Forssberg K. S. E. and Forsling W. Shear flocculation and flotation of galena and synthetic PbS. In: International Symposium on Advances in Fine Particles Processing: Fine Particles Society, Boston, Massachussetts, 22-26 August, 1989.

19. Forsling W. Subrahmanyam T. V. and Molin E. Surface reactions of PbS in aqueous solutions: Electrokinetic and Potentiometric Studies (submitted to the International Symposium on Surface and Colloid Engineering, San Diego, California, 1990.

Role of chloride hydrometallurgy in processing of complex (massive) sulphide ores

D.N. Collins
D.S. Flett
Minerals and Metals Division, Warren Spring Laboratory, Stevenage, Hertfordshire, England

SYNOPSIS

The paper critically examines the current problems in the physical beneficiation of complex sulphide ores, the state of the technology and its likely future. New concepts on the way this field may progress are discussed and illustrated by means of recent WSL test data. The alternative scenario of bulk flotation/hydrometallurgical extraction is then assessed with specific reference to chloride extraction processes. A brief account of the essential technical differences between the processes is given together with their likely impact on process costs and metal recoveries. The likely deportment of toxic metals is examined as is the recovery of precious metals.

The economic benefits-deficiencies of the bulk float/chloride hydrometallurgical route are finally considered against the conventional differential flotation/smelting process. The implications of metal prices, smelter terms, leach recoveries etc on the relative economics are broadly discussed.

INTRODUCTION

Despite considerable research and development effort in the field of hydrometallurgical treatment of sulphide ores and the emergence of roast leach/electrowinning and pressure leaching for the recovery of Zn from primary sources, the commercial use of hydrometallurgy in the broader field of base metal recovery from primary sulphide ores has not developed to any degree. The reasons for this are numerous and were the subject of a recent study undertaken by Warren Spring Laboratory for the Minerals Industry Research Organisation (MIRO)[1] on Chloride Hydrometallurgy. This paper attempts to draw some of the conclusions of this study together and look in more depth at some aspects of the problem with the aid of practical examples.

In evaluating the potential of hydrometallurgy in the treatment of complex sulphide ores one has to consider current practice and this is largely based on differential flotation and the production of concentrates for direct smelting and refining. Despite the rather poor metallurgy generally achieved in the flotation of complex ores, technology is also moving in this area and it is essential to consider the implications of changes here before looking at the broader concepts of bulk flotation linked with hydrometallurgy.

As a final introductory remark complex sulphides have been defined[2] as

(a) having a very high pyrite content

(b) having very fine mineral intergrowth (often down to 0.03-0.02 mm)

(c) usually containing minor elements (both precious and penalty)

(d) showing alteration and oxidation affects in some cases

The first two points are normally considered essential in the categorisation of complex sulphide ores.

FLOTATION UPGRADING

Differential flotation

Since flotation is, and is likely to remain, the principal upgrading method for base-metal sulphide ores in general it is important to understand how the flotation of complex ores differs from standard base metal processing. Much data exists in the literature on the differential flotation of complex sulphide ores and three symposia[3-5] have recently been held on this subject and a further one is planned[6]. From information taken from these sources coupled with our own experiences the following differences emerge.

(i) Comminution

For adequate liberation of the individual minerals very fine grinding is necessary for complex sulphide ores (frequently down to 80% passing 0.4 to 0.2 mm). At these product sizes energy usage is frequently doubled with respect to conventional fine grinding and the efficiency limits of a standard tumbling mill are approached. One advantage is that the high density of the associated gangue (pyrite) will permit higher pulp densities to be achieved, thus leading to energy savings.

(ii) Flowsheets

From data abstracted from the WSL Flotation Reagent Database (FREDA)[7] the most common circuit used in conventional Cu-Pb-Zn differential flotation involves the bulk flotation of Cu + Pb followed by the differential flotation of zinc on the Cu/Pb tail. The Cu + Pb concentrate is then separated by depression of the Pb (favoured) or the Cu. For massive sulphide ores however the direct differential flotation of the values is equally popular despite being the most cost intensive option as far as circuitry is concerned. It does however permit the highest technical efficiency since one is selecting flotation conditions to suit each commodity rather than compromising selectivity by the flotation of two or more commodities. For complex ores this extra selectivity can be essential if acceptable concentrate grades are to be produced. For this reason bulk Cu + Pb + Zn flotation followed by subsequent differential flotation is not used at all for massive sulphide ores and even in the application of bulk Cu + Pb flotation the complication of retreating the subsequent Cu (or Pb) concentrate for residual pyrite or zinc removal is a common feature.

Reagent selection

Although analysis of FREDA has indicated a greater use of the more selective short chain xanthates, alkyl dithiophosphates and thionocarbamates (frequently in combination) for copper flotation together with short chain xanthates and aryl dithiophosphates for lead flotation, there is little commercial evidence to identify any one reagent as being specific for a given mineral system and one suspects that the combined use of reagents is done more with cost considerations in mind than for the desired improvement in metallurgy. Analysis of the massive sulphide ores in particular[8-24] suggests that the major innovations have been on the depressant systems used. The trend in copper flotation has been to a lower pH operation (around neutral) utilising SO_2 and the use of sulphite or bisulphite ion to depress pyrite, galena and sphalerite (sometimes in conjunction with $ZnSO_4$) and occasionally using aeration. NaCN and starches are more rarely used. Certainly a greater emphasis has been on Eh-Ph control and generally selectivity has improved substantially over the more traditional lime /NaCN system. Similar depressants have been used in bulk Cu-Pb flotation although the pH has generally been higher (in the 8-10 range). NaCN is an alternative reagent for pyrite depression for this system - often used in combination with $ZnSO_4$. Zn flotation has normally been based upon very conventional lines with the use of most of the common collectors at an operating pH of 9-12 (lime usually but occasionally NH_3 being used as pH modifier). Residual depressants from stage 1 may have influenced metallurgy to some degree as they will have done in lead flotation which was operated under similar conditions to the bulk Cu-Pb flotation. Most effort has been centred on the Cu-Pb separation which normally has involved depression of the lead with $Na_2Cr_2O_7$ (pH 8-9.5) or with SO_2/starch (dextrin) in the pH 4-5.5 range. Activated C has been used for collector desorption while lead depression has sometimes been accelerated using elevated temperatures. Less frequently NaCN has been used for Cu depression, again using activated carbon for collector desorption.

The bulk Cu-Pb float followed by Cu-Pb separation frequently does not result in the production of clean concentrates for complex sulphide ores and often the lead concentrates have to be subsequently upgraded - by reverse flotation of either sphalerite or pyrite. High temperature treatment is a feature of reverse flotation procedures.

Future developments

Despite the more sophisticated depressant systems used in complex sulphide flotation, there remains an inherent problem with the selective flotation of the lead. Basically the conditions under which Pb floats selectively (high NaCN additions) cannot be used for Cu/Pb bulk flotation due to the depression of Cu and, for both Cu/Pb bulk flotation or for the Pb stage of a Cu-Pb-Zn direct differential float, the use of SO_2 to depress pyrite is less effective at the pH necessary for good Pb recovery (8-8.5). Lime depression of pyrite also affects Pb recovery although Chinese workers[25] claim to have floated galena selectively at pH 11-11.5 using the thionocarbamate collector from a reground bulk concentrate.

The selection of the best collectors cannot be emphasised too much and thionocarbamates have generally proved their selectivity for copper flotation (and therefore by inference for activated sphalerite) while dithiophosphinates we have found selective for galena although not extensively used commercially. The trend should be towards special purpose built reagents [26] [27]

for specific mineral systems. Of recent interest has been the use of sodium sulphide for Eh control[28] [29]. This reagent also has the advantage of precipitating most pulp cations in a very insoluble form, and also of sulphidising any oxidised surfaces, but its usage has been more accepted in the USSR than in the West. Recent work at Warren Spring Laboratory has shown its effectiveness in counteracting activated sphalerite and in depressing pyrite. In Table 1 some tests were undertaken on a Cu/Zn/Sn ore on which the plant was having significant problems in producing a Cu concentrate due to the presence of highly activated sphalerite. Even with SO_2 depressant and a selective copper collector the zinc was floating as strongly as the copper. Grinding in the presence of Na_2S reduced the zinc flotation substantially and further improvement in performance was obtained when the collector was omitted (ie collectorless flotation) when Cu grade improved to 14.1% Cu and Zn distribution to the concentrate reduced to 10.3%. Under these conditions copper recovery to the cleaner concentrate was still satisfactory at 72.7% from this low grade feed. The technique has now been used on a number of ores and shows substantial advantages where there is any tendency for the zinc to float. The application to Cu-Pb-Zn ore flotation is shown in Table 2, the ore sample again has very fine mineral dissemination and the zinc again showing a high degree of natural floatability. In addition to improving overall grades and recoveries of the principal metals, depression of pyrite, in the first two concentrates particularly, has been very effective by this method. Other points of interest were that the change of collector at the Pb stage was essential (since the thionocarbamate will not float the galena after SO_2 depression unless it is activated with $CuSO_4$) and the low pH's used at both the Pb and Zn stages were achieved by virtue of using the Na_2S/SO_2 combination and the more selective collectors.

The fact that Zn can be floated selectively from pyrite at near neutral pH was demonstrated in work carried out on the bulk flotation of Pb/Zn from a complex Peruvian ore (Table 3).

235

Table 1   Effect of Na$_2$S on naturally activated sphalerite in Cu/Zn ore (UK)

		Test 1					Test 2			

Conditions: Grind 80% minus 0.044 mm   80% minus 0.044 mm

Condition 500 g/t SO$_2$ 500 g/t Na$_2$S to mill

           25 g/t thionocarbamate   500 g/t SO$_2$

           30 g/t Frother AF65   25 g/t Iso propyl ethyl thionocarbamate

                            30 g/t Frother A65

Float   pH 5.0   pH 5.0

Clean 1   50 g/t Na$_2$SO$_3$ 50 g/t Na$_2$SO$_3$

Clean 2   Naturally   50 g/t Na$_2$SO$_3$

Metallurgical Results:

	wt %	Cu		Zn		wt %	Cu		Zn	
		Assay %	Distbn %	Assay %	Distbn %		Assay %	Distbn %	Assay%	Distbn %
Cl Conc	3.27	3.98	85.5	46.0	86.3	1.01	10.9	77.2	25.9	14.8
Cl Tail	2.64	0.48	8.3	6.5	9.9	2.85	0.47	9.4	11.4	18.3
Tailing	94.09	0.01	6.2	0.07	3.8	96.14	0.02	13.4	1.23	66.9
Head (calculated)	100.0	(0.15)	100.0	(1.74)	100.0	100.0	(0.14)	100.0	(1.77)	100.0

Table 2   Use of Na$_2$S on finely disseminated Cu/Pb/Zn ore (UK)

Conditions Grind   80% passing 0.038 mm   Pb Float Condition 125 g/t Na$_2$S
                   with 250 g/t Na$_2$S   500 g/t SO$_2$
                                                      15 g/t AF65

Cu Float Condition 1000 g/t SO$_2$ 10 g/t di iso butyl
                     30 g/t iso propyl   dithiophosphinate
                          ethyl   Float pH 5.7-6.2
                        thionocarbamate   Clean x 2 with 50 g/t SO$_2$
                   15 g/t AF65
                   500 g/t Na$_2$S   Zn Float Condition 250 g/t CuSO$_4$
                   Float pH 5.7-6.4   50 g/t iso propyl ethyl
                   Clean x 2 Naturally   thionocarbamate
                                                     50 g/t AF65
                                                     Float pH 6.4-7.0
                                                   Clean x 2 with 10 g/t AF65

Metallurgical Balance

Product	% Wt	Cu		Pb		Zn	
		Assay	% Dist	Assay	% Dist	Assay	% Dist
CuCl Conc	3.22	25.00	78.1	5.80	8.4	8.10	6.4
PbCl Conc	3.32	2.80	9.0	50.10	74.5	17.30	14.1
ZnCl Conc	5.81	1.10	6.2	3.00	7.8	53.00	75.1
ZnCl Tails	3.75	0.49	1.8	0.83	1.4	1.20	1.1
ZnRo Tails	83.90	0.06	4.9	0.21	7.9	0.13	2.7
Head (calculated)	100.0	(1.03)	100.0	(2.23)	100.0	(4.07)	100.0

The final grade of the concentrate (51.1% combined Pb + Zn) may be further improved with Na$_2$S depression of the pyrite. Recovery increases beyond the levels quoted in Table 3 necessitated the regrinding of a second bulk concentrate.

Commercially the use of bulk Pb/Zn flotation has not been used on Cu, Pb, Zn ores despite its usage on Pb/Zn ores. The reason for this is not clear particularly in the case of complex ores where the difficulty in floating the lead would be improved by copper activation. Pyrite depression could also be maximised. Bulk Pb/Zn flotation offers a further advantage in that, should there be difficulty in producing a clean Pb concentrate subsequently, the Zn contaminated Pb product could still go forward to ISF smelting.

The foregoing serves to highlight areas in which complex sulphide flotation could improve and although metallurgy achieved on complex sulphides is always likely to be inferior to that achieved on other base metal ores there is certainly scope for improvement on current practices.

Bulk sulphide flotation

The economic advantages of applying hydro-metallurgy to complex sulphide ores primarily lie in the potential extra payable metal recovery achievable. This in turn hinges on maximising the recovery at the bulk flotation stage.

The principal recovery improvement during bulk flotation will be the recovery of base metals which are misplaced to the wrong concentrates during differential flotation. Precious metals associated with zinc concentrates are frequently not payable and this will also reflect in additional recovery to a bulk concentrate. It has also been argued that by opting for bulk flotation it is possible to increase overall recovery of base and precious metals by accepting greater overall dilution with pyrite with the view to recovering more of the composite grains. While this is undoubtedly true there is a trade off in subsequent processing costs and in practice most recoverable Cu and Pb composites are usually recovered with subsequent concentrates (ie Pb and Zn) during differential flotation and would therefore be recovered to a selective bulk concentrate. Some improvement in precious metal recovery could however be achieved particularly where the metals are associated with arsenic and antimony minerals. Since zinc is the major valuable mineral phase present in complex sulphides non selective flotation of this mineral is not recommended due to the heavy

associated dilution with pyrite.

In terms of operating costs bulk flotation offers the simplest circuity and reagent requirements, with potential energy savings on comminution also. One could envisage a circuit as shown in Fig 1 offering the maximum scope for energy savings as well as metallurgical flexibility.

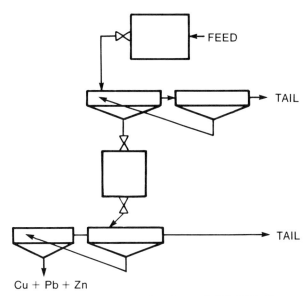

Cu + Pb + Zn

FIG. 1 TYPICAL CIRCUIT FOR BULK SULPHIDE FLOTATION

To maximise base metal recovery a neutral pH float could be used with the more selective collectors (eg thionocarbamate or dithio-phosphinate) using the SO_2 depressant system and $CuSO_4$ activation. Sodium sulphide additives could be considered on either of the grinding stages as an additional pyrite depressant. It is arguable that one should be selecting collectors on strength rather than selectivity for this type of circuit but selective collectors will have less affinity for pyrite and collector utilisation will be greater on these massive sulphide ores leading to significantly reduced collector consumption.

Recovery of precious metals may however be another element in the equation and could also dictate choice and level of depressants.

Table 3 Bulk Pb/Zn flotation of Peruvian ore

Conditions	Grind	68% passing 0.075 mm
	Conditions	200 g/ℓ CuSO$_4$
		125 g/t SO$_2$
		20 g/t iso propyl ethyl thionocarbamate
		30 g/t AF65
	Float	pH 7

Metallurgical Balance

Product	Weight %	Assay Pb + Zn %	Pb Assay %	Pb Dist %	Zn Assay %	Zn Dist %	Ag Assay oz/t	Ag Dist %
Prim Cl Conc 2	9.93	51.10	13.89	70.9	37.21	88.8	64.25	55.1
Prim Cl Tails	4.03		1.94	4.0	0.79	0.8	7.28	2.5
Prim Tails	86.04		0.57	25.1	0.50	10.4	5.71	42.4
Head (Calculated)	100.00		(1.95)	100.0	(4.16)	100.0	(11.58)	100.0

Metallurgical data

On the basis of the average of 13 massive
sulphide operations taken from literature
sources[8-24] a "typical" metallurgical balance
for the differential flotation of a complex
sulphide ore is shown in Table 4.

In calculating the bulk flotation data (also
Table 4) it has been assumed that a 20% increase
in weight of total concentrate has floated
relative to differential flotation, that base
metal recoveries are the sum of those reporting
to individual concentrates while additionally
for Ag and Au an extra 5% (absolute) recovery
has been achieved. Although the assumptions are
somewhat subjective they are not unreasonable on
the basis of observations made earlier in the
text.

HYDROMETALLURGY

Hydrometallurgical processes for treating
sulphide concentrates, in general, have received
considerable attention over the last three
decades. Many flowsheets have been developed,
many pilot-plant studies have been undertaken
and several processes are now successfully
operated. For example, direct ammoniacal
pressure leaching has been used by Sherritt
Gordon Mines since the 1950's to extract nickel
from pentlandite concentrates[30], the
roast-leach-electrowin route is widely used to
recover zinc from sphalerite concentrates and
recently direct sulphuric acid pressure leaching
of sphalerite concentrates has been successfully
commercialised[31][32]; also Duval have operated a
direct cupric/ferric chloride leaching process
with electrowinning to recover copper from
chalcopyrite concentrates[33] this latter process

Table 4 Average metallurgical data for complex sulphide ore

	Tonnes/Day	Assay %			g/t		% Distribution				
		Cu	Pb	Zn	Ag	Au	Cu	Pb	Zn	Ag	Au
Head	3000.0	0.80	2.53	7.68	112	0.25					
Cu Conc	75.0	21.8	9.1	9.2	1657	3.2	68	9	3	37	32
Pb Conc	94.2	2.6	49.2	9.8	927	0.24	10	61	4	26	3
Zn Conc	359.1	0.7	2.9	52.6	122	0.21	10	14	82	13	10
Bulk Conc	600.0	3.52	10.6	34.2	426	0.56	88	84	89	81	50

(now closed) being the only chloride process which has been commercialised for a sulphide flotation concentrate. Despite these developments, however, it is now considered, particularly for copper, that direct hydrometallurgical processes are in the main unlikely to be able to compete successfully with smelting for large-scale operations for simple high grade single metal concentrates, and whilst such concentrates can be produced commercially, smelting is likely to remain the principal processing route[34]. Whether the CUPREX Metal Extraction Process now currently undergoing extensive and highly successful pilot scale testing in Spain[35] will prove to be the exception is not yet clear. It will be noted however that this is a chloride based process. However, complex sulphide ores offer greater opportunities for reasons given earlier and much interest continues to be shown world-wide in the hydrometallurgical processing of these materials. Many of these studies have concentrated on chloride hydrometallurgical routes, and it does seem at present that this may be the preferred medium if commercialisation comes. Indeed, in a review of chloride leaching of copper concentrates McLean[36] concludes that chloride leaching is a viable alternative technology to smelting, although no single flowsheet so far devised will be suitable for every type of concentrate. Particularly, however, he sees the lack of viable automatic control sensors for use in the highly corrosive

acid chloride medium for process optimisation as a problem area.

When complex sulphides are treated with chloride lixiviants most of the metal sulphides are likely to be taken into solution and attempts to achieve selectivity through the different ease of solubility of different sulphides have met with little success. The implication of this is that subsequent solution treatment is required to separate the metals in solution either to recover the values or to obtain them in a form suitable for disposal, depending on quantities and economics. Thus the economics of solution treatment and metal recovery will be very important, and little can be done at the leaching stage to lessen these problems. The leaching stage has to be considered in terms of the efficient dissolution of the metal values with maximisation of yield and minimisation of the energy consumption.

Why should chloride based processes be more favoured than those based on other media of which sulphate is dominant? A comparison of sulphate and chloride based leaching processes is given in Table 5. The table shows that there are distinct advantages in chloride hydrometallurgy particularly in the leaching process. Further advantages arise in the subsequent metal separation and purification steps wherein the chemistry of chloride based separations provides more flexibility than sulphate based alternatives. Major difficulties with chloride based processes may rest with

Table 5 Comparison of sulphate and chloride based leaching process

---

Factors favouring sulphate processes

Sulphate	Chloride
1. Materials selection is well understood.	1. More expensive materials are required. Pumps etc may cause particular problems.
2. Copper and zinc electrowinning from sulphate electroytes are proven commercial processes.	2. Experience of Cu, Zn and Pb electrowinning from chloride solutions is limited. Cu and Zn chlorides can be converted to sulphates for electrowinning but oxidation of cuprous chloride is required.

Factors favouring chloride processes

Chloride	Sulphate
1. Leaching can be carried out at ambient pressure.	1. Pressurised reactors for gas/liquid/solid contact are required.
2. Attack on pyrite is limited and recovery of elemental sulphur is high.	2. More difficult to limit attack on pyrite and obtain high recovery of elemental sulphur.
3. Copper metal can be obtained by reducing a cuprous salt or solution with a potential resultant saving in energy.	3. Copper metal is obtained by reducing cupric salt or solution.
4. Leaching can be carried out to obtain lead and silver in the leach solution.	4. Lead and silver report in the leach residue which requires further treatment for Pb and Ag recovery.
5. Faster leaching rates and more concentrated leach liquors.	
6. Lixiviant is readily recycled.	

---

deportment of impurities, an area less fully explored to date and with the final metal winning steps, if the metals, particularly copper and zinc, are to be won from chloride electrolytes.

While some sulphides can be leached directly by HCl, production of H_2S would normally be considered a drawback. Oxidative leaching which solubilises the values of interest and produces elemental sulphur is clearly the preferred option. Most favoured oxidative lixiviants are $FeCl_3$, $CuCl_2$ or a combination of the two.

Dutrizac[37] has noted that cupric chloride leaching processes offer some advantages over ferric chloride leaching in terms of reaction rates, subsequent solution purification operations and ease of lixiviant regeneration.

A wide range of processes have been developed for chloride hydrometallurgical treatment of complex sulphide ores and these have been reviewed by several authors[34][36][38][39]. The individual flowsheets of these processes can be combined into a general process flowsheet as devised by ICI[40] (Fig 2). Often at times two stages of leaching are used, one of which may be combined with the lixiviant regeneration step thus providing for one main solid residue from the process. Lead removal by $PbCl_2$ crystallisation is a common first stage. Thereafter copper recovery choices obviously relate to copper content in the bulk concentrate and the relative economics of producing cement copper or high grade cathode. None of the flowsheets as currently devised recover copper directly by electrowinning, but those which do employ copper electrolysis, interface copper winning to the main process flowsheet via solvent extraction. With one exception, eg CUPREX Metal Extraction Process adapted for complex sulphide ore processing, copper is transferred via solvent extraction to a sulphate solution for conventional electrowinning. Zinc

is always treated by solvent extraction and depending upon the reagent used is then electrowon from either chloride or sulphate solution. A comprehensive review of lead electrowinning and cell design has been given recently by Ozberk et al[41]. Iron elimination from the process is generally achieved via goethite precipitation but as sulphate is produced in greater or lesser extent in the oxidative leaching then some jarosite precipitation can occur. Sulphate control in all circuits is an important consideration. While considerable information has been published mostly from laboratory studies, but in some cases also form quite extensive pilot scale testwork, much less information is available on impurity deportment and on precious metal recovery. Some further comments on the process flowsheet steps are however considered appropriate.

(a) Leaching

Ferric or cupric chloride leaching of Cu/Pb/Zn sulphide concentrates will produce leach liquors containing either cuprous and ferrous chlorides or cupric and ferrous/ferric chlorides depending on whether the primary leach is carried out in reducing or oxidising conditions viz:-

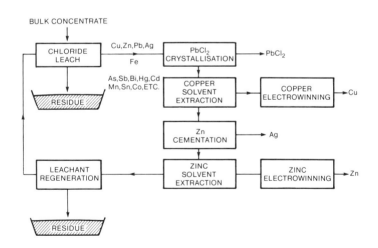

FIG. 2 OUTLINE CIRCUIT FOR CHLORIDE HYDROMETALLURGICAL TREATMENT OF
BULK CONCENTRATES (REF 40)

$$CuFeS_2 + 3FeCl_3 \rightarrow 4FeCl_2 + CuCl + 2S^\circ \text{ (reducing)}$$

$$CuFeS_2 + 3CuCl_2 \rightarrow 4CuCl + FeCl_2 + 2S^\circ \text{ (reducing)}$$

or

$$CuFeS_2 + 4FeCl_3 \rightarrow 5FeCl_2 + CuCl_2 + 2S^\circ$$
$$\text{(oxidising)}$$

Thus for oxidising leaching an excess of $FeCl_3$ is required.

When CuCl is the desired leaching product then excess chloride is required to render the CuCl soluble through the formation of complex anions such as $CuCl_2^-$. Leaching yields are poor and the circuit must therefore accommodate a further oxidative leaching step to achieve a high recovery of values. If the copper recovery option is from cuprous chloride then the oxidative leaching step is carried out after the copper recovery step as in the Duval CLEAR process or its variations as devised for complex sulphide concentrates or Red Sea muds[42]. However if cupric chloride is the desired product then either one or two consecutive oxidative leaching steps can be performed at the head of the flowsheet. The consensus of opinion would appear to favour 2-stage leaching for high copper recovery although high lead and zinc recoveries are readily achieved in one stage. Choice of leaching has implications for iron elimination and its sequence within the flowsheet.

(b) Lead Recovery

The only process reported in the literature to use zinc cementation for lead recovery directly from the leach liquor is Tecnicas Reunidas's Zinchlor process[43a]. However this has been superceded' by the Ledchlor process[43b] wherein lead is electrowon from aqueous $PbCl_2$ solutions after recovery from the leach liquor and purification via a crystallisation step. This $PbCl_2$ crystallisation step is employed in all other processes for treatment of complex sulphide ores, with lead recovery thereafter by electrowinning either from aqueous or fused salt media. Ozberk et al[41] has comprehensively reviewed lead chloride electrolysis and a schematic (Fig 3) for technology options has been given. From this review it was concluded

that technical feasibility of lead electrowinning from either media has been proved at pilot scale. Economic viability however is not so far proven and thus confident selection of the best technology is difficult. The purity of the lead so produced is dependent on the purity of the lead chloride produced at the crystallisation stage and thus this stage is of considerable importance.

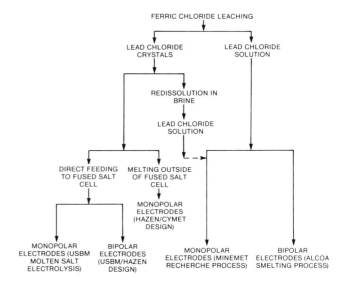

FIG. 3 OPTIONS FOR LEAD CHLORIDE ELECTROLYSIS
(REPRODUCED FROM EXTRACTION METALLURGY '89,
OZBERK ET AL REF 41)

Lead chloride is much less soluble than cuprous chloride particularly at ambient temperature and will report in the lead residue unless leaching conditions are chosen carefully. The solubility of lead chloride in a solution of NaCl and HCl has been reported by Muir et al[44] who showed that the solubility of lead rose sharply only in solutions of high ionic strength.(>3 M) but that NaCl was much more effective than HCl in dissolving $PbCl_2$, Fig 4, presumably because the activity of water and hence that of the chloride ion varies according to the counter-ion as well as the ionic strength of the solution. The proton is hydrated more strongly and has much greater activity than Na^+ at the same ionic strength and thus it would associate more strongly with Cl^- thereby decreasing the single ion activity of Cl^- in HCl solution relative to NaCl solution.

More recently the solubility of aqueous lead chloride solutions has been studied by Holdich and Lawson[45], who present data for the solubility of lead in acidic solutions of between 5 and 10 M chloride concentrations, present as the salts of Cu^{2+} and of either sodium or calcium. Cooling these solutions from boiling point to 20°C precipitates approximately two thirds of the lead as $PbCl_2$ crystals of >99% purity. The solubility of lead is shown to depend upon the activity of the lead and chloride ions in solution and can be predicted from known thermodynamic constants.

Although crystallisation of lead chloride seems easy because of its low solubility in low chloride tenor liquors its solubility is in fact quite complex as shown above and varies considerably with solution composition and temperatures. Thus relatively large concentrations of lead may remain in leach liquors so treated and the effect of lead at such concentrations on downstream processes such as solvent extraction of other metals needs to be considered carefully when designing these processes.

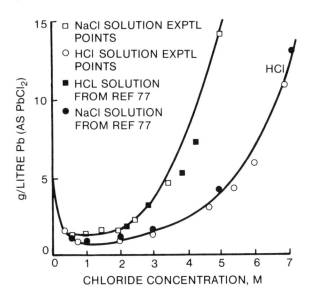

FIG. 4 SOLUBILITY OF $PbCl_2$ IN HCl AND NaCl SOLUTIONS AT 25°C (REF 44)

(c) Copper Recovery

If the level of copper in the leach liquor is too low to warrant recovery via solvent extraction and electrowinning then cementation would be the preferred option. Zn dust is the preferred reagent for Red Sea muds processing[46]

and this step would also recovery silver with the copper. Lead has been proposed for copper and silver removal prior to lead crystallisation in the CANMET process but was later abandoned because of unacceptably high lead addition requirements and lead contamination of the cement product[47][48]. In the current version of the flowsheet copper and silver are now recovered after $PbCl_2$ crystallisation by cementation using iron granules or shredded steel[49].

As already noted, copper recovery by electrolysis has to be interfaced with the leach liquor by solvent extraction. At concentrations between 1-5 g l^{-1} any of the modern hydroxyoxime reagents currently used for copper extraction from sulphate leach liquors can be used to transfer copper from the chloride leach liquor to a sulphate electrolyte, although reagents modified with α-hydroxyoximes as accelerators must be avoided as they transfer chloride ion. Di 2-ethylhexyl phosphoric acid (DEHPA) can also be used for this purpose provided the leach liquor contains little or no ferric iron. Control of pH can be achieved using Minemet Recherche Technology[50] whereby cuprous copper is oxidatively extracted according to the following reaction.

$$2CuCl + \overline{2RH} + \tfrac{1}{2}O_2 \rightarrow \overline{R_2Cu} + CuCl_2 + H_2O$$

At concentrations much above 5 g l^{-1} neither the hydroxyoxime reagents nor DEHPA are suitable extractants. However the new ICI reagent DS5443 (or Acorga CLX20) has been tailor-made for extraction of $CuCl_2$ from strong copper bearing leach liquors via the following reaction.

$$Cu^{2+} + 2Cl^- + \overline{2L} \quad \overline{L_2 CuCl_2}$$

This reagent which is a pyridine dicarboxylate[51] is highly selective for copper and thus can produce very pure cupric chloride strip liquors from which a high purity granular copper product can be produced. Special cell design is necessary for successful electrowinning of copper from chloride solutions.

Direct electrolysis of cuprous chloride has been carried out by Duval in their CLEAR process plant[52] using a diaphragm cell designed whereby the copper product was continuously removed from the cell on an endless belt which ran below the cell cathodes and onto which the cathode product fell. Cuprous copper was oxidised to cupric at the anode to regenerate lixiviant and to minimise energy consumption. However this plant has closed down and Fletcher[42] has commented that it is now evident that CLEAR technology is not able to compete with the smelting/electrorefining route when clean copper concentrates are to be processed.

The CUPREX Metal Extraction Process[35] on the other hand uses the Metchlor cell to electrowin copper from the cupric chloride electrolyte produced via cupric chloride solvent extraction with DS5443 (Acorga CLX20) from the ferric chloride leach liquor. The Metchlor cell, developed in Spain by Tecnicas Reunidas is of a two compartment design with the cathode and anode compartments separated by a cation-selective ion-exchange membrane such as NAFION 117. The cathodes are titanium and dimensionally stable anodes are employed.

The feed to the cathode compartment is a pure, >90 g l^{-1} Cu, cupric chloride solution containing added sodium chloride to improve its conductivity and to avoid precipitation of cuprous chloride. As copper deposition takes place sodium ions move through the membrane from the anode compartment. The spent catholyte is a sodium chloride solution containing approximately 10 g l^{-1} copper, roughly 50% Cu^{2+} and 50% Cu$^+$. This liquor is treated with Cl$_2$ to reoxidise the Cu$^+$ and then treated by solvent extraction with DS5443 to completely remove all the copper. The essentially copper free raffinate is fed to the anode compartment of the Metchlor cell where chlorine gas is produced at the anode. The cell voltage is reported as 2.3 to 2.5 V and current efficiencies have been given as between 97-99% at a current density of 1500 Am^{-2}. The copper is deposited in a compact granular form which facilitates separation and washing and avoids the problem of oxidation of more powdery forms of copper deposits. Membrane

lifetime is predicted to be 5 years.

Technical viability of this process has been demonstrated in a 1 tonne per day pilot plant near Madrid. The economics of the process are not yet available in any detail.

A recent paper by Tailoka and Fray[53] has described an improved system for increasing mass transfer in copper chloride electrowinning systems. Using microporous membranes to create a very fine dispersion of bubbles has permitted electrowinning of copper from chloride electrolytes as a coherent planar deposit at high current densities provided the cupric chloride concentration was not greater than 0.4 M. Replenishment of the cupric chloride was found necessary and thus interface with a solvent extraction step such as used in the CUPREX Metal Extraction Process would seem ideal. The use of such membranes for gas sparging in the Metchlor cell could therefore further enhance the CUPREX Metal Extraction Process.

(d) Zinc Recovery

Zinc forms relatively weak complexes with chloride ions but the existance of all chloro species up to ZnCl$_4$$^{2-}$ has been demonstrated. Thus zinc is extractable in some form of chloro complex by alkyl amines, tributyl phosphate (TBP), alkyl phosphonates, trioctyl phosphine oxide (TOPO) and acid extractants. TBP is the preferred extractant for zinc in several processes where the ability to extract zinc as the ZnCl$_2$ species provides a precise materials balance between the solvent extraction feed, strip and tankhouse circuit where Cl$_2$ is the anode product. If a complex anionic species was extracted a chloride imbalance in the circuit would result.

A full comparison of the extraction behaviour of trialkyl phosphates, dialkyl alkyl phosphonates, alkyl dialkyl phosphonates and trialkyl phosphine oxides for zinc, iron and cadmium from chloride solution has been carried out by Preston and du Preez[54]. The study showed moderately good selectivity for zinc over ferric iron and excellent selectivities for zinc over iron (II), copper (II), lead (ii) and cadmium (II).

The CANMET process[55], the Elkem[38] and the Elkem/ICI[56] process, and the Zinchlor process[43] all employ solvent extraction to recover $ZnCl_2$ from the process liquors and transfer the $ZnCl_2$ to an advance tankhouse electrolyte. Extensive work at CANMET has fully defined conditions for zinc electrowinning from chloride electrolytes and a full-height cell (3" x 4" x 48") was used to test the electrolysis conditions developed in the laboratory scale testwork on a mini-plant scale[57]. The zinc deposits were smooth, compact and dendrite free. The average cell voltage was 5.5 V but addition of NaCl to the electrolyte decreased the cell voltage. These additions did not significantly affect the current efficiencies but at concentrations >1 M tended to promote nodular edge growth. In the Elkem and Elkem/ICI processes, TBP is used as the extractant for $ZnCl_2$ and thus iron and copper are the two contaminants with highest concentrations in the advance electrolyte. Iron is removed by precipitation as goethite while zinc dust is used to cement out copper. Electrowinning is carried out in diaphragm cells at a current density of 200-400 A m^{-2}. An organic additive, unspecified, is used to control the growth of dendrites. The current efficiency is said to be about 90%. Chlorine gas produced at the anodes is recycled for use in the process flowsheet.

In the Zinchlor process dipentyl pentaphosphonate is used as the extractant for $ZnCl_2$ and the Metchlor cell is used for zinc electrowinning. Results of electrowinning zinc in the Metchlor cell together with operating conditions in association with the Leadex process have been published and are reproduced in Table 6[58].

It is interesting to note that the chosen current density is at the top end of the range used by Elkem. The cell voltage is also much less than that reported by MacKinnon in the CANMET work.

e) Iron Control and Removal

Iron control in chloride systems has been discussed by Raudsepp and Beattie[59]. The predominant route which has been adopted for the removal of iron from chloride leach liquors involves the oxidation of the iron to the ferric state accompanied by the precipitation of basic iron compounds ie goethite or jarosite compounds. The variations in iron removal strategies which have been adopted are primarily in which product is formed and where in the flowsheet the iron control is implemented. According to the conditions used, iron can be precipitated in many forms and the reactions can be represented as follows:

$$Fe^{3+} + 3H_2O = Fe(OH)_3 + 3H^+ \qquad <60°C$$
$$\text{ferric hydroxide}$$

$$Fe^{3+} + 2H_2O = FeOOH + 3H^+ \qquad 60-150°C$$
$$\text{goethite}$$

$$2Fe^{3+} + 3H_2O = Fe_2O_3 + 6H^+ \qquad >150°C$$
$$\text{haematite}$$

$$3Fe^{3+} + 2SO_4^{2-} + M^+ + 6H_2O = MFe_3(SO_4)_2 (OH)_6 + 6H^+$$
$$\text{jarosite}$$

where M = H_3O, NH_4, Na, K or 0.5 Pb

In hydrometallurgical processes for the treatment of metal sulphides, iron is usually converted to goethite and/or jarosite because ferric hydroxide is difficult to filter and precipitation of haematite requires high temperatures.

Thus in chloride solution jarosite precipitation will remove any sulphate formed in leaching and could thus potentially be useful in sulphate control. Iron precipitation may also control other impurities eg arsenic by precipitation of ferric arsenate. Coprecipitation or adsorption of metal values on the iron precipitates may also occur. For this reason it is usually desirable to precipitate iron from solutions which have been previously depleted of metal values.

In several of the proposed processes[39] iron is rejected during the leaching step by

Table 6   Electrowinning Results for a Leadex Pilot Plant[58]

Operating Conditions		Results		Product quality
No of cathodes	= 6	Cathodic C Ef	= 94-96%	Pb = 0,0026
No of anodes	= 7	Anodic C Ef	= 92-94%	Fe = 0,0004
Cathode	= titanium	Cell voltage	= 2.6-2.7 V	Cd ≤ 0,0001
Anode	= DSA			Cu ≤ 0,0001
Membrane	= NAFION 117			Sn ≤ 0,0005
Catholyte temp	= 35°C			Zn ≥ 99,99
Cath current density	= 400 A/m²	Energy Consumption		
Organic additives	= Gum arabic	2.33 kwh/kg of Zn SHG		
Zinc production	= 8 kg/d			

introduction of air eg Dextec, Cymet and CLEAR. While this has the advantage of achieving leaching and iron control in the same vessel, it is disadvantageous to have iron products in the leach residue when it is to be treated for precious metal recovery. Iron removal by hydrolysis from recycle liquors is incorporated into the flowsheets of the Elkem[38], CANMET[55] and UBC-Cominco[36] processes.

In the Minemet process[38] however, iron is eliminated as goethite from the cuprous chloride leach liquor itself by injection of air or oxygen in the solution[60]. Low residual levels of iron were achieved in the copper liquor (~10 mg l^{-1}) and copper entrainment was limited to 0.7% for pH values between 2 and 2.5. Filtration was improved by operating the process near boiling point. Additionally information on the kinetics of oxidation of cuprous ion in this system is also provided.

(f) Sulphate Control

Some sulphate is always produced by oxidation of the elemental sulphur produced during the leaching of sulphide minerals. Dutrizac[61] has studied the deportment of sulphate during ferric chloride leaching of a pyritic Zn-Pb-Cu-Ag bulk concentrate and investigated several control options. Although sulphate can be limited by jarosite formation this is not recommended as

significant contamination of the $PbCl_2$ crystallisation product by KPb_2Cl_5 occurs. The presence of calcium chloride is recommended by Smyres[62] to control sulphate build up but Dutrizac[61] notes that the presence of calcium chloride in the leaching medium gives a final sulphate concentration of 1-3 g l^{-1}. Calcium sulphate contamination of the $PbCl_2$ product is also a concern. Barium chloride on the other hand effectively controls sulphate but excess $BaCl_2$ results in the precipitation of $BaCl_2H_2O$ which impedes residue filtration and contaminates the $PbCl_2$ intermediate product. Based on the technical data of the various sulphate control options, the relative costs of the reagents involved, and the fact that rapid crystallisation and filtration of the $PbCl_2$ intermediate product minimises contamination, sulphate control by $CaSO_4$ precipitation is concluded to be the preferred option.

(g) Impurity Control and Removal

A major advantage claimed for chloride/oxygen leaching systems[62] is the selective extraction of the metal values of interest and rejection of impurities such as Al, As, Cr, Fe, Sb and silicates to the leach residues. However this is dependent on solution acidity and chloride level.

In acid ferric and cupric chloride leaching

of complex sulphide ores considerable dissolution of impurities such as Sb, As, Hg, Mn, Bi, Cd, Sn etc can be expected and data are provided by Andersen et al[63]. Iron precipitation can be expected to scavenge some of these impurities and significant quantities also report to the copper product when copper removal is carried out by cementation. In an experimental study at CANMET[64] it was found that the rate of cementation of metals from acid solutions by zinc was in the order Cu>Pb>Cd>Bi>Ni>Co. The rate of cementation of lead was decreased by the addition of sodium chloride as was that of Bi and Cd. Mercury if present in the leach liquor can also be expected to co-cement and of course silver will also co-cement with copper. Further information on coprecipitation of arsenic and antimony in the iron elimination step is provided by Raudsepp and Beattie[59], while Beutier et al[60] report partition of elements between solution and precipitate for goethite precipitation in the Minemet Recherche process. Here greater than 97% of soluble arsenic and antimony are coprecipitated with iron but the removal of bismuth was low.

The use of solvent extraction with Acorga CLX 20 for copper ensures an impurity free copper chloride strip liquor for copper electrowinning. Iron precipitation prior to zinc solvent extraction minimises iron contamination of the zinc electrolyte if dialkyl alkyl phosphonate is used as the zinc extractant. Further iron removal if necessary can be achieved by precipitation on addition of zinc oxide for pH adjust. Other impurity deportment should not present problems in this solvent extraction step provided any copper present in the feed to zinc solvent extraction is in the cupric form otherwise cuprous copper carryover must be eliminated by cementation with zinc dust.

(h) Precious Metal Deportment and Recovery Oxidative leaching in the presence of high chloride levels will ensure that silver reports to the leach liquor. Gold is not leached and remains in the leach residue.

Silver can be recovered by cementation with copper if copper is subsequently to be recovered via solvent extraction and electrowinning. If copper is recovered by cementation then silver will report to the copper cement. Most of any bismuth and mercury present in the leach liquor will also be cemented out at this stage.

Recent work by Abe and Flett[65] has shown that silver could be recovered by solvent extraction with Cyanex 471. If copper is present in the cupric state then good selectivity is achieved over copper, lead and zinc. Although no information is available it is expected that bismuth might extract with the silver while mercury certainly will[66]. Thus significant further processing would seem to be required before a high grade silver product could be recovered.

For most processes gold recovery is dismissed as being readily achieved by conventional processing of the leach residue. By this is presumably meant cyanidation. Gold recovery from leach residues has been discussed by Wilson[67] who comments that the inability of most hydrometallurgical processes to solubilise gold is an important short coming as the presence of sulphur in the residue renders the normal process for gold recovery from such materials uneconomic. Oxidative dissolution of gold using a strong saline solution of cupric chloride with chlorine sparging can dissolve up to 95% of contained gold which is then readily recovered by electrowinning and copper can be reoxidised at the anode. No information is given as to the extent of attack on pyrite or on elemental sulphur oxidation. A more general examination of the fate of gold in cupric chloride hydrometallurgy has been given by McDonald et al[68] whose results compliment the data given by Wilson.

ECONOMIC CONSIDERATIONS

Turning now to economic considerations of the bulk sulphide flotation/chloride hydrometallurgical option; on the basis of the examples given in Table 4 it is possible to evaluate the net smelter return (NSR) - see Table 7 - for concentrates obtained by differential flotation and the intrinsic value (IV) of the bulk flotation concentrate (this was approximately double the NSR). Further the costs of operating both the bulk and differential circuits could also be evaluated (mining costs were excluded as being common to both circuits). Cost data was based on literature sources[69-72] and the rate of discounting cash was 15% over 8 years.

The net additional value (NAV) of the bulk concentrate produced was therefore given by the following:

$$NAV \ (Bulk) = IV \ (Bulk) - NSR \ (Differential)$$
$$+ \ (Costs \ Differential - Costs \ Bulk)$$

This equated to a value of 285 $US/tonne of concentrate at periods of low metal prices (refer Table 8) and 406 $US/tonne concentrate at periods of high metal prices. Accepting that the mining operation would be profitable by the

normal recovery methods, it therefore follows that the hydrometallurgical operation has to be justified solely on the basis of this net additional value.

The methodology used for the cost evaluation of the chloride leach route utilised a computer model developed using a Lotus 123 spreadsheet system and cost data was again based on literature sources[70 73-76] and from private communications also. The process costed was based on the flowsheet as given in Fig 5 which is similar to that described in Fig 2. The rate of discounting cash was as for the flotation plant.

The cost of the chloride leach (ferric chloride) process (229 US $/tonne concentrate) excluded recovery of Ag and Au and the cost of recovering the silver has subsequently been added. It was decided however not to take account of the Au recovery from the residue. Since the net available value (NAV) of the bulk concentrate assumes total metal recovery the metal loss on chloride leach treatment becomes

Table 7   Smelter Returns

	Smelter Return (% Intrinsic Value)		
	Copper Conc*	Lead Conc+	Zinc Conc+
Prime Metal Return[70]	67	50	50.5
Silver Return[71]	93.0	89.9	–
Gold Return[70]	65.3	–	–

*   Curve 1 p 766 of reference 70

+   Curve 2 p 766 of reference 70

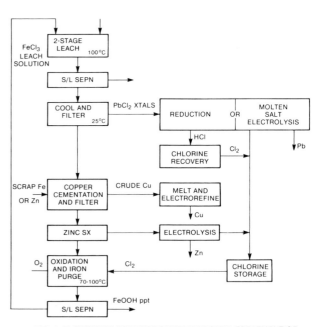

FIG. 5 FLOWSHEET FOR HYDROMETALLURGICAL TREATMENT OF A BULK SULPHIDE CONCENTRATE

Table 8 Metal prices		
	Low (Jan 87)	High (June 1988)
	US $/tonne (kg*)	
Cu	1400	2510
Zn	610	1400
Pb	950	660
Ag	172	225*
Au	12530	14420*

Table 9 Metal losses FCL	
Zn	1.5%
Pb	1.5%
Cu	3 %
Ag	5 %
Au	100 %

as

1) Concessions by smelters at periods of high metal prices.

2) Achieving high leaching and extraction recoveries of all metals (apart from gold) and high elimination of impurity elements.

3) The potential for improving recoveries and grades to differential concentrates.

an additional cost factor.

  Thus

Profit = NAV - Costs [Ferric Chloride Leach
        (FCL) + Ag Smelting + Metal Losses]

Figures obtained from private sources indicated metal losses as in Table 9.

  On the basis of the above it was calculated that the rate of return on an FCL installation treating the bulk concentrate from this 'average' mine would be 9% at the low metal prices and 44% at the high metal prices – illustrating the high dependence of profitability on metal price. Further it was calculated that if the NSR on intrinsic metal value increased by 5% for Cu and Zn during periods of high metal prices (a not unreasonable attitude by the smelters) then the rate of return on the FCL operation, even at the high metal prices, would drop to 34%.

CONCLUSIONS

On the basis of cost calculations undertaken on a ferric chloride leach (FCL) process, as applied to a typical bulk flotation concentrate produced from a complex sulphide ore, it would appear that the process could be economically justified at periods of high metal prices although it would be barely competitive against the traditional differential flotation/refining route when low metal prices prevail.

  In addition to metal prices the degree of profitability will be influenced by such factors

With respect to the latter item our discussion on the current state of the art for direct differential flotation has indicated that there is a margin for metallurgical improvement requiring only relatively minor modifications to existing technology. On the other hand many of the steps in the chloride hydrometallurgy route still have to be integrated and commercially exploited demanding new technology in the areas of equipment and sensors. Additionally the fate of certain impurities in the bulk concentrates (particularly Bi and Hg) cannot as yet be quantified or costed for. Although the basic metal content of a bulk concentrate will not be totally dissimilar to our example, sufficient variation in levels (particularly Cu and Pb) could occur to modify the process route and the methodology for recovering both the principal metal phases and the impurities/silver also. Consequently a global solution to this problem may not be available. Investors are likely to want assurances on metal recoveries and process costs at sufficiently large scale to justify the risk capital in a fluctuating metal market and on a case by case basis another factor that will enter into the equation will be the relative position of the leaching plant and smelter to the mill.

As a final discussion point it is debatable whether from an environmental standpoint one transfers an air pollution problem to a landfill problem or whether for example Sb and As are more stable as minerals in tailings streams or as coprecipitates with geothite.

ACKNOWLEDGEMENTS

The authors would like to acknowledge the Minerals Industry Research Organisation for permission to use certain extracts from the final report on the MIRO RC44 project and to Dr R Spencer (consultant) who was responsible for economic assessment data on the Ferric Chloride Leach Process.

References

1. Chloride Hydrometallurgy - A Technicoeconomic Assessment. MIRO Project RC44.

2. G. Barbery. Complex Sulphide Ores. Processing Options. Mineral Processing at the Cross Roads. Ed. B.A. Wills, R.W. Barley. NATO ASI Series 1986.

3. Complex Sulphide Ores. IMM London 1980. Ed. M.J. Jones.

4. Complex Sulphides, Processing of Ores, Concentrates and By-products. TMS-AIME. New York 1985. Ed. D. Zunkel, R.S. Boorman, A.E. Morris and R.J. Wesley.

5. Flotation of Sulphide Minerals. Elsevier, Amsterdam 1985. Ed. K.S.E. Forssberg.

6. 2nd Workshop on Flotation of Sulphide Minerals June 18-21 1990. Lulea Sweden.

7. D.N. Collins, D.S. Flett, J. Melling and S.P. Barber. Reagent Databases for the Metallurgical Industry. Advances in Mineral Processing Ed. P. Somasundaran, SME-AIME 1986 pp 380-394.

8. C.R. Johnson. Lead Zinc Ore Concentration by Electrolytic Zinc Co of Australasia Ltd. Mining and Metallurgical Practices in Australasia, 1980.

9. C.J. Burns, P.J. Duke and S.R. Williams. Process Development and Control at Woodlawn Mines. Miner Process Congr. Tech. Pap. 14th 1982.

10. Tveter and Mcquiston. Plant Practice in Sulfide Flotation. Froth Flotation 50th Anniversary Volume (AIME), 1962.

11. Thompson, Amsden and Chapman. Texas Gulf Canada - Kidd Creek Concentrator. Milling Practice in Canada, 1977.

12. Anon. Zinc Mining and Concentrating at Rosebury. Mining Engineering, 1964.

13. T.R. Twidle, P.C. Engelbrecht and J.W.S. Koel. Optimising Control of Lead Flotation at Black Mountain. Int. Miner Process Congr. Tech. Pap. 15th 1985 Cannes.

14. T.W. Twidle, P.G. Engelbrecht and J.W.S. Koel. Improvements in Stabilising Control at Black Mountain. J S African Inst. Mining Metallurgy V86 N1 P15-24, 1986.

15. Neumann and Schnarr. Concentrator Operation at Brunswick Mining and Smelting Corporation, No 12 Mine. AIME World Symposium. Lead and Zinc, 1970.

16. Schnarr. Brunswick Mining and Smelting Corporation. Milling Practice in Canada, 1977.

17. Allan and Bourke. Mattabi Mines Ltd. Milling Practice in Canada, 1977.

18. I. Maeshiro. Recovery of Valuable metals from 'Black Ore'. Proc. Hong K Meeting 11 Commonw. Min. Metall. Congr., 1978.

19. Anon. Rammelsberg Mine. Mining Magazine, 1979.

20. Anon. Kosaka Mine and Smelter. Mining Magazine V151 N5, 1984.

21. Anon. Bleikvassli and Mofzell. Mining Magazine V143 N5, 1980.

22. Anon. Expansion at New Brunswick. Mining Magazine V137 N3, 1977.

23. Brooks and Barnett. Noranda Mines Ltd - GECO Division. Milling Practice in Canada, 1977.

24. S.R. Williams and J.M. Phelan. Process Development at Woodlawn Mines. (As Ref. 3).

25. Zhang Hui-wen. New flotation flowsheet for treatment of Fankon complex lead zinc ore. Mineral Processing and Extractive Metallurgy Oct/Nov 1984 Kunming. Ed. M.J. Jones and P. Gill IMM.

26. R. Klimpel, R.D. Hansen, B.S. Fee. Recent advances in new frother and collector chemistry for sulphide mineral flotation. 16th IMPC Stockholm 1988. Elsevier.

27. D.R. Nagaraj, M.E. Lewellyn, S.S. Wang, P.A. Mingiona, M.J. Scanton. New sulphide and precious metal collectors: for acid, metal and mildly alkaline circuits. 16th IMPC Stockholm 1988. Elsevier.

28. A.A. Abramov, G.G. Shtoik, Yu I Filshin, Et Al. Improvement for Beneficiation of Polymetallic Ores of Variable Composition. Int. Miner. Process Congr. Tech. Pap. 15th 1985 Cannes.

29. A.P. Vargas. Process for the Selective Separation of Base Metal Sulphides and Oxides Contained in an Ore. Patent No. WO 84/00704 (World Intellectual Property Organisation), Aug. 1983.

30. J.R. Boldt and P. Queneau, The Winning of Nickel. London: Methuen and Co. Ltd, 1967.

31. E.G. Parker and S. Romanchuk. Pilot plant demonstration of zinc sulphide pressure leaching. Lead-Zinc-Tin '80. J.M Cigan, T.S. Mackay and T.J. O'Keefe, eds. Warrendale, Pa: Met. Soc. AIME, 1979, pp 407-425.

32. R.L. White. Sherritt commercialises zinc pressure leaching. Engng. Min. J., 1981, 182(8), 76-79.

33. G. Thorsen. Extractive metallurgy of copper. Topics in Non-ferrous Extractive Metallurgy. Critical Reports on Applied Chemistry Volume 1. London: Society of Chemical Industry, 1980, pp 1-41.

34. D.S. Flett. The Role of Hydrometallurgy in Extractive Metallurgy. Chemistry and Industry, 20 June, 1981.

35. R.F. Dalton, E. Hermana and B. Hoffmann. The CUPREX process - a new chloride-based hydrometallurgical process for the recovery of copper from sulphidic ores. Separation Processes in Hydrometallurgy, ed. G.A. Davies, Ellis Horwood Ltd, 1987, pp 466-476.

36. D.C. McLean. Chloride Leaching of Copper Concentrates. Practical Operational Aspects. Paper presented at the 111th AIME Ann. Mtg. Dallas, Texas, USA, Feb 13-18 1982.

37. J.E. Dutrizac. The leaching of galena in cupric chloride media, Metall. Trans. B. 1989, 20B 475-483.

38. D.S. Flett, J. Melling and R. Derry. Chloride hydrometallurgy for the treatment of complex sulphide ores, Warren Spring Laboratory Report, 1983, No LR 461 (ME).

39. J.E. Dutrizac. Recent advances in the leaching of sulphides and the precipitation of iron. MINTEK 50 ed. L.F. Haughton. The Council for Mineral Technology, Johannesburg RSA 1985, pp 39-62.

media - their application in the processing of complex base metal concentrates. Paper presented at the Joint EEC-Canada Seminar on the Treatment of Complex Minerals. Ottawa, Canada Oct 12-14, 1982.

41. E. Ozberk, J.E. Dutrizac and R. Minto. Lead-chloride electrolysis and the conceptual design of a cell. Extraction Metallurgy '89 London, IMM 1989 pp 861-884.

42. A.W. Fletcher. Future potential for chloride hydrometallurgy. Advances in Mineral Processing ed. P. Somasundaran SME/AIME 1986, pp 495-508.

43(a). E.D. Nogueira, L.A. Suarez and P. Cosmen. The Zinchlor process: simultaneous productions of zinc and chlorine, Zinc '83. 13th Annual Hydrometallurgical Meeting, CIM Edmonton, Alberta, Canada, 1983, Paper No 7.

43(b). Tecnicas Reunidas. Private Communication.

44. D.M. Muir, D.C. Gale, A.J. Parker and D.E. Giles. Leaching of the MacArthur River lead zinc sulphide concentrate in aqueous chloride and chlorine system. Proc. Australia. Inst. Min. Metall. 1976, 251 23-35.

45. R.G. Holdich and G.J. Lawson. The solubility of aqueous lead chloride solution. Hydrometallurgy, 1987, 19, 199-208.

46. H. Weber, K. Pretzch, G. Barbery and A.W. Fletcher. Metallurgical treatment of Red Sea Concentrates. Inter Ocean '81 Dusseldorf, 1981, 116-122.

47. D.J. MacKinnon, J.E. Dutrizac and D.J. Hardy. Cementation of silver and copper on lead dust and chemical upgrading of the cement product. Ottawa: CANMET, Report No MRP/MSL 85-46 (TR).

48. J.E. Dutrizac. Private communication.

49. D.J. MacKinnon and J.M. Brannan. Cementation of copper and silver on granular iron from simulated iron chloride leach solutions, Ottawa: CANMET, Report No MRP/ML 86-107 (TR).

50. J.M. Demarthe and A. Georgeaux. Hydrometallurgical treatment of complex sulphides, Complex Metallurgy '78 ed. M.J. Jones, London, IMM 1978, pp 113-120.

51. K.H. Soldenhoff. Solvent extraction of copper (II) from chloride solution by some pyridine carboxylate esters. Solvent Extr. and Ion Exch. 1987 5 (5) 833-851.

52. F.W. Schweitzer and R.W. Livingston. Duval's CLEAR hydrometallurgical process. Chloride Electrometallurgy. ed. P.D. Parker, TMS-AIME 1982, pp 221-227.

53. F. Tailoka and D.J. Fray. Increasing mass transfer in aqueous solution by use of microporous membranes, Paper presented at TMS Annual Meeting, Anaheim, California February 1990.

54. J.S. Preston and A.C. du Preez. The solvent extraction of zinc, iron and indium from chloride solutions by neutral organophosphorus compounds MINTEK Report 1985, No M 228.

55. M.C. Campbell and G.M. Ritcey. Application of chloride metallurgy to base-metal sulphide and uranium ores at CANMET. Extraction Metallurgy '81, IMM, London 1981, pp 76-90.

56. G.H. Danielsson, G.M. Boe and P.M. Finne. Extraction of cupric chloride from iron chloride leach liquors. Proc. Int. Conf. ISEC '83 Denver, Colorado, USA 1983, 536.

57. D.J. MacKinnon, J.M. Brannnen and D.M. Morrison. Zinc electrowinning from aqueous chloride electrolytes. J. Appl. Electrochem. 1982, 12 39-53.

58. E.D. Nogueira. Recent advances in the development of hydrometallurgical processes for the treatment of base-metal sulphides MINTEK 50 Ed. L.F. Haughton. The Council for Mineral Technology, Johannesburg RSA, 1985 pp 677-693.

59. R. Raudsepp and M.J.V. Beattie. Iron control in chloride systems. Iron Control in Hydrometallurgy, eds J.E. Dutrizac and A.J. Monhemius, Ellis Horwood Ltd 1986 pp 163-182.

60. D. Beutier and J.P. Bunzynski. Iron elimination by oxygen in acid cuprous chloride solutions: the case of the Minemet process. Iron Control in Hydrometallurgy eds. J.E. Dutrizac and A.J. Monhemius. Ellis Horwood Ltd 1986 pp 640-656.

61. J.E. Dutrizac. Sulphate control in chloride leaching process, Hydrometallurgy, 1989, 23 1-22.

62. G.A. Smyres, Chloride-oxygen leaching of sulphide, oxide and scrap metal feeds, Extraction Metallurgy '89, IMM London 1989, 839-860.

63. E. Anderson, G.H. Boe, T. Danielssen and P.M. Finne. Production of base metals from complex sulphide concentrates by the ferric chloride route in a small, continuous pilot plant. Complex sulphide ores, IMM, London 1980, pp 186-192.

64. D.J. MacKinnon and J.M. Brannen. Zinc dust cementation of bismuth, cadmium, cobalt, copper, lead and nickel from chloride solution, Ottawa, CANMET, Report No MRP/MSL 85-136 (TR).

65. Y. Abe and D.S. Flett. Solvent extraction of silver from chloride solution by Cyanex 471X, paper submitted to ISEC '90, Kyoto, Japan, July 1990.

66. Y. Baba, Y. Umezaki and K. Inoue. Extraction equilibrium of mercury (II) with trisobutyl phosphine sulfide. Solvent Extr. and Ion Exch. 1986, 4(1), 15-26.

67. A.T. Wilson. An economical method for the recovery of gold from the sulphur containing residue of a hydrometallurgical process. Complex sulfides: Processing of ores, concentrates and by-products eds. A.D. Zunkel, R.S. Boorman, A.E. Morris and R.J. Wesley. TMS-AIME 1985 pp 143-148.

68. G.W. McDonald, A. Saua, M.S. Barger, J.A. Koutsky and S.H. Langer. The fate of gold in cupric chloride hydrometallurgy. Hydrometallurgy 1987 18 321-336.

69. A.A. Matthews. Capital and Operating Cost Estimating System Handbook. Mining and Beneficiation of Metallic and Non Metallic Minerals in the US and Canada. USBM Open File Report 10-78 Dec 1977.

70. Marshall and Swift. Annual Cost Indices - Mining and Milling. Chemical Engineer. Jan 1987.

71. P.J. Lewis and C.G. Streets. An analysis of base metal smelter terms. Complex sulphides. IMM London 1980 Ed. M.J. Jones. pp 753-767.

72. De Cuyper and P.D. Ondenne. Experience in the Treatment of Sulphide Ores containing Precious Metals. Flotation of Sulphide Minerals Ed K.S.E. Forssberg. Lulea Sweden 1985.

73. T.A. Phillips. Economic Evaluation of a Process for Ferric chloride Leaching of Chalcopyrite Concentrate.

74. Bureau of Mines Cost Estimating System Handbook. Part 2 Mineral Processing. USBM Information Circular 9143 (1987).

75. H.C. Bauman. Fundamentals of Cost Engineering in the Chemical Industry. Reinhold Publishing Corp. New York pp 364 (1964).

76. T.A. Phillips. Economic Evaluation of an Electrolyte Process to recover lead from scrap batteries. USBM Information Circular 9071 (1986).

77. W.F. Linke and A. Seidell. Solubility of
inorganic and metal organic compounds, New York:
Amer. Chem. Soc. 1965.

Evaluation of the CANMET Ferric Chloride Leach (FCL) process for treatment of complex base-metal sulphide ores

W.J.S. Craigen
F.J. Kelly
D.H. Bell
Energy, Mines and Resources Canada, CANMET, Mineral Sciences Laboratories, Ottawa, Canada
J.A. Wells
Kilborn, Ltd., Toronto, Canada

SYNOPSIS

Significant deposits of complex fine-grained base metal sulphide ores occur in many parts of the world. To obtain high metal recoveries from such ores requires the production of bulk rather than selective concentrates. CANMET has developed a Ferric Chloride Leach (FCL) process to treat these relatively low grade bulk concentrates and to obtain very high recoveries of the contained metals. Typical extractions in CANMET's two-stage leach process are 99% for zinc and lead, 97% for copper and 96% for silver. Reactive sulphides are converted to elemental sulphur, which along with inert sulphides (mainly pyrite), remains in the leach residue and can be recovered as a valuable by-product.

The CANMET FCL process is capable of obtaining very high extractions of Zn, Pb, Cu and Ag, regardless of concentrate grade. Only the economic viability is affected by the combination of concentrate grade and throughput. This paper provides a comparative evaluation of the FCL process, versus conventional technology, for the processing of these complex ores. Specific examples are given to relate ore and concentrate grades and throughputs, to capital and operating costs, cash flow and return on investment. In all cases where a substantial increase in recovery (>20%) is achievable, the bulk concentrate/FCL process option is considerably more economically attractive than the conventional processing route. In addition the process has the potential to be economically viable at zinc production rates as low as 30,000 tpy.

INTRODUCTION

Significant deposits of complex fine-grained base metal sulphide ores occur in many parts of the world. Although many of these deposits are relatively high grade, their economic exploitability has been severely hampered by the metallurgical difficulties encountered in producing readily saleable selective concentrates with high metal recoveries.

For the majority of these complex sulphide deposits it has been adequately demonstrated that metal recoveries from the ore can be very significantly improved by producing bulk rather than selective concentrates, with overall recoveries generally being inversely proportional to the bulk concentrate grade. For the New Brunswick deposits in Canada, CANMET has demonstrated that the increase in recovery for bulk versus selective concentrates will be in the range 28 to 37% of ore value (1,2). This percentage increase is not significantly affected by metal prices. If the bulk concentrates are produced directly from the ore, grades of 30 to 40% (combined metals) are possible. If one or more selective concentrates is also produced along with the bulk concentrate, overall metal recoveries can remain about the same, but the grade of bulk concentrate will be proportionately reduced.

Regardless of the method of production, the bulk concentrates, because of their complexity and relatively low grade, are not amenable to conventional extraction/refining processes. To overcome these constraints to the processing of low grade bulk concentrates, CANMET has developed a Ferric Chloride Leach (FCL) process which can obtain very high recoveries of zinc, lead, copper and silver from bulk base metal sulphide concentrates, and is relatively insensitive to concentrate grade. The process produces refined zinc and lead and a high grade copper/silver cement product, equivalent to blister copper. Cadmium is also recovered. Reactive sulphides are converted primarily to elemental sulphur, which can be recovered as a by-product. No SO_2 is generated and there are therefore no gaseous emission problems. The process maximizes lixiviant recycle and the iron bearing residues for disposal to tailings are relatively free of environmentally deleterious impurities.

This paper summarizes the present state of development of CANMET's FCL process, and provides an estimate of its economic potential for the treatment of complex base metal sulphide ores and concentrates.

THE CANMET FERRIC CHLORIDE LEACH (FCL) PROCESS[2]

General Description

A simplified flowsheet for the FCL process is shown in Figure 1. The process consists of a two-stage ferric chloride leach, with inter-stage thickening. Thickener overflow becomes the pregnant solution for metals recovery while thickener underflow goes to the second stage leach. Second stage leach liquor, after solids/liquid separation, is regenerated with chlorine from electrowinning and recycled to the first stage leach. Sulphate build-up in the leach circuit is controlled by $CaCl_2$ (which is produced in the oxyhydrolysis circuit) to precipitate $CaSO_4$. Final leach residue, which contains primarily pyrite and sulphur, is thickened, filtered, washed and sent to tailings, or, if warranted, to sulphur and gold recovery. Lead chloride is recovered by crystallization from the pregnant solution, while copper and silver are recovered as a high grade metallic product by cementation with iron. Lead chloride crystals are washed, dried and converted to high purity lead by fused salt electrolysis. Zinc chloride is separated by a selective solvent extraction step, and is further purified by a second solvent extraction step to remove residual iron. Any other residual impurities are removed by zinc dust cementation prior to aqueous chloride electrolysis to produce cathode quality zinc. Spent electrolyte from zinc electrowinning is recycled to the zinc SX stripping circuit.

The raffinate from solvent extraction is regenerated with chlorine from zinc and lead electrowinning, prior to recycling to the leach circuit. Iron and impurity build-up in the process are controlled by a bleed stream from the SX raffinate. The bleed stream is treated with oxygen and lime in an oxyhydrolysis reactor to remove iron as goethite or hematite, after which the solution is recycled to the leach circuit to provide the $CaCl_2$ requirements for sulphate control and ensure recovery of other con-

tained chlorides and residual metal values (primarily Zn, Pb, Cd). Cadmium is allowed to build up in the process and is eventually recovered from the bleed stream by a separate processing step. To guard against build-up of magnesium, or other detrimental minor elements that may not be removed in oxyhydrolysis or cementation, a small portion of the raffinate bleed stream is sent directly to effluent treatment and subsequent tailings disposal with the iron oxide residue.

Application and Incentives for Use

The FCL process is primarily applicable to complex sulphide ore deposits, where high grade selective concentrates are difficult or impossible to produce, and where the production of a bulk concentrate is essentially the only alternative to ensure high recoveries from the resource base. The FCL process is readily adaptable to bulk concentrates of different compositions and grades. Typically bulk concentrates in the range 25 to 40% combined metals can be readily treated. This is well below the normally accepted 45% minimum grade (Zn and Pb) required for the Imperial Smelting Process (ISP), and in addition the FCL process has no upper limit on copper content. CANMET has shown that for typical Canadian complex sulphide ores, the bulk concentrate/FCL process option results in a net increase in recovered metal values (inclusive of FCL process recoveries) of 26 to 38% over that obtainable using the selective concentrate/conventional technology route. When compared to the production and sale of selective and/or high grade bulk concentrates to conventional processing plants (based on present day smelter terms), an on-site FCL process plant can result in a net increase in cash flow to the mine of 33% to 72% (after all FCL process capital and operating costs have been paid for), for throughputs in the range 40,000 to 100,000 tpy of zinc (see Table 7). This percent increase is relatively independent of metal prices.

In general therefore, the bulk concentrate/FCL process option should be seriously considered over conventional technology where production of a relatively low grade bulk concentrate can increase recoveries from the ore by at least 20%. The minimum grade and throughput for which the FCL process is viable will be determined primarily by economic rather than technical considerations. This is further discussed later in this report.

Typical Ores Amenable to the FCL Process

Typical analyses of Canadian complex sulphide ores and concentrates, which CANMET would consider amenable to FCL process treatment, are shown in Table 1. Typical chemical and mineralogical analyses of FCL bulk concentrates are shown in Tables 2 and 3 respectively. Typical ore to concentrate recoveries, based on present practice and testwork carried out at CANMET, are shown in Table 4. Because the FCL process bulk con-centrates have no stringent grade constraints, the recoveries of all valuable major metals are substantially higher than those obtained in selective concentrates and/or ISP bulk concentrates. The value of this additional recovery is shown in Table 5, based on the metal prices shown. The figures indicate that total metal values recovered in the FCL bulk concentrate are 29 to 37% higher than those obtainable if only selective concentrates are produced, and are at least 23% higher than those obtainable if

Table 1   Typical Ore and Concentrate Grades

	Zn (%)	Pb (%)	Cu (%)	Ag (g/t)	Au (g/t)
Ore	4.3-8.9	1.6-3.9	0.4-1.0	60-110	0-1.7
Selective Concentrates	49-51	29-34	22-24	500-2700	2-7
ISP Bulk Concentrates	32-34	13-18	0-1.5[1]	300-350	0-2.5
FCL Bulk Concentrates	25-31	7-11	0.8-6	145-300	0-2.4

[1]   Copper content is normally limited to 1.5% maximum

Table 2   Chemical Analysis of Bulk Concentrate (%)

Zn	Pb	Cu	Ag[1]	Fe	S_t	SO_4	oxides	As	Cd	Sb	Bi
29	8.7	0.8	235	19.5	35	4.5	1.5	0.4	0.05	0.1	0.04

[1]   g/tonne

Table 3   Calculated Mineralogical Analysis of Bulk Concentrates[1]

Mineral	Formula	Wt %
Sphalerite	ZnS	49
Pyrite	FeS_2	33
Galena	PbS	9.4
Anglesite	$PbSO_4$	2.0
Chalcopyrite	$CuFeS_2$	1.9
Siderite	$FeCO_3$	1.8
Quartz	SiO_2	0.90
Calcite	$CaCO_3$	0.54
Arsenopyrite	FeAsS	0.46
Dolomite	$MgCO_3 \cdot CaCO_3$	0.38
Tetrahedrite	$[Cu,Ag]_{12}Sb_4S_{13}$	0.25
Pyrrhotite	FeS	0.01

[1]   Based on mineralogical examination and chemical analysis

Table 4   Typical Recoveries – Ore to Concentrate (%)

	Zn	Pb	Cu	Ag	Au
Selective (Zn, Pb & Cu) Concentrates	70–75	50–60	50–60	50[2]	20–29[2]
ISP Bulk Concentrates	73–80	67–71	–[1]	55–57	25–34
FCL Bulk Concentrates	93–96	70–84	63–85	58–77	34–42

[1] Because of limits on copper content, copper normally must be rejected from ISP bulk concentrates

[2] In Cu + Pb concentrate only

Table 5   Typical Ore and Concentrate Values and Recoveries

	Value[1] $C/tonne of Ore Mined	% of Ore Value Recovered	% Increase (over selective) in Ore Value Recovered
Ore	166–278	100	–
Selective Concentrates Only	108–173	60–67	–
Selective + ISP Bulk Concentrate	193	75	12
ISP Bulk Concentrates Only	136–173	62–65[2]	3–8
FCL Bulk Concentrates	139–237	82–92	29–37

[1] Based on Metal Prices as follows ($C)
Zn   1.00/lb
Pb   0.32/lb
Cu   1.60/lb
Ag   7.20/oz
Au   435/oz

[2] Recovery does not include copper because of Cu limit in ISP bulk concentrate

an ISP bulk concentrate is produced. This percent increase is relatively independent of metal prices.

The net increase in recovery for the FCL process, versus conventional technology, is of course dependent upon the recovery in the process itself. Table 6 shows expected recoveries from concentrate in the FCL process, versus conventional processes for zinc, lead and copper. The recovery of major metals (Zn, Pb, Cu) is expected to be equal to or greater than with conventional technology. However, silver recovery is expected to be slightly lower and gold recovery is assumed to be zero, since the FCL process, at its present state of development, does not recover gold. Any gold present in the FCL bulk concentrate will end up in the leach residue, along with unreacted pyrite and the elemental sulphur produced in the process. Recovery of this gold will require additional processing of the leach residue in an economically and environmentally attractive manner, and preferably should include recovery of saleable elemental sulphur. Alternatives to achieve these objectives are presently being investigated.

Table 6   Processing Recoveries Assumed for Economic Assessment (%)

	FCL Process	Conventional Processes			
	Bulk Conc.	Zn Conc.	Pb Conc.	Cu Conc.	ISP Bulk Conc.
Zn	96	95	–	–	95
Pb	96	–	95	–	95
Cu	95	–	–	95	–
Ag	95	–	96	96	96
Au	0	–	96	96	96

Mine Revenues for Selective and Bulk Concentrates

Expected mine revenues for four different ores, based on "paid for" metal values using conventional smelter charges (Table 14) for selective and/or ISP bulk concentrates, are shown in Table 7 for comparison with "paid for" metals from the FCL process. Charges for the latter are based on the capital and operating costs documented in Table 10 and Figure 2. The figures indicate that by using the FCL process to treat a low grade (30 to 40% combined metals) bulk concentrate, mine revenues can be increased by 33 to 72% over that obtained for the selective concentrate/conventional technology route. If the latter route involves the production of an ISP bulk concentrate the mine revenue increase for the FCL process is still expected to be in the range 47-66%. Again, this percentage increase for FCL is relatively insensitive to metal prices.

It should be noted that the increased mine revenues in Table 7 do not include savings in the mill from producing bulk rather than selective concentrates. Kilborn (3) has estimated this savings at nearly $2.50/tonne of ore for operating costs alone. In addition the FCL process costs include the cost of flotation for producing the bulk concentrate. To compare directly to the conventional technology route this cost should be omitted and would result in an additional increase in mine revenue for the FCL option of about 2%.

FCL Process Capital and Operating Costs

Capital and operating costs for a FCL plant capable of producing 100,000 tpy zinc from 351,000 tpy of bulk concentrate have been estimated by Kilborn Ltd. (3) and are shown in Tables 8 and 9 respectively. Costs include a flotation plant to produce the bulk concentrate but do not include sulphur recovery. Capital costs assume a "greenfields" plant and

include a 25% contingency. Operating costs are based on labour, energy and reagent costs in eastern Canada (New Brunswick).

CANMET estimates of FCL process capital and operating costs, based on up-dating and in some cases modifying the original Kilborn estimates, are also shown in Tables 8 and 9 for comparison. The increased operating cost is primarily due to higher labour burden, maintenance and plant overhead used by CANMET. The lower capital cost is primarily due to lower equipment costs in the SX circuits (based on a quotation from Krebs), and to lower building and insulation costs in the leach and crystallization circuits. CANMET costs are for January 1990, while Kilborn costs are January, 1989.

Effect of Throughput on Capital and Operating Costs

To determine the effect of throughput on FCL process economics, Kilborn and CANMET have calculated capital and operating costs for zinc production rates down to 25,000 tpy. The effect of zinc throughput on capital and operating costs is shown in Table 10 and plotted graphically (in $/annual tonne of zinc) in Figure 2. These FCL process costs are used in Table 7 to determine mine revenues for ore deposits where throughputs are less than 100,000 tpy zinc.

Although zinc throughput is very important to FCL process economics, the effect of by-product credits also has to be considered. To evaluate this effect, the relationship of FCL process costs to total revenue for different throughputs is plotted in Figures 3 and 4 (CANMET estimates only). Decreasing total revenue and/or throughput by a factor of four increases operating costs from 20% of annual metal values recovered to 34%. This percentage range is considerably lower than the 26% to 44% range obtained if only zinc values are considered and indicates that FCL operating costs can be maintained at less than 25% of metal

259

Table 7   Net Mine Revenues for Selective and Bulk Concentrates
($C/tonne of ore mined)

	166[1]		257		220[1]		278[1]	
Ore Value	166[1]		257		220[1]		278[1]	
CONVENTIONAL PROCESS								
Concentrate Value — Selective	108[1]		173	173	−		−	
Concentrate Value — ISP Bulk	−		−	20	145[1]		176[1]	
Total Concentrate Value	108[1]		173	193	145[1]		176[1]	
Smelter Charges	42		73	83	73		88	
Transportation Charges	2		4	4	3		3	
Net Mine Revenue	64[1]		96	106	69[1]		85[1]	
FCL PROCESS								
Concentrate Value — FCL Bulk Concentrate	139[1]		237	237	181[1]		227[1]	
Net Value[2]	130		227	227	165		208	
	Kilborn	CANMET	Kilborn	CANMET	Kilborn	CANMET	Kilborn	CANMET
Cu/Ag Smelting/Refining Charges[3]	13.6	9.2	4.3	2.9	3.7	2.5	3.7	2.5
FCL Process Operating Costs	21.9	26.0	41.5	46.9	37.1	45.7	44.7	53.0
FCL Process Capital Costs[4]	9.1	8.7	15.9	14.8	16.2	15.6	18.8	18.0
Total FCL Process Costs	44.6	43.9	61.7 to 64.6		57.0	63.8	67.2	73.5
Net Mine Revenue	85.4	86.1	165.3 − 162.4		108.0	101.2	140.8	134.5
% Increase in Revenue for FCL	33 to 35		69–72	53–56	56 to 47		66 to 58	
Mining Rate $(10^6$ t/y)	1.5		1.24	1.24	0.7		0.7	
Zinc Production $(10^3$ t/y) — Conventional Process	46		75	83	34		44	
Zinc Production — FCL process	57		100	100	41		53	

[1] Includes gold values
[2] Based on recoveries in FCL process
[3] Includes all treatment and transportation costs. Note: Kilborn costs based on 50% copper product; CANMET costs based on 90% copper product
[4] Based on 15 year capital write-off

revenues at zinc throughputs down to at least 50,000 tpy. The higher the by-product credits, the lower the zinc production required for economic viability. This is exemplified by the point in Figure 3, located below the "total metal value" curve, where operating costs at 57,000 tpy zinc are the same as at 100,000 tpy zinc (20% of metal revenue). The reason for this is that this particular bulk concentrate contains considerably more copper, which results in a higher by-product credit and reduces the economic dependence on zinc production.

Figure 4 shows the relationship of capital costs to metal values recovered. The trends are similar to the operating cost curves, with capital requirements increasing from 98% of annual metal values (125% of zinc value) to 169% of annual metal values (218% of zinc value) for a factor of four decrease in total revenue and/or zinc throughput. The beneficial effect of the high copper bulk concentrate is again indicated by the point below the general curve, showing capital requirements per unit of revenue remaining relatively constant over the range 100,000 to 57,000 tpy of zinc (yearly revenues of $282M to $195M). Figures 3 and 4 therefore aptly demonstrate the importance of by-product credits to FCL process economic viability, especially at zinc production rates of 50,000 tpy or less.

Table 8   FCL Process Capital Costs ($)

	Kilborn	CANMET
Bulk Flotation	6,100,803	5,873,817
FCL Leach	15,966,175	10,700,219
Oxyhydrolysis	967,532	1,342,991
Zinc SX	17,790,070	12,697,163
Iron SX	6,790,969	5,033,201
Pb Crystallization	10,256,290	7,490,144
Fe Cementation	2,516,040	1,584,929
Zn Purification	2,591,066	2,011,222
Pb Electrolysis/Casting	9,984,404	9,780,462
Zn Electrolysis	62,428,488	62,137,709
Zn Casting	11,587,488	14,349,979
Plant Subtotal	146,979,325	133,001,836
Facilities	15,301,000	10,810,483
Utilities	3,668,096	8,107,862
Oxygen, Chlorine	8,609,600	8,976,197
Tailings and Environment	3,652,600	3,619,695
Total Direct Costs	178,210,621	164,516,073
Total Indirect Costs	43,822,700	41,129,018
Contingency	56,256,400	51,411,273
Total Capital	278,289,721	257,056,365
Working Capital	17,423,201	17,993,946
Total Investment	295,712,922	275,050,310

Table 9   FCL Process Operating Costs

	Kilborn		CANMET	
	$C/y	$/t ore	$C/y	$/t ore
Bulk Flotation	2,011,406	1.62	2,662,262	2.15
Leach	8,990,234	7.26	9,291,795	7.50
Crystallization	969,965	0.78	989,422	0.80
Lead Electrowinning	3,075,010	2.48	3,178,325	2.57
Cementation	868,675	0.70	971,155	0.78
Solvent Extraction	2,131,044	1.72	1,535,193	1.24
Oxyhydrolysis	3,221,844	2.60	4,428,181	3.57
Zn Electrolyte Purification	1,819,000	1.47	2,176,120	1.76
Zinc Electrowinning	23,074,600	18.63	25,782,645	20.81
Zinc Casting	2,273,975	1.83	2,237,057	1.81
Water	72,000	0.06	72,000	0.06
General Administration	1,888,000	1.52	2,766,550	2.23
Laboratory	1,025,400	0.83	2,000,000	1.61
TOTAL	51,421,153	41.50	58,090,705	46.89

261

Table 10   FCL Process Capital and Operating Costs versus Throughput

Zinc Production (t/y)	Capital Costs ($Cx10^6)		Operating Costs ($Cx10^6/y)	
	Kilborn	CANMET	Kilborn	CANMET
25,000	122	120	19.2	23.6
50,000	190	181	30.1	35.9
75,000	–	231	–	47.2
100,000	296	275	51.4	58.1

Financial Analysis

A financial analysis was carried out by Kilborn (3) for an FCL plant produc-ing 100,000 tpy of zinc from a bulk concentrate containing 30% Zn, 10% Pb, 1% Cu and 278 g/t Ag. The financial analysis was based on a 15 year plant life and 100% equity financing and included a comparison with conventional technology, assuming feed from the same ore body and a mining rate of 1.24 million tonnes/y. The conventional plant produced zinc, lead and copper concentrates, and an ISP bulk concen-trate. The copper and ISP bulk concen-trates were sold to existing smelters. The zinc and lead were processed on-site; the zinc in a conventional roast-leach-electrowin plant and the lead in a Kaldo TBRC and conventional thermal refinery. A common acid plant captured SO_2 gases from both zinc and lead operations.

Physical throughputs, grades and recoveries assumed for the financial analysis are shown in Table 11. A summary of the capital and operating costs used for the financial analysis is shown in Table 12. Mining and milling costs are common to both FCL and con-ventional plants, and are typical of eastern Canadian underground base metal mines. Costs for the conventional zinc and lead processing plants were provided by Kilborn.

A summary of the financial analysis is shown in Table 13. Net smelter revenues were based on the metal prices document-ed in Table 5. Payments for the ISP and copper concentrates in the conventional plant and for the Cu/Ag cement product in the FCL plant were calculated using present day smelter contract terms (Table 14). The financial analysis indicates a DCF ROI before taxes of 36.3% for the FCL plant versus 22.8% for the conventional plant.

Sensitivity Analysis

The effect of variances in capital and operating costs and total revenue on return on investment for the FCL plant is shown in Figure 5. As expected, revenue has the greatest effect and zinc price is the dominant factor, as shown in Figure 6. However the influence of zinc price on ROI will depend upon the composition of the bulk concentrate, and will not be as pronounced if higher value by-products (such as Cu and Ag) are present in greater amounts.

Effect of Throughput on Economic Viability

The effect of concentrate throughput and/or zinc production rate on DCF ROI is shown graphically in Figure 7 for production rates down to 25,000 tpy of zinc, using both Kilborn and CANMET FCL process cost estimates. Because the higher CANMET operating costs are off-set by lower capital costs the DCF ROI's are very similar regardless of whether Kilborn or CANMET cost estimates are used. The figures indicate a decrease in ROI from 37% at 100,000 tpy zinc to about 15% at 25,000 tpy zinc. It would appear that to maintain an ROI of at least 20%, zinc production should be at least 35,000 tpy, or equivalent to a net smelter revenue of at least $96M/y. This would indicate that the gross value of FCL bulk concentrate should be about $100M/y for the FCL process to be econ-omically attractive, with zinc accounting for about 75% of this value. If by-product credits amount to more than 25% of total value, then economic viability will be less dependent on zinc throughput and the process could be economically attractive at lower zinc production rates. However, since at least 70% of the capital costs and 55% of the operating costs are zinc related, zinc throughput rates must be high enough to justify these costs. Under these circumstances it is unlikely that the FCL process will be economically attractive at zinc production rates below about 30,000 tpy even if by-product credits are very high.

Table 11   Physical Throughputs, Grades and Recoveries for Financial Analysis

Ore Production: 1,239,000
Ore Grade: 8.85% zinc
3.58% lead
0.36% copper
105 g/t silver

	FCL Process Bulk Conc.	Conventional Process				
		Zn Conc.	Pb Conc.	Cu Conc.	ISP Bulk Conc.	TOTAL
Concentrate Production (t/y)	351,120	158,760	66,360	10,920	26,040	262,000
Concentrate Grade						
Zn (%)	30	50	4.5	3	33	
Pb (%)	10.4	2	34	7	18	
Cu (%)	1.0	0.3	0.6	23	0.7	
Ag (g/t)	278	90	495	2950	350	
Flotation Recovery						
(ore to conc.) Zn (%)	96	72	3	0.3	8	83.3
Pb (%)	82	7	50	2	11	70
Cu (%)	78	10	9	58	4	81
Ag (%)	75	11	25	25	7	68
Metal in Concentrate						
Zn (t/y)	105,230	79,253	3,100	300	8,552	91,206
Pb (t/y)	36,376	3,100	22,397	750	4,700	30,962
Cu (t/y)	3,476	400	400	2,574	200	3,587
Ag (kg/y)	97,646	14,500	32,941	32,159	9,114	88,662
Process Plant Recoveries						
(physical only) Zn (%)	96	95	–	–	95	
Pb (%)	96	–	95	–	95	
Cu (%)	95	–	–	95	–	
Ag (%)	95	–	97	97	97	
Metal Produced						
Zn (t/y)	101,020	75,290	–	–	8,124	83,414
Pb (t/y)	34,920	–	21,277	–	4,465	25,752
Cu (t/y)	3,300	–	–	2,445	–	2,445
Ag (kg/y)	92,765	–	31,953	31,195	8,841	71,988
Overall Physical Recovery from Ore						
Zn (%)	92.1					76.1
Pb (%)	78.7					58.0
Cu (%)	74.0					54.8
Ag (%)	71.3					55.3
Gross Metal Value Recovered ($Cx1000/y)	282,000					224,000

CONCLUSIONS

The CANMET FCL process is a technically viable and economically attractive alternative for the treatment of low grade base metal sulphide bulk concentrates, and can achieve very high recoveries of the valuable metals (Zn, Pb, Cu, Ag) in the form of high value products. An economic evaluation based on engineering feasibility studies and capital and operating costs generated by Kilborn and CANMET indicates that, at present day metal prices, the return on investment for an FCL plant producing 100,000 tpy of zinc is about 37% before taxes. This compares to about 23% for a conventional plant treating the same amount of ore over the same mine life. The evaluation also indicates that the FCL process has the potential to be economically attractive at zinc production rates as low as 30,000 to 35,000 tpy.

263

Table 12   Capital and Operating Cost Summary for Financial Analysis

CAPITAL COSTS ($Cx1000)	FCL Plant	Conventional Plant
Common Costs		
Mine Development	20,000	20,000
Mine Equipment	16,000	16,000
Replacement Equipment[1]	16,000	16,000
Crushing and Milling	16,000	16,000
Total Common Costs	68,000	68,000
Processing Costs		
Concentrator	6,000	26,000
Processing Plant	172,000	203,000
Acid Plant	–	27,000
Total Direct Costs	178,000	256,000
Total Indirect Costs	44,000	59,000
Contingency	57,000	47,000
Working Capital	17,000	17,000
Total Processing Cost	296,000	379,000
Total Investment	364,000	447,000

OPERATING COSTS ($Cx1000/y)			
Common Costs	($C/t)		
Mine	25.0	31,000	31,000
Mill	8.0	10,000	10,000
Admin.	0.5	500	500
Total Common Costs	33.5	41,500	41,500
Processing Costs			
Concentrator		2,000	5,000
Processing Plant		49,500	38,100
Acid Plant		–	3,200
Total Processing Costs		51,500	46,300
TOTAL COSTS		93,000	87,800

[1] In years 7 and 8

Table 13   Summary of Financial Analysis

	FCL Plant	Conventional Plant
Ore Mined (t/y)	1,239,000	1,239,000
Concentrate Produced (t/y)	351,000	262,000
Zinc Produced[1] (t/y)	101,000	82,000
Capital Cost ($C)	364 M	447 M
Net Smelter Revenue ($C/y)	275 M	219 M
Operating Costs ($C/y): Mining and Milling Processing	93 M (41.5 M) (51.5 M)	87.8 M (41.5 M) (46.3 M)
Cash Flow Before Taxes ($C/y)	182 M	131 M
DCF ROI (%)	36.3	22.8

[1] Zinc metal paid for

Table 14   Smelter Contract Terms

COPPER CONCENTRATE AND COPPER/SILVER CEMENT		
Concentrate:	Treatment Charge ($US/t)	107.00
Copper Payment:	Metal Paid Minimum Deduction (units) Refining Costs ($US/lb)	100% 1% 0.078
Silver Payment:	Metal Paid Minimum Deduction (g/t) Refining Costs ($US/kg)	93% 30 8.40
LEAD CONCENTRATE		
Concentrate:	Treatment Charge ($US/t)	157.00
Lead Payment:	Metal Paid Minimum Deduction (units)	95% 3%
Silver Payment:	Metal Paid Minimum Deduction (g/t) Refining Costs ($US/kg)	95% 31 8.40
ZINC CONCENTRATE		
Concentrate:	Treatment Charge ($US/t)	234.00
Zinc Payment:	Metal Paid Minimum Deduction (units)	85% 8%
ISP BULK CONCENTRATE		
Concentrate:	Treatment Charge ($US/t)	231.00
Zinc Payment:	Metal Paid Minimum Deduction (units)	80% 7%
Lead Payment:	Metal Paid Minimum Deduction (units)	95% 3%
Silver Payment:	Metal Paid Minimum Deduction (g/t) Refining Costs ($US/kg)	75% 90 8.40

265

In addition, because of higher recoveries from the resource base, the bulk concentrate/FCL process option can increase mine revenues by 33 to 72% over those obtainable using conventional technology. This percentage increase is relatively independent of metal prices. In all cases where a substantial increase in recovery (>20%) is achievable, the bulk concentrate/FCL process option is considerably more economically attractive than conventional technology.

To verify the potential attractiveness of the FCL process, CANMET is actively seeking industry partners to build and operate a demonstration plant to confirm the engineering criteria and economic feasibility, and to resolve any operating problems that arise from continuous operation of a fully integrated process flowsheet. Initial estimates indicate that the cost of such an undertaking will be in the range of $5 M to $10 M, depending upon the size of the demonstration plant.

ACKNOWLEDGEMENTS

The assistance of Lorna Paquette and Ron Molnar in the preparation of this publication is gratefully acknowledged.

REFERENCES

1.   W.J.S. Craigen and J.R. Schnarr, "Economic Incentives for the Production of Bulk Base Metal Sulphide Concentrates";   Complex Sulphides TMS-AIME-CIM San Diego, California, November 10-13, 1985;   CANMET MSL 85-130(OP).

2.   W.J.S. Craigen and CANMET/MSL Staff, "The CANMET Ferric Chloride Leach Process for the Treatment of Bulk Base Metal Sulphide Concentrates";   CANMET MSL 89-67(OP), June 1989.

3.   Kilborn Ltd., "Marketability Assessment of the CANMET Ferric Chloride Leach Process";   November 1989.

Figure 1
CANMET FCL PROCESS FLOW DIAGRAM

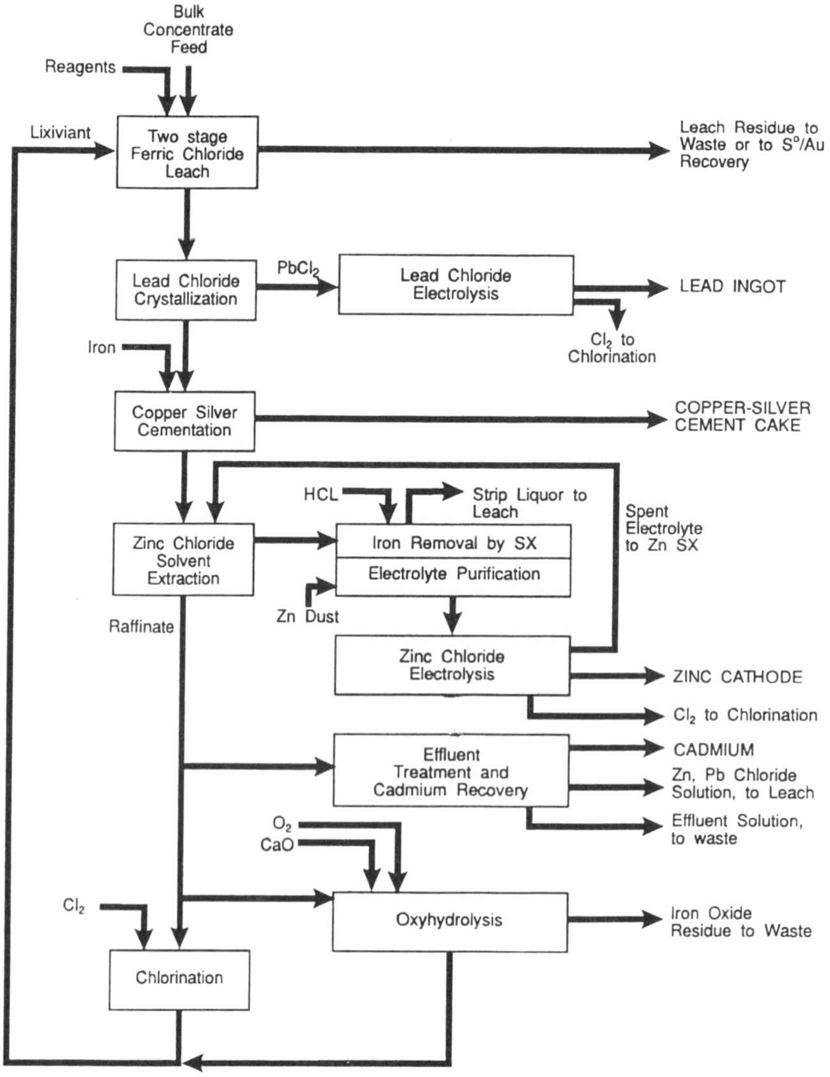

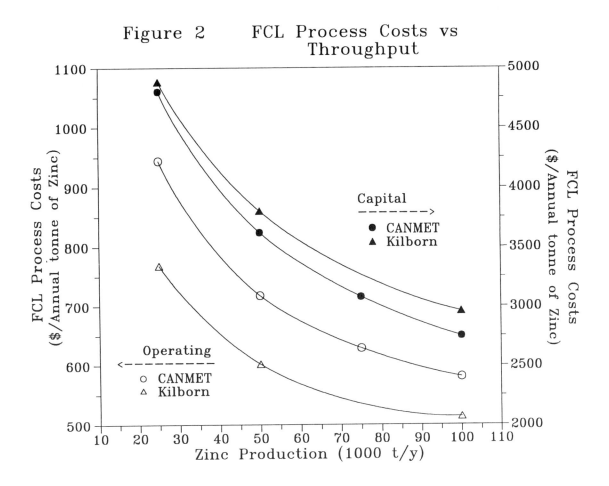

Figure 2   FCL Process Costs vs Throughput

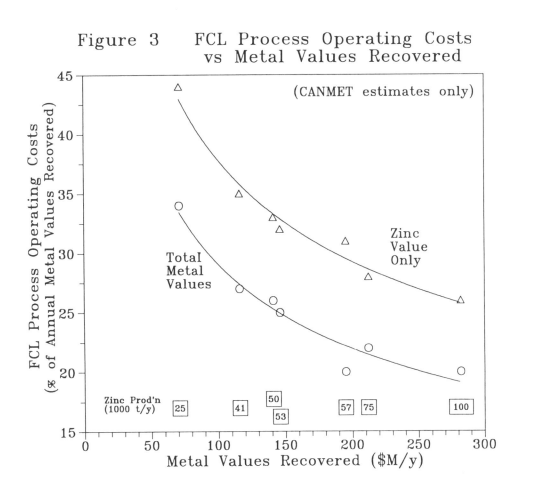

Figure 3   FCL Process Operating Costs vs Metal Values Recovered

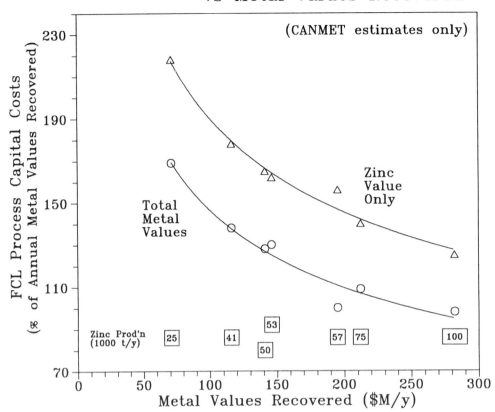

Figure 4　FCL Process Capital Costs
vs Metal Values Recovered

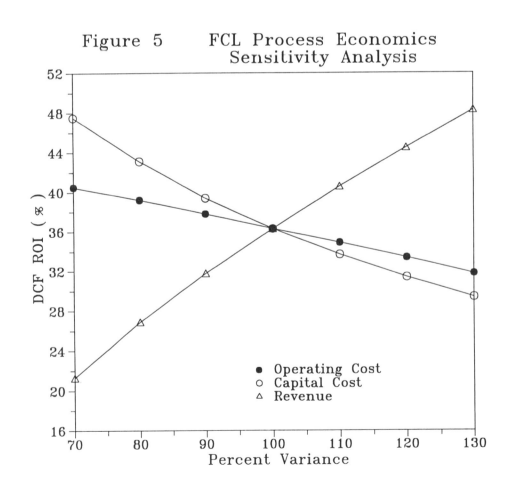

Figure 5　FCL Process Economics
Sensitivity Analysis

268

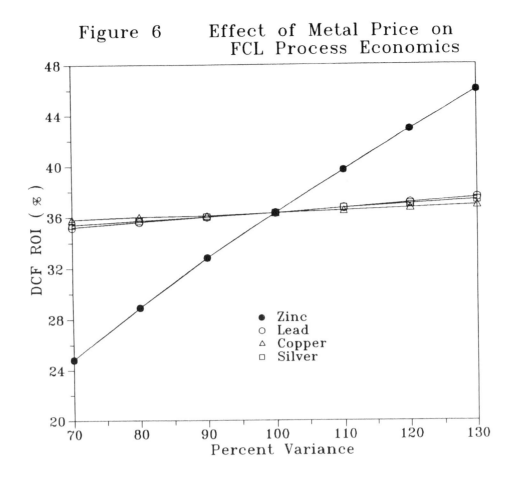

Figure 6   Effect of Metal Price on
            FCL Process Economics

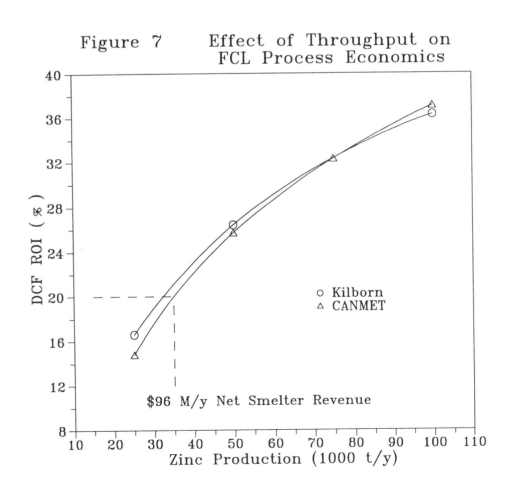

Figure 7   Effect of Throughput on
            FCL Process Economics

Lead production from high-grade galena concentrates by ferric chloride leaching and molten-salt electrolysis

J.E. Murphy
Reno Research Center, Bureau of Mines, U.S. Department of the Interior, Reno, Nevada, U.S.A.
M.M. Wong
UNOCAL, Science and Technology, Brea, California, U.S.A.

SYNOPSIS

The U.S. Bureau of Mines investigated an alternative process to smelting for the production of lead metal. The process eliminates sulfur dioxide generation and minimizes particulate lead emissions. The process involves ferric chloride leaching of galena concentrate to produce lead chloride, followed by molten-salt electrolysis of the lead chloride to yield lead metal and chlorine. The chlorine is used to regenerate ferric chloride in the leaching solution.

The process was tested on both a bench scale and a larger scale which had a capacity of 225 kg of lead metal per day. The large-scale leaching and electrolysis tests were performed in three 10-day and six 5-day campaigns, during a period spanning 28 mo. The study was carried out using a southeastern Missouri lead concentrate. Typical lead recovery was 98%. Process problems related to impurity buildup and control were studied. When copper and silver impurities were removed from the leaching solution, lead metal purity increased to 99.999%. Lead-in-air levels and lead-in-blood levels of operating personnel were monitored.

Large-scale electrolysis tests were performed in a 3,000-A cell with horizontal, flat plate electrodes. Electrolysis proved to be the most expensive step in the process because the energy requirement in the large cell was 1.32 $kW \cdot h/kg$. Subsequently, the cell was operated with a more efficient electrode design, and the energy requirement was decreased to 0.66 $kW \cdot h/kg$.

INTRODUCTION

Primary lead is commercially produced from lead sulfide concentrates by a smelting process consisting of sintering, blast furnace reduction, and refining. The pyrometallurgical method is low cost and requires relatively little energy, but generates sulfur dioxide and particulate lead, which must be controlled to prevent air pollution. Because of concern about acid rain and difficulties in meeting regulations for lead exposure, there has been considerable interest in hydrometallurgical alternatives to the smelting process. A major part of the research effort has involved chloride leaching, much of which was reviewed by Dutrizac.[1-2]

The application of ferric chloride leaching to treating galena concentrates has been the subject of several investigations.[3-7] James and Bounds reported on St. Joe Minerals Corp. operation of a 2-ton-per-day pilot plant which employed continuous ferric chloride leaching.[6] High-grade galena concentrate was continuously fed to the leaching tanks as a slurry in the operation. Very pure $PbCl_2$ was produced, and the operation proved to be safe for the lead worker and for the environment. However, the process "did not prove to be economically competitive" with smelting. The investigators concluded that the most promising application for ferric chloride leaching would be in treating complex concentrates. Demarthe and Georgeaux of Minemet Recherche discussed the application of ferric chloride leaching to both high-grade and complex concentrates.[7] The economic advantages of treating a bulk concentrate produced from complex ores was discussed by Craigen and Schnarr.[8]

The U.S. Bureau of Mines investigated ferric chloride leaching of a high-grade concentrate obtained from St. Joe Mineral Corp. The concentrate was from south-eastern Missouri, a region which accounts for most of the U.S. lead production and which represents one of the world's major lead deposits. The Bureau's process consists of leaching galena concentrate with ferric chloride-sodium chloride solution,

$$PbS + 2FeCl_3 \xrightarrow{NaCl} PbCl_2 + 2FeCl_2 + S°,$$

cooling the solution to crystallize $PbCl_2$, electrolyzing the lead chloride in a molten-salt bath to produce lead metal and chlorine,

$$PbCl_2 \xrightarrow{Electrolysis} Pb + Cl_2,$$

and using the chlorine to regenerate ferric chloride in the leaching solution,

$$2FeCl_2 + Cl_2 \rightarrow 2FeCl_3.$$

After a preliminary study of the method[3], bench-scale investigations were performed on the leaching[4] and electrolysis steps[9], followed by a preliminary evaluation on the

process economics[10]. The process was then
tested on a larger-scale (225 kg of lead per
day) with integrated leaching and electrolysis
operations. A 3,000-A cell was used for
electrolysis. Four lead producers joined the
Bureau in this effort on a cost-sharing
basis.[5]

The objective of the investigation was to
identify possible problems in an integrated
operation of the process. Lead levels in the
workplace and worker health hazards were
monitored closely during the operation.

After the larger scale operation was
terminated, the leaching step was performed on
a continuous basis on a bench scale to further
streamline the process.[11] Subsequently, the
3,000-A cell was operated with an improved
electrode design which dramatically decreased
the energy required for electrolysis.[12] The
emphasis of this paper is to update the
earlier larger scale tests especially with the
improvements in the molten-salt electrolysis.

MATERIALS, EQUIPMENT, AND OPERATING PROCEDURE

Lead concentrate

Analyses of two lots of the concentrates are
given in Table I, and a wet screen analysis is
presented in Table II. The lead concentrate
contained galena, chalcopyrite, sphalerite,
marcasite, silver associated with zinc,
cadmium in sphalerite, and nickel and cobalt

Table I Analysis of lead concentrates, percent

	Lot 1	Lot 2[1]
Ag..........................	0.006	0.012
As..........................	.01	.08
Ca..........................	.35	.37
Cd..........................	.02	.02
Co..........................	.04	.04
Cu..........................	1.5	1.2
Fe..........................	3.7	3.9
Mg..........................	.2	.3
Ni..........................	.07	.06
Pb..........................	73.5	74.2
Sb..........................	.002	([2])
Zn..........................	1.6	1.6
SO$_4$.......................	2.4	1.2
Total S.................	16.1	15.7

[1]Used in campaigns 7 and 8.
[2]Not determined.

Table II Screen analysis of lead concentrate[1]

Size fraction, mesh[2]	Percent
Plus 65................................	1.4
65 by 100..............................	2.1
100 by 200.............................	9.8
200 by 325.............................	20.0
Minus 325..............................	66.7

[1]Wet screen analysis of lot 1 lead concentrate.
[2]Tyler standard sieves.

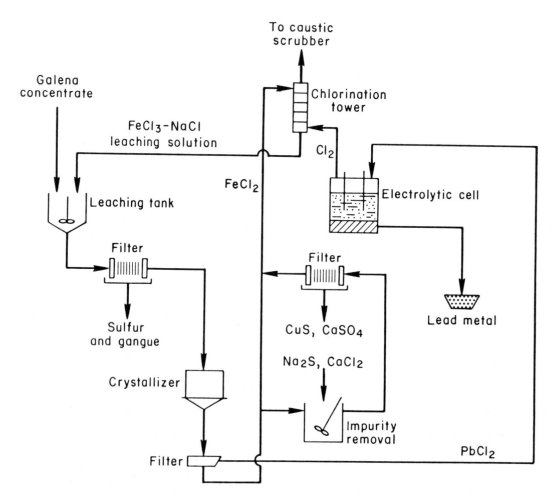

Fig. 1 Flow diagram of ferric chloride leaching, molten-salt electrolysis procedure.

272

in siegenite. The gangue material contains a small quantity of dolomite. All analyses in this paper were by atomic adsorption, inductively coupled plasma optical emission spectroscopy, or wet chemical methods.

Equipment

A schematic flow diagram for the process is shown in Figure 1. In the large-scale tests, the leaching circuit was sized to treat six 57-kg batches of flotation concentrate per day. The six leaching cycles produced enough lead chloride to yield about 225 kg of lead metal. An overall view of the experimental unit is shown in Figure 2. The leaching section, which occupied two levels, is located on the right and the electrolytic cell on the lower left.

The leaching vessel was a 1,900 L, coned-bottom steel tank lined with polyvinylidene fluoride, thermally insulated and equipped with a steam-heated titanium coil and a titanium stirrer. The concentrate was fed from a hopper by a screw feeder into the tank. The slurry was discharged through an air-activated, straight-through, pinch-type chlorobutyl-lined cast iron valve and was transferred by a spring-activated, fluorinated-rubber-lined diaphragm pump to a

filter press. A Teflon* expansion joint and a flexible stainless steel pipe section with a polyvinylidene fluoride liner were also installed in the line to absorb the vibrations of the spring-activated pump.

The polypropylene plate and frame filter press were dressed with polypropylene felt cloths. Hot pregnant solution from the filter was transferred through a chlorinated polyvinyl chloride (CPVC) pipe to a crystallizer, which was a 1,500 L, coned-bottom polypropylene-lined steel tank equipped with a water-cooled titanium coil and a titanium stirrer. The slurry was discharged through an air-activated, pinch-type fluorinated-rubber-lined valve. The line connecting the leaching tank, filter press, and crystallizer was thermally insulated.

Lead chloride was recovered in a fiber-reinforced plastic (FRP) vacuum pan filter dressed with a polypropylene cloth. The spent solution was transferred to a preparation tank (not shown in Fig. 1) by a magnetically driven, polypropylene centrifugal pump. The preparation tank was a 1,500-L, coned-bottom, polypropylene-lined, thermally insulated steel

---

*Reference to specific products does not imply endorsement by the Bureau of Mines.

Fig. 2 Integrated leaching-electrolysis plant.

tank, equipped with a steam-heated titanium coil and a titanium stirrer. The crystallizer, pan filter, and preparation tank were connected by CPVC pipes, fittings, and ball valves. Liquor from the preparation tank was discharged into a 2,300-L FRP surge tank (not shown in Fig. 1) by gravity through an air-activated, polypropylene-lined, Teflon diaphragm valve.

The surge tank was equipped with a steam-heated titanium coil and a titanium stirrer. A stream of solution from the surge tank was circulated by a neoprene flexiliner pump through a chlorination tower. The lines between the surge tank and the chlorinator were made of CPVC, and the solution flow rate was measured by an ultrasonic Doppler flowmeter clamped on the line.

The chlorinator was a 30-cm-diam by 3-m-high FRP tower, packed with 35 cm of 7.6-cm polypropylene saddles and 1.2 m of 2.5-cm Glitch butterflies and equipped with a polypropylene mist eliminator and a polypropylene spray nozzle. Chlorine adsorption by the stripped solution was usually more than 98%. To ensure that the gas discharged to the atmosphere was free of chlorine, another tower of identical construction to the chlorinator was connected in series with the gas stream from the chlorinator and used caustic solution for scrubbing. The gas stream was drawn through both towers by a 0.75 kW blower.

All the tanks in the leaching section were covered and the fumes were vented through a water scrubber consisting of a 30-cm-diam by 3-m-high FRP tower packed with 1.5 m of 7.6-cm polypropylene saddles.

The $PbCl_2$ from the pan filter was dried in sulfon-X trays in an electric oven with a stainless steel interior and a horizontal air flow of 5 L/sec. A Burr crusher was used to break up the lead chloride lumps and a coned-bottom, 60-cm-diam stainless steel container with a clamp-on lid and a polyvinyl chloride (PVC) ball valve at the bottom was used to transport the lead chloride to the feed hopper of the electrolytic cell.

The electrolytic cell construction is shown in Figure 3. The cell exterior dimensions were 135 cm long, 112 cm wide, and 76 cm high. The cavity was 86 cm long, 64 cm wide, and 46 cm deep. The inside walls were constructed of silica bricks, and the lid was constructed of cement and low-density aggregate. Two graphite plate anodes, 35.6 by 61 by 7.6 cm, were threaded and were attached to two 15-cm-diam graphite rods which were connected to a busbar. The graphite rods above the anode plates were protected from air oxidation by mullite sleeves. Another graphite plate, 74 by 61 by 5.1 cm, was supported by four graphite blocks which were partially immersed in a pool of molten lead metal. Under the molten lead metal pool, a steel bar protruding through the bottom of the cell was cathodically connected to a 5,000-A rectifier. The anode plates and the cathode plate were separated by 1.9-cm silicon oxynitride spacers. The surfaces of the anode and cathode plates were grooved with six 0.64-cm-wide slanted channels to guide the flow of chlorine to one side of the cell and lead metal to the opposite side.

The electrodes were subsequently modified as shown in Figures 4 and 5 to decrease the energy required for electrolysis.[12] The new

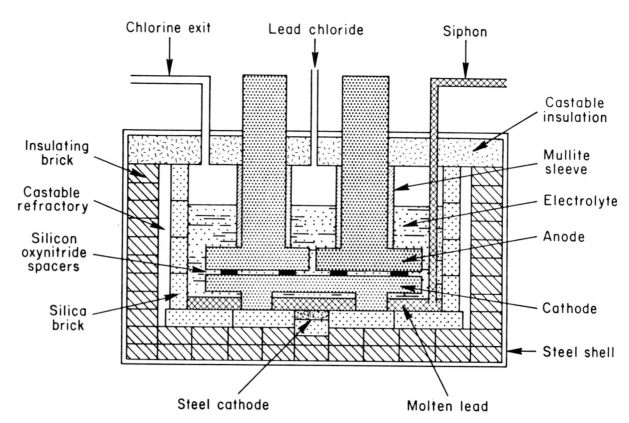

Fig. 3 Electrolytic cell with horizontal, flat-plate electrodes.

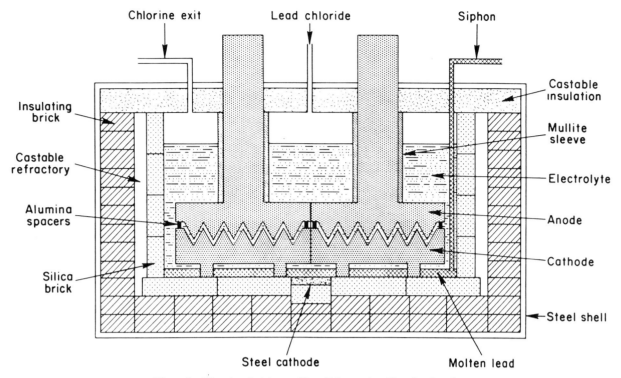

Fig. 4 Electrolytic cell with sawtooth electrodes.

electrode assembly, termed the sawtooth design, had triangular grooves with 60 degree anglcs. The depth of the grooves was 5.1 cm. The anodes had 1.27-cm-diam holes cut in the top of the trough and spaced 3.9 cm apart to permit chlorine escape.

The cathodes were similar to the anodes, but without holes. The lead metal formed on the cathode, moved down the side of the sloping well to the bottom of the trough, and flowed along the trough to the end of the cathode.

Lead chloride was added to the cell from the feed hopper by a screw feeder. Chlorine from the exit port was diluted to 30 vol% with air through an adjustable draft outside the cell and drawn through an FRP pipe to the bottom of the chlorination tower. Chlorine from a bottle provided a backup source when needed and compensated for the losses of chlorine in the exhaust from the chlorination tower and extraction of metals other than lead in the circuit. Lead metal produced in the electrolytic cell was siphoned through a removable Pyrex tube connecting the cell bottom to a vacuum chamber in which a mold was used to collect the metal. Two removable graphite ac heating electrodes (not shown in Figs. 3 and 4) were used to keep the electrolyte and lead metal molten when the cell was idle.

Operating procedure

A 57-kg batch of lead concentrate (55-kg on dry basis) was leached at 95°C in 1,135 L of solution initially containing 73 g/L $FeCl_3$, 254 g/L NaCl, and sufficient HCl to give a pH of about 0.3. After a 15-min leach, the slurry was filtered. Air was blown through the filter cake after each filtration. The residues accumulated after three to four batches were washed with a steam-water mixture

Fig. 5 Sawtooth electrode assembly.

before being removed from the filter press. Each load of residue was washed with 75 to 115 L of water at 4 L/min.

Hot pregnant solution from the filter was transferred to a crystallizer in which lead chloride precipitated when the solution was cooled to about 20°C. The lead chloride crystals were separated from the solution in a vacuum pan filter and washed with about 75 L of water followed by drying in an electric oven at 150°C for 6 h. The dry $PbCl_2$ was crushed and placed in a portable, closed container inside a fume hood. The container was connected to the feed hopper of the electrolytic cell.

The spent solution from the pan filter was transferred to a preparation tank, heated to 70°C, and then transferred to a surge tank. The solution was maintained at 70° to 75°C in the surge tank. Makeup water besides the water used in washing the residue and $PbCl_2$ was added to the circuit to compensate for the water losses such as entrainments in the residues and $PbCl_2$ and vapors from the tanks and chlorination tower.

The spent solution from the pan filter was periodically transferred to the leaching tank for copper and sulfate removal. Copper and silver were removed from the spent leaching solution with Na_2S. Enough Na_2S was added to decrease the oxidation-reduction potential to 260 mV when measured with a Ag-AgCl reference electrode (Eh = 482 mV). Copper and silver sulfides, precipitated as fine black particles, were removed from the solution by filtration in the filter press. The sulfides were washed with steam and water prior to being emptied from the filter press. Sulfate was removed from the spent solution with $CaCl_2$. About 5 kg of $CaCl_2$ was added for every 1,135 L of solution at 25°C. The precipitated $CaSO_4$ was removed from the solution by filtration in the filter press. This procedure was repeated until the sulfate was lowered to the desired level. Prior to emptying the filter press, the $CaSO_4$ residue was washed with steam and water. Following precipitation of copper and sulfate, the filtered solution was returned to the preparation tank.

A stream of about 30 L/min of solution from the surge tank was constantly circulated through the chlorination tower to convert ferrous chloride to ferric chloride. A total of 1,700 L of solution was kept in the surge tank. The extent of chlorination was determined by monitoring the oxidation-reduction potential and by determining the Fe^{2+} in solution with a dichromate titration. The oxidation-reduction potential was measured with platinum and Ag-AgCl combination electrodes, using Ag-AgCl as the reference electrode. The desired chlorination was reached at a measured redox potential of 580 mV (Eh = 802 mV). When the desired chlorination was obtained, 1,135 L of the solution was transferred to the leaching tank. When needed, additions of NaCl, $FeCl_3$, and HCl were made to adjust the solution composition pH before the solution was heated to temperature for another cycle of leaching.

The lead chloride feeder, which was controlled by a timer, added lead chloride automatically to the electrolytic cell. The initial bath was composed of 380 kg of salt mixture containing, in mole %, 25 LiCl, 32 KCl, and 43 $PbCl_2$. The lead metal produced was periodically siphoned through a removable

Pyrex tube connecting the lead metal pool in the cell and the vacuum chamber. Lead metal was collected in a mold in the vacuum chamber.

RESULTS AND DISCUSSION

Leaching

The leaching unit was operated on a 8-h basis and in 5- and 10-day continuous campaigns. Altogether, 419 leaching cycles were made. After some initial modifications no serious problems were encountered with corrosion of construction materials. A variety of commercially available materials, such as fiberglass-reinforced plastic, chlorinated polyvinyl chloride, polyvinylidene fluoride, polypropylene, fluorinated rubber, and titanium, withstood the corrosiveness of the chloride solutions.

Table III shows the analysis of the leaching solution. The major impurities introduced into the leaching circuit were copper, zinc, and sulfate. Copper and zinc originated from the leaching of chalcopyrite and sphalerite, whereas nearly all of the sulfate came from $PbSO_4$ formed by oxidation of PbS during storage of the concentrate. For example, lot 1 concentrate contained 2.4% sulfate when it was obtained and this increased to 4.5% over a 2-year period.

Buildup of copper and silver in the leaching solution did not adversely affect the leaching efficiency, but resulted in the contamination of the lead chloride and consequently the lead metal. The correlations of copper and silver in the leaching solution and the lead metal product are shown in Figures 6 and 7.

A laboratory investigation to remove copper from the leaching solution indicated several possible methods. The results of these tests are given in Table IV. Although all methods were successful, sodium sulfide precipitation was selected because of the efficiency and ease of using this method. With a 4.9-g addition of $Na_2S \cdot 9H_2O$ per liter of leaching solution, the copper decreased from 19 to about 3 g/L in 10 min.

Zinc concentration built up slowly and leveled off at about 15 g/L. Magnesium and cadmium also showed a moderate increase. At the concentrations of zinc, magnesium, and cadmium shown, there was no apparent effect on leaching.

Table III Analysis of leaching solution, grams per liter

Campaign	Ag	As	Ca	Cd	Co	Cu	Fe	Mg	Na	Ni	Pb	Zn	SO_4
1........	0.01	–	1.6	0.03	–	2.6	32	0.9	96	0.03	14	3.8	8.9
2[1].......	.03	–	.8	–	0.03	13.0	27	2.3	66	–	6	5.4	16.7
3........	.06	<0.025	.5	.19	<.025	19.0	21	3.4	110	–	17	12.0	25.6
4[2].......	.04	<.025	.6	.22	<.025	4.3	15	7.0	100	.14	16	14.0	23.0
5[3].......	.04	–	1.1	.21	–	5.9	22	–	94	–	13	13.0	14.0
6........	.05	–	.4	.24	–	9.1	20	–	93	–	–	15.0	18.0
7[4].......	.01	<.025	3.5	–	–	.8	22	6.3	106	–	–	21.0	3.0
8........	–	–	3.8	.24	.05	1.3	23	4.6	15	.14	15	15.0	2.7

– Not analyzed.
[1] A 300-gal bleed was made, and NaCl concentration adjusted thereafter.
[2] Copper and silver were removed by precipitation with $Na_2S \cdot 9H_2O$.
[3] Sulfate was removed by precipitation with $CaCl_2$.
[4] Removal of copper and sulfate with $Na_2S \cdot 9H_2O$ and $CaCl_2$. Use of fresh (lot 2) concentrate was started.

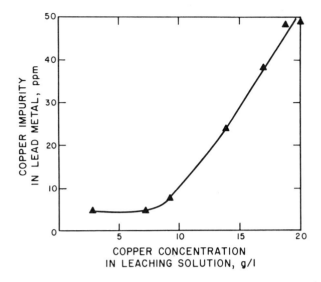

Fig. 6 Correlation of copper impurity in leaching solution and lead metal product.

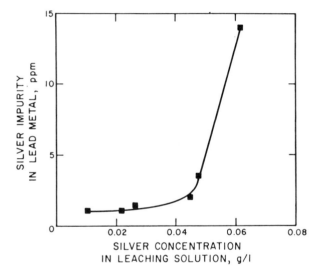

Fig. 7 Correlation of silver impurity in leaching solution and lead metal product.

Sulfate buildup in the leaching solution decreased the lead extraction, contaminated the electrolyte, and formed a precipitate in the chlorination tower that obstructed the gas and solution flows. An analysis of a sample of the precipitate removed from the chlorination tower showed, in percent, 20 Fe, 9.0 Ca, 5.2 Na, 3.0 Pb, 0.48 Zn, 0.19 Cu, 36 SO_4, 3.3 Cl, and 0.50 S°. When the sulfate content of the leaching solution was about 34 g/L, iron sulfate precipitation occurred during the crystallization of lead chloride.

Precipitation of sulfate with calcium chloride was effective in removing sulfate from the leaching solution. The laboratory study shown in Table V indicates that the sulfate can be removed to <0.1 g/L with calcium chloride. The precipitate after a water wash contained, in percent, 26 Ca, 62 SO_4, 0.3 Cu, 0.3 Na, 5 Pb, 0.02 Zn, and 0.8 Fe. In the large-scale tests, the sulfate level in the leaching solution was controlled to about 3 g/L by maintaining the calcium content at 3 to 4 g/L as was done in campaigns 7 and 8 (Table III). Dutrizac investigated sulfate control during ferric chloride leaching and concluded that $CaSO_4$ precipitation "is the preferred option."[13]

To determine the effect of sulfate buildup, aged lead concentrate with a high sulfate content was used during the early part of the investigation to accelerate the sulfate buildup. The sulfate in lot 1 concentrate increased from 2.4 to 4.5% by aging for 2 years. Fresh concentrate was used for campaigns 7 and 8. In an industrial operation using a fresh concentrate, the sulfate buildup would not be as serious.

Table VI shows the analysis of the leach residues. Prior to analysis, the residue samples from the filter press were rewashed with hot water in the laboratory. Thorough washing of the residues in the filter press was difficult. Generally the residue samples that were not rewashed in the laboratory contained up to 5% more lead sulfate and chloride and up to 1% more copper and zinc. Fourteen kilograms (dry basis) of residue was produced from leaching 55 kg (dry basis) of concentrate. The residue contained 40 to 50% elemental sulfur. The remainder of the sulfur was present as insoluble sulfates and unreacted sulfides of iron, copper, zinc, and lead. Microscopic and X-ray diffraction examinations of the residues showed a major amount of alpha-sulfur and minor amounts of

Table IV Laboratory copper precipitation[1]

Reagent	Weight, g	Time, min	Analysis, g/L[2]			
			Cu	Fe	Pb	Zn
Fe.........	20	60	3.2	60	10	17
	20	20	3.2	53	12	15
	20	5	3.1	50	12	14
Pb.........	20	60	3.0	26	19	17
	20	20	10.6	22	20	15
	20	5	17.9	23	16	14
$Na_2S\cdot9H_2O$..	3.9	10	4.4	27	18	17
	4.3	10	3.6	22	15	14
	4.9	10	3.1	22	15	14
	5.9	10	2.8	23	15	14

[1] 1,000 mL leaching solution (19.3 g/L Cu, 23 g/L Fe^{2+}, 15 g/L Pb, and 14 g/L Zn.
[2] Solution analysis after precipitation.

Table V Laboratory sulfate precipitation[1]

$CaCl_2$ addition, g	Analysis of leaching solution, g/L[2]						
	Ca	Cu	Fe	Pb	Na	Zn	SO_4
5.4........	1.0	5.9	18	5.9	75	14	15
44.........	8.5	6.0	18	6.8	74	14	1.5
89.........	24	5.1	18	6.7	74	14	.08

[1] Initial solution (g/L): 0.5 Ca, 19 SO_4, 80 Na, 6.7 Cu, 20 Fe, 7.0 Pb, 15 Zn.
[2] Solution analysis after precipitation.

Table VI Analysis of leach residues,[1] pct

Campaign	Ag	Cd	Cu	Fe	Pb	Zn	SO$_4$	Total sulfur
1.......	0.007	0.03	3.5	15	4.6	2.4	-	58
2.......	.010	.08	1.8	15	11.0	5.1	-	-
3.......	.005	-	1.8	10	7.6	2.5	5.7	60
4.......		.04	4.0	9	37.0	3.7	-	-
5.......	-	.05	1.9	10	13	3.0	7.8	51
6.......	-	.07	4.6	9	17	4.0	6.6	-
7.......	-	-	2.4	12	32	1.6	-	-
8.......	.008	.02	-	13	14	1.6	5.4	57

- Not analyzed.
[1]All samples taken from filter press and
rewashed with hot water in laboratory except
sample from cycle 1, which was not rewashed.
The residues also contained, in percent,
0.08 As, 0.14-.54 Ca, 0.13-.20 Co, 0.14-.30 Ni,
and 0.03 Sb. Fresh lot 2 concentrate was used
after cycle 6.

Table VII Metal extraction,[1] percent

Campaign	Ag	Cu	Pb	Zn
1.......	64	32	98	56
2.......	32	58	95	0
3.......	80	73	98	65
4.......	-	0	82	13
5.......	-	59	96	51
6.......	-	27	94	17
7.......	-	3	87	15
8.......	64	-	95	59

- Not analyzed.
[1]Based on laboratory washed
leach residue.

Table VIII Analysis of lead chloride,
parts-per-million

Campaign	Ag	Ca	Cu	Fe	Mg	Na	Zn	SO$_4$
1.......	<10	-	<20	120	-	180	60	-
2.......	<10	170	<20	130	-	120	<20	2,800
3.......	<10	<60	63	92	-	370	34	800
4[1].....	<3	<60	5	34	-	110	10	170
5[2].....	<3	1,110	7	95	4	180	15	7,700
6.......	-	<60	12	44	-	85	14	730
7.......	<3	<60	<3	80	-	150	9	350
8.......	<3	<60	<3	24	5.5	97	9	350

- Not analyzed.
[1]Start of periodic removals of copper from
leaching solution.
[2]Start of periodic removals of sulfate from
leaching solution.

pyrite, sphalerite, chalcopyrite, covellite,
pyrrhotite, galena, and gypsum. Sulfur grains
averaging 5 to 10 μm in diameter occurred in
the matrix of the leach residue. Covellite,
which was not detected in the galena
concentrate, occurred as a coating on galena
particles and galena pseudomorphs.

Elemental sulfur could be recovered from the
residue by steam filtration at temperatures
above the melting point of sulfur and with
steam or organic solutions can be used in
place of steam if a suitable method for
recovering the organics from the residue is
developed.[14] A mixed sulfide concentrate for
subsequent recovery of the metal values would
be obtained. Another possible method to

recover the sulfur value would be to burn the
residue to produce sulfur dioxide to
manufacture sulfuric acid and a mixed oxide
for subsequent recovery of the metal values.
Even without sulfur removal, long-term
stabilization of the residue appears
feasible.[15]

Metal extractions based on the analysis of the
laboratory-washed leach residues are shown in
table VII. The lead extraction averaged 93%
but there was considerable variation in the
data. In subsequent work on continuous ferric
chloride leaching, lead extractions
consistently exceeded 99% in 15 min of
leaching if the redox potential of the spent
leaching solution was 480 mV or above.[12]
Extractions over 99 pct were also obtained in
previous bench-scale tests.[4] Extractions of
copper, zinc, and silver were higher than
those obtained in bench-scale research,[4,12] in
which better control of leaching time was
possible. Additional leaching time increased
solubilization of chalcopyrite and sphalerite
in the bench-scale tests.[4] Because leaching
and filtration were performed batchwise in the
larger-scale unit, the contact time between
solution and concentrate was at least 30 min,
which may account for the increased extraction
of copper, silver, and zinc.

Lead chloride solubility was affected by
sodium chloride concentration and temperature.
During leaching, the lead chloride was kept in
solution by the sodium concentration and the
high temperature. After leaching, lead
chloride crystallization occurred as the
temperature was decreased in the crystallizer.
The lead chloride produced was granular and
had a bulk density of approximately 2.7 g/cm^3.

As shown in Table VIII, the impurities
detected in the lead chloride were sodium,
iron, copper, zinc, magnesium, calcium,
silver, and sulfate. Copper and magnesium
corresponded to their concentrations in the
leaching solution. Sulfate and calcium
occurred together in the lead chloride either
as anhydrite or gypsum. Removal of sulfate
from the leaching solution by precipitation
with calcium chloride decreased, but did not
eliminate, sulfate in the lead chloride. In
the subsequent bench-scale continuous leaching
tests, sulfate was controlled to 100 ppm in
the lead chloride product by maintaining at
least 3 g/L calcium in the leaching
solution.[12] The lack of correlation between
the concentrations of iron, sodium, and zinc
in the lead chloride and their respective
concentrations in the leaching solution was
probably caused by variations in washing the
lead chloride.

Electrolysis

The liquidus surface diagram of the PbCl$_2$-KCl-
LiCl system (Fig. 8) shows that operation at
about 450°C permitted a wide variation of
electrolyte compositions, particularly the
PbCl$_2$.[16] Within the ranges of variation of
electrolyte temperature and compositions
encountered during the operation of the
electrolytic cell, the electrolyte was always
fluid.

Table IX shows the operating data of the
electrolyte cell with horizontal, flat-plate
electrodes shown in Fig. 3. Heat from
electrolysis maintained the cell at the
desired temperature. The cathode current
efficiency was lower, and the cell voltage was

278

higher than those obtained in previous bench-scale research.[3,9] Variations in cell resistance (Table IX) resulted primarily from changes in melt composition.

The analysis of the electrolyte is shown in Table X. The impurity elements that are less noble than lead, such as Ca, Na, Fe, Zn, and Mg, increased in the electrolyte. The zinc and magnesium increased very slowly. Copper in the electrolyte codeposited with the lead metal and had to be removed from the leaching solution to prevent contamination of the lead chloride. Sulfate in the lead chloride feed accumulated in the electrolyte. During the last campaign, a layer of brownish foam, which was high in sulfate, was observed on the bath surface. Iron in the electrolyte was present as Fe_2O_3 (black salt) and at the levels encountered had no discernible effect on cell performance. The fluctuation in Pb, K, and Li concentrations was caused by periodic addition of KCl and LiCl to the cell.

Copper and silver were the only detectable metallic impurities in the lead metal product (Table XI). When copper built up in the leaching solution, a corresponding increase was observed in the lead chloride, electrolyte, and lead metal product. Silver behaved in a similar manner to copper. The content of copper and silver in the lead chloride must be controlled to maintain a high-purity lead product. When copper was removed from the leaching solution, 99.99% pure lead metal was produced.

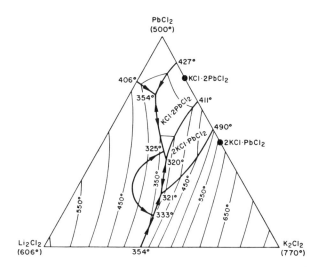

Fig. 8 Liquidus diagram of the $PbCl_2$-LiCl-KCl system (temperatures in degrees Celsius and compositions in mole %).

Table IX Operating data for electrolytic cell with horizontal, flat plate electrodes

Campaign........	1	2	3	4	5	6	7	8
Av. current, A..	3,000	2,860	3,350	2,948	2,868	2,365	3,100	3,050
Av. voltage, V..	5.1	4.8	4.3	4.4	4.5	5.2	4.7	4.5
Av. electrolyte temp., °C......	439	453	437	450	457	454	455	454
Av. anode current density, A/cm^2.........	0.69	0.66	0.77	0.68	0.66	0.54	0.72	0.70
Av. cathode current density, A/cm^2.........	0.66	0.63	0.74	0.65	0.63	0.52	0.69	0.67
Ampere-hours used..........	483,000	603,500	532,500	597,500	266,353	244,496	277,715	307,145
$PbCl_2$ added, kg.	3,108	2,993	2,443	2,864	1,432	1,148	1,309	1,326
Lead metal produced, kg......	1,709	2,140	1,930	2,157	974	856	987	1,018
Cathode current efficiency, pct...........	92	92	94	93	95	91	92	86
Energy requirement, kW·h/kg lead..........	1.44	1.35	1.19	1.22	1.23	1.49	1.32	1.36

Table X Analysis of electrolyte[1]

Campaign	K	Li	Pb	SO_4	Ca	Cu	Fe	Mg	Na	Zn
1....	7.6	2.0	49	0.90	1,500	16	310	–	700	<20
2....	6.4	1.3	56	.36	1,100	42	1,100	23	1,700	80
3....	11.0	3.3	40	2.49	3,400	90	2,100	48	1,400	280
4....	9.0	2.7	48	2.06	2,900	39	2,100	49	1,700	400
5....	16.8	4.4	23	6.50	7,500	34	2,400	55	26,000	400
6[2]...	–	–	–	–	–	–	–	–	–	–
7....	14.0	2.1	49	5.60	9,300	12	110	34	–	–
8....	11.9	4.1	40	6.60	6,000	11	1,800	51	12,000	320

– Not analyzed.
[1]Ag was <10 ppm for all periods.
[2]114 kg of electrolyte was replaced.

279

Table XI Analysis of lead metal,[1]
parts-per-million

Campaign	Ag	Cu
1.............	1	8
2.............	2	24
3.............	14	49
4.............	<1	14
5.............	<1	14
6.............	<1	10
7.............	<1	14
8.............	2	7

[1]Other impurities below
detection limits include,
in parts-per-million,
<15 Bi, <0.7 Ca, <08 Cd,
<2.5 Co, <3 Fe, <2.5 Ni,
<25 Sb, and <1 Zn.

TABLE XII Operating data for electrolytic cell
with sawtooth electrodes

Average cell current.......A..	3,000	4,000
Average cell voltage.......A..	2.5	3.0
Average electrolyte temp..°C..	450	450
Average electrode current density.......A/cm^2..	0.36	0.48
Ampere-hours used.............	426,527	230,640
PbCl$_2$ added...............kg..	2,446	1,182
Lead metal produced.......kg..	1,633	887
Current efficiency.......pct..	98.7	99.5
Energy requirement...kW·h/kg..	0.66	0.77

After more than 19 months of electrolytic cell intermittent operation, one of the alternating current heating electrodes failed and caused the electrolyte to freeze. At this point, the cell was taken apart to determine how much deterioration of the electrodes had taken place. Other than slight erosion of the anodes, the electrodes appeared to be in excellent condition. However, they disintegrated after being exposed to air for a week or so. Intercalation of alkali metal chlorides into the graphite interstices was presumed to be responsible.

As mentioned previously, the current efficiency was lower and the cell voltage higher than had been predicted from bench scale tests. In an effort to improve the efficiency of the electrowinning, it was decided to rebuild the cell and test a more efficient electrode design.[12] After bench scale tests with various electrode designs, the sawtooth design shown in figures 4 and 5 was chosen for incorporation into the rebuilt cell.

With the sawtooth electrodes, the relationship between groove angle, depth of groove, and electrode spacing had to be considered to maximize the effective electrolyte area. A mathematical analysis was used to determine the ideal groove angle after selecting the depth of groove and the electrode spacing. A groove depth of 5.1 cm and an electrode spacing of 1.27 cm were selected for the electrode assembly. The optimum groove angle was then calculated to be 60 degrees. Smaller angles would increase the total electrode surface area but the effective bath area between the electrodes would decrease unless electrode spacing was also decreased. The effective bath area was defined as the area

which was covered by parallel faces of both the anode and cathode. It was assumed that the electrolysis that occurred between all other electrode surfaces was insignificant. While this assumption was not strictly true, it greatly simplified the mathematical analysis.

The sawtooth cell was operated continuously at 3,000 A for 6 days. The results are shown in table XII. Current efficiency was 98.7% which was similar to the current efficiencies obtained in the bench-scale tests. The energy requirement of 0.66 kW·h/kg of lead metal produced was a significant improvement over the 1.22 to 1.49 kW·h/kg obtained in the earlier tests. The voltage of 2.5 V given in table XII was measured, as previously, from the anode connecting rod to the cathode steel busbar. Measuring the voltage directly between the anode and cathode gave an average voltage of 1.9 V, and thus, the voltage drop in the electrode leads was 0.6 V. The voltage losses in the electrode leads could be decreased by using larger leads and by minimizing contact resistance. Lowering the voltage drop in the electrode leads is a simple means of further decreasing the energy required for lead electrowinning.

Since the potential drop through the electrolyte was dramatically decreased by the sawtooth electrodes, much less heat was generated in electrolysis. As a result, 6 kVA had to be furnished by the ac power supply during electrolysis to keep the electrolyte at 450°C.

Following cell operation at 3,000 A, the cell was operated at 4,000 A for 58 h. The operational data at 4,000 A are given in table XII. Current efficiency was excellent at 99.5%. Cell voltage increased to 3.0 V and resulted in an energy requirement of 0.77 kW·h/kg. Voltage losses in the leads at 4,000 A increased to 0.9 V. Cell potential from anode to cathode was 2.1 V and showed little influence of chlorine gas on the electrolyte conductivity. At 4,000 A, the ac power requirement to keep the electrolyte at 450°C was decreased to 4 kVA.

Lead Monitoring

Exposure of personnel to lead was monitored during the larger scale operation. Air samples taken by samplers carried by operating personnel were used to determine lead-in-air levels. Lead analyzers used to determine the concentration of lead in the air were a Bendix model 44, and an MSA Drager-type Multi-Gas Detector model 21/21. Blood tests for lead were administered to monitor lead-in-blood levels of operating personnel. Figure 9 shows that most of the lead-in-air results were <30 µg Pb/m^3 of air. Figure 10 shows that the majority of the blood samples contained <40 µg Pb/100 g of whole blood. Lead monitoring data indicated that lead exposures were within Occupational Safety and Health Administration standards. However, lead monitoring data were obtained from an intermittent operation, and a small amount of lead was involved compared to the quantities handled in a commercial production plant.

The results of lead monitoring in different locations of the workplace are given in Table XIII. Air samples taken near the covered tanks, filter press, electric oven,

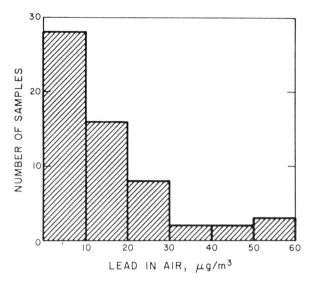

Fig. 9 Lead-in-air monitoring.

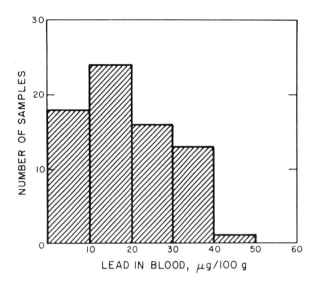

Fig. 10 Lead-in-blood monitoring.

Table XIII Lead monitoring in work areas

Sample location	Instrument	Lead in air, $\mu g/m^3$	Number of samples
2 m above floor........	Bendix 44	2- 28	4
Deck..................	MSA	<6	5
Leaching tank, steam...	MSA	100-180	3
Leaching tank, covered.	MSA	ND	1
Crystallizer, steam....	MSA	400	1
Crystallizer, covered..	MSA	ND	4
Surge tank, steam......	MSA, Bendix	100-150	2
Surge tank, covered....	Bendix 44	14	1
Above opened filter press................	Bendix 44	18	1
Addition of concentrate to the feed hopper....	MSA	100	1
$PbCl_2$ handling.........	MSA	ND	3
Electrolytic cell......	MSA	ND	6
In scrubber air stream near roof.............	MSA	ND	2
Ambient air 200 m downwind.............	Bendix 44	<2- 2	2
Ambient air 1 km downwind.............	Bendix 400	<0.3- .33	9

ND Not detected.

electrolytic cell, and the fume hood in which the dry lead chloride was handled showed that the lead concentration was not significantly higher than in ambient air. The highest lead concentrations were in the vapors inside the leaching and crystallizer tanks, indicating that all tanks should be covered and that vapors from hot lead-containing solutions should be vented and scrubbed.

DISCUSSION AND CONCLUSIONS

The ferric chloride leaching-molten salt electrolysis process produces high-purity lead from galena concentrates with minimum pollution of the environment and with low risk to worker health. Leaching is extremely fast and can be performed on a continuous basis with excellent lead recoveries.[12] Control of impurities such as copper, silver, and sulfate in the leach liquor is relatively simple. Control of other impurities such as zinc was not investigated, but will be necessary in a plant operation. Lead losses to the residue are low which simplifies residue treatment.

The electrolytic recovery of lead metal works well and the energy required is low, especially with electrodes of sawtooth design. Cell construction is simple and the cell materials withstood long-term tests without serious degradation. The $LiCl-KCl-PbCl_2$ electrolyte is conductive and is fluid at 450°C. The electrolyte is reasonably tolerant of most impurities but will require a bleed to control some impurities such as sulfate. With the $LiCl-KCl-PbCl_2$ electrolyte, the bleed will have to be purified and returned to the cell. A $NaCl-PbCl_2$ electrolyte which was tested on bench-scale experiments may be advantageous because the electrolyte bleed could be returned to the leaching solution.[3,9] Long-term electrolysis tests would be necessary with the $NaCl-PbCl_2$ electrolyte to determine if the electrolyte is suitable.

The economics of the process appear to be close to pyrometallurgical processes.[10] Moreover, application of the technology to complex concentrates should improve the economics of the process.

REFERENCES

1. Dutrizac, J. E. The Chloride Processing of Lead Concentrates. CANMET Division Report MRP/MSL 84-41, 1984, 44 pp.

2. Dutrizac, J. E. MINTEK 50, L. F. Haughton, ed., Johannesburg, S.A., vol. 1, 1985, pp. 39-62.

3. Murphy, J. E., Haver, F. P., and Wong, M. M. Recovery of Lead From Galena by a Leach-Electrolysis Procedure. Bureau of Mines Report of Investigations 7913, 1974, 8 pp.

4. Haver, F. P., and Wong, M. M. Ferric Chloride-Brine Leaching of Galena Concentrate. Bureau of Mines Report of Investigations 8105, 1976, 17 pp.

5. Wong, M. M., Sandberg, R. G., Elges, C. H., and Fleck, D. C. Integrated Operation of Ferric Chloride Leaching, Molten Salt Electrolysis Process for Production of Lead. Bureau of Mines Report of Investigations 8770, 1983, 21 pp.

6. James, S. E., and Bounds, C. O. High Purity Lead From the Ferric Chloride Leaching of Complex Sulfides. Complex Sulfide Processing of Ores, Concentrates and By-Products, ed. by A. D. Zunkel, R. S. Boorman, A. E. Morris, and R. J. Wesely, Metallurgical Society of AIME, 1985, pp. 441-453.

7. Demarthe, J. M., and Georgeaux, A. Hydrometallurgical Treatment of Lead Concentrates. Lead-Zinc-Tin '80, ed. by J. M. Cigan, T. S. Mackey, and T. J. O'Keefe, TMS-AIME, Warrendale, Pennsylvania, 1979, pp. 426-444.

8. Craigen, W. J. S., and Schnarr, J. R. Economic Incentives for the Production of Bulk Base Metal Sulfide Concentrates. CANMET Division Report MRP/MSL 85-130, 1985, 13 pp.

9. Haver, F. P., Elges, C. H., Bixby, D. L., and Wong, M. M. Recovery of Lead From Lead Chloride by Fused-Salt Electrolysis. Bureau of Mines Report of Investigations 8166, 1976, 18 pp.

10. Phillips, T. A. Economic Evaluation of a Leach-Electrolysis Process for Recovering Lead From Galena. Bureau of Mines Information Circular 8773, 1978, 23 pp.

11. Murphy, J. E., Eichbaum, B. R., and Eisele, J. A. Continuous Ferric Chloride Leaching of Galena. Mineral and Metallurgical Processing, vol. 3, No. 2, Feb. 1985, pp. 38-42.

12. Murphy, J. E., Chambers, M. F., and Eisele, J. A. Molten-Salt Electrolysis of Lead Chloride in a 3,000-Ampere Cell With an Improved Electrode Design. Paper in Molten Salts, ed. by M.-L. Saboungi, D. S. Newman, K. Johnson, and D. Inman, Electrochemical Society, 1985, 612 pp.

13. Dutrizac, J. E. Sulfate Control in Chloride Leaching Processes. Hydrometallurgy, vol. 23, 1989, pp. 1-22.

14. Herve, B. P., Eichbaum, B. R., Murphy, J. E., and Sandberg, R. G. Sulfur Extraction From Elemental Sulfur-Bearing Materials. Bureau of Mines Open File Report No. 93-85, 1985, 7 pp.

15. Carnahan, T. G., and Lucas, M. A. Weathering of a Base-Metal Sulfide Leaching Residue. Bureau of Mines Report of Investigations 8667, 1982, 11 pp.

16. Levin, E. M., Robbins, C. R., and McMurdie, H. F. Phase Diagrams for Ceramists, 1969 Supplement. American Ceramic Society, Columbus, Ohio, 1969, p. 332 (fig. 3254).

Mercury production from sulphide concentrates by cupric chloride leaching and aqueous electrolysis

J.E. Murphy
H.G. Henry
J.A. Eisele
Reno Research Center, Bureau of Mines, U.S. Department of the Interior, Reno, Nevada, U.S.A.

SYNOPSIS

The Bureau of Mines developed a hydrometallurgical method for treating mercury sulfide
concentrates to recover mercury. Sulfide
flotation concentrate from the McDermitt Mine
(Nevada) was leached in a cupric chloride
solution at 80° C. The cupric chloride
concentration of the solution was maintained
during leaching by chlorine sparging. Mercury
extractions exceeded 99 pct in 3 h of
leaching. After leaching, the pH of the
solution was increased from 1 to 4.5 to
precipitate copper as atacamite. The pregnant
solution, which typically contained 100 g/L
Hg, was sent to electrolysis to produce high-
purity mercury metal, and chlorine for
recycle. In a 200-A electrolytic cell
operated for 24 h, current efficiency was 99
pct, and the energy requirement was 2 kW·h/kg
of mercury produced. Mercury removal from the
waste stream was also investigated. Iron or
zinc cementation, sulfide precipitation with
H_2S, and activated carbon adsorption all
decreased the mercury concentration from 10
g/L to 0.01 mg/L.

INTRODUCTION

The U.S. Bureau of Mines investigated a cupric
chloride leaching-aqueous electrolysis process
for recovering mercury metal from sulfide
concentrates.[1,2] Other researchers have also
investigated hydrometallurgical methods for
treating mercury concentrates.[3-9]

The concentrate used in the present
investigation was obtained from the McDermitt
Mine, the major mercury producer in the United
States. The impetus for developing a
hydrometallurgical method for recovering
mercury from concentrates was to eliminate
from the workplace mercury vapors that may be
difficult to control at very low levels in a
conventional furnacing operation.

The flow diagram for the process is shown in
figure 1. The process involves leaching as-
received mercury flotation concentrate with
cupric chloride solution to obtain mercuric
chloride solution according to equation 1.

$$HgS + 2CuCl_2 \rightarrow HgCl_2 + 2CuCl + S° \qquad (1)$$

Chlorine is sparged into the solution to
maintain the redox potential at 850 mV (SHE)
during leaching.

$$2CuCl + Cl_2 \rightarrow 2CuCl_2 \qquad (2)$$

The solution is regenerated with chlorine
during leaching to decrease the concentration
of copper required for leaching. After
leaching is complete, the residue is filtered
from solution, washed, and sent to waste.
Copper is precipitated from the solution as
atacamite, $Cu_2Cl(OH)_3$, by adding $CaCO_3$ to
increase the pH to 4.5.

$$8CuCl_2 + 6CaCO_3 + 6H_2O$$
$$\rightarrow 4Cu_2Cl(OH)_3\downarrow + 6CaCl_2 + 6CO_2 \qquad (3)$$

Copper must be removed from the leaching
solution because it causes a mercurous
chloride sludge to form on the cathode during
subsequent electrolysis. Atacamite is
filtered from solution, redissolved in
sulfuric acid, and returned to leaching.
Sulfuric acid is used to dissolve the
atacamite in order to control the
concentration of $CaCl_2$ in solution which is
produced by reaction 3. During leaching,
insoluble $CaSO_4$, either anhydrite or gypsum,
is formed. The insoluble $CaSO_4$ reports to the
residue and is sent to waste.

The nearly copper-free mercury solution is fed
to the electrolytic cell to produce mercury
metal and chlorine for recycle.

$$\begin{array}{c} \text{Electrolysis} \\ HgCl_2 \quad \rightarrow \quad Hg + Cl_2\uparrow \end{array} \qquad (4)$$

The residue from leaching must be thoroughly
washed to remove toxic mercuric chloride.
Wash water can be treated with H_2S to recover
HgS for recycling to leaching. Alternatively,
wash water can be treated by iron cementation
to produce a relatively impure mercury
product. Salable mercury metal could be
obtained from the impure mercury by filtering
and acid washing.

EQUIPMENT AND MATERIALS

Mercury sulfide flotation concentrate was
provided by the McDermitt Mine. Analysis of
the concentrate is shown in Table I.
Approximately 70 pct of the mercury was
present as cinnabar (HgS) and 30 pct as
corderoite ($Hg_3S_2Cl_2$).

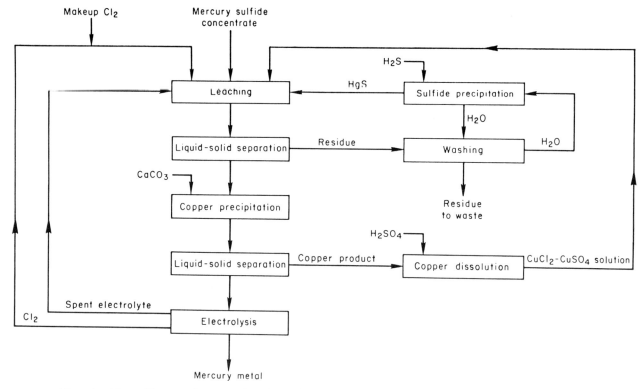

Fig. 1 Flow diagram for cupric chloride-chlorine leaching of mercury concentrate.

Table I   Analysis of flotation concentrate, percent

Mercury...	77.7	Antimony...	0.08
Sulfur....	11.9	Magnesium..	.07
Chlorine..	2.7	Calcium....	.03
Silicon...	1.8	Arsenic....	.01
Aluminum..	.9	Elemental	
Iron......	.8	sulfur....	.05
Titanium..	.08	Sulfate....	.07

Chemicals used in the experiments were reagent grade. Leaching was performed in a 1- or 2-L glass beaker on a hotplate with a thermocouple and controller to maintain the leaching temperature. A Teflon[5] paddle-type stirrer was used to keep the concentrate suspended in the solution. Chlorine was sparged into the slurry through a capillary glass tube with a 1.5-mm-diam bore. A redox probe with a platinum electrode and a silver/silver chloride reference was used to measure the redox potentials of the slurries. All redox potentials given in the paper are versus the standard hydrogen electrode. Filtering was done with Whatman No. 5 filter paper on a Buchner funnel with slight vacuum.

The bench-scale electrolytic cell, which is shown in Fig. 2, consisted of a 1-L glass beaker with a 9 cm-diam, 1.3-cm-thick graphite plate for the anode and a 1-cm-deep mercury pool as the cathode. A 0.5-cm-diam iron rod, which was protected from the electrolyte by a glass sleeve, served as electrical contact to the cathode. A 1.3-cm-diam graphite rod, which was threaded into the anode plate, served as the anode connection. The cell was operated at 7A except when the current density was varied. Cell temperature, typically 50° C, was maintained by a hotplate that was

controlled by a glass-encased thermocouple in the electrolyte. A paddle-type polyethylene stirrer was used for electrolyte agitation during electrolysis.

The 200-A cell is shown in Fig. 3. The cell container was a 29-cm-ID by 30-cm-high Pyrex glass vessel. A 5-cm-diam graphite rod was connected to the anode bus and was threaded into a 2.5-cm-thick by 23-cm-diam graphite plate. A 1-cm-deep mercury pool was electrically connected to the cathode bus by a 2.5-cm-diam mild steel rod encased in glass tubing. Mercury metal was removed from the cell by siphoning through a 1.3-cm-ID Pyrex glass tube. The cell operated at a cathode and anode current density of 0.30 and 0.48 A/cm[2], respectively, and required cooling to maintain the temperature at 60° C. The electrolyte was circulated between the cell and a 95-L reservoir tank which was equipped with a water cooling coil. Chlorine generated during electrolysis was scrubbed by passing the cell offgas through a Na_2CO_3 solution.

EXPERIMENTAL PROCEDURES AND RESULTS

Leaching Tests

The leaching solution normally contained, in grams per liter: 50 Cu, 100 Ca, 3.8 HCl, and from 100 to 300 Hg concentrate. Leaching was performed at 80° C for 3 h with enough stirring to keep the concentrate in suspension. Chlorine was slowly sparged into the solution to maintain the copper in the cupric state. A stable redox potential of the solution of 850 mV (SHE) indicated that leaching was essentially complete. The leaching solutions filtered easily.

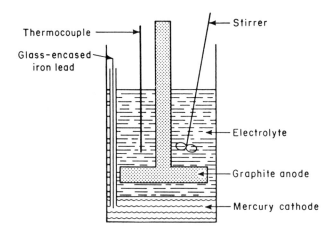

Fig. 2 Bench-scale electrolytic cell.

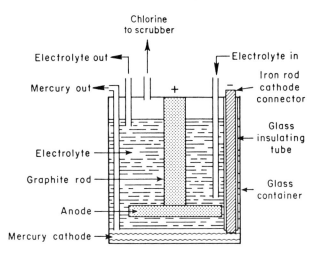

Fig. 3 200-A electrolytic cell.

A series of leaching experiments was performed with 300 g/L of mercury concentrate. The results of these experiments, which are given in Table II, showed that mercury extractions were more than 99 pct if the redox potential of the solution was 850 mV (SHE). The mercury concentration of the pregnant solution was more than 200 g/L. Analysis of the residue showed that from 2 to 7 pct of the sulfur in the concentrate was converted to sulfate when mercury extraction was 99 pct. An analysis of a typical residue is given in Table III.

The pulp density during leaching was quite flexible because $HgCl_2$ is very soluble. The advantage of using higher pulp densities for leaching solution is that more concentrate can be leached before copper has to be removed from solution.

Increasing the pH of the pregnant solution with calcium or sodium carbonate precipitated a copper oxychloride which was identified by X-ray diffraction as atacamite. The carbonate was added slowly to the solution over a 15-min period with constant stirring. Na_2CO_3 worked as well as $CaCO_3$, but calcium is advantageous because the calcium concentration of the solution can be controlled by adding sulfate as H_2SO_4 when atacamite is dissolved to reclaim the copper.

Copper was removed from the pregnant solution containing over 200 g/L mercury. Calcium carbonate was added very slowly to the stirred slurry until the pH was raised to a selected level. The slurry became very thick after the $CaCO_3$ was added, but filtered easily. Results of the tests are presented in Table IV and demonstrate that excellent separation of the copper was possible with $CaCO_3$. In some tests an equal volume of water was added to the filtrate from leaching. Dilution of the filtrate made the copper precipitate easier to filter and decreased the mercury concentration in solution to about 100 g/L, which was a good concentration for electrowinning. In practice, spent electrolyte could be used to dilute the filtrate which would help in maintaining the water balance in the process.

Table II Effect of leaching solution redox potential on mercury extraction

Final redox potential (SHE)	Mercury extraction, pct	Sulfur converted to sulfate, pct
868	99.6	6.8
857	99.2	6.1
852	98.9	2.1
840	99.0	5.0
832	99.1	5.5
815	98.1	.37

Table III Analysis of the residue from leaching, percent

Mercury...	0.7	Antimony...	0.2
Copper....	.2	Magnesium..	.02
Chlorine..	.3	Calcium....	5.3
Silicon...	8.7	Arsenic....	.02
Aluminum..	4.3	Elemental	
Iron......	3.5	sulfur....	58.0
Titanium..	.4	Sulfate....	8.8

Table IV Removal of copper from simulated pregnant solution containing 200 g/L Hg by precipitation with $CaCO_3$

Final pH	$CaCO_3$ added, g/L	Cu remaining in solution, g/L	Hg in Cu precipitate, pct
4.0	16.1	1.2	0.6
4.5	20.5	.13	1.0
5.0	25.7	.005	.9

Analyses of the pregnant solutions which had been diluted and the copper precipitated from them are given in Table V. Adding 100 g $CaCO_3$ per liter of solution, to obtain a pH of 4.6, decreased the copper to less than 0.1 g/L, which is lower than necessary for electrowinning. Mercury concentration in the precipitates averaged less than 0.3 pct. Even this small amount of mercury would not be lost, but would be redissolved in H_2SO_4 along with the copper and returned to leaching.

Acid consumption can be minimized by keeping the concentration of HCl in the leaching solution low and by maintaining a high ratio of mercury to copper in solution with high pulp densities. Leaching tests with 2 g/L HCl were successful, but the pH must be kept low enough to prevent the precipitation of copper and iron.

Electrolysis Tests in 7-A Cell

Preliminary electrolysis experiments showed that a low copper concentration in the electrolyte was required to avoid the formation of sludge on the liquid mercury cathode. The sludge consisted mainly of mercurous chloride and finely divided mercury metal, and contained a few percent copper. Decreasing the copper concentration in the electrolyte dramatically decreased sludge production and increased current efficiency as shown in Fig. 4. Below 1 g/L Cu, the mercury cathode remained clean and bright during electrolysis and current efficiency was more than 99 pct.

Experiments were made with solutions from which the copper had been precipitated and containing 100 g/L Hg, 100 g/L Ca, and 3.8 g/L HCl. Data from a typical experiment are shown in Table VI. Current efficiency was more than 99 pct and the energy requirement was 0.8 kW·h/kg. No sludge formed on the mercury metal cathode which remained bright and shiny until electrolysis was stopped.

Mercury metal electrowon from the solutions from which copper had been removed was very pure. In Table VII, analysis of the electrowon mercury is compared with analyses of mercury produced at the McDermitt Mine by conventional retorting procedures and triple-distilled mercury from the J. T. Baker Company.

In a series of experiments, electrolysis was continued on depleted solutions to determine the effect of low mercury concentration in the electrolyte. At 5 g/L Hg the cell voltage gradually increased, and at 2 g/L Hg, hydrogen production was apparent on the cathode.

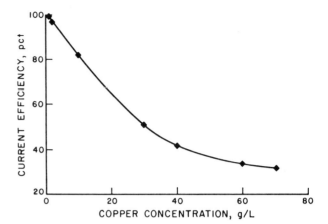

Fig. 4 Dependence of current efficiency on copper concentration in electrolyte.

Table VI Operating data for 7-A electro-winning cell[1]

Cell current................A..	7	
Cell voltage................V..	3.0	
Anode current density....A/cm^2..	0.11	
Cathode current density..A/cm^2..	0.09	
Electrolyte temperature.....°C..	50	
Electricity used..........A·h..	10	
Metal produced..............g..	37.3	
Current efficiency........pct..	99.7	
Energy required.....kW·h/kg Hg..	0.80	

[1]Initial electrolyte composition, in grams per liter: 100 Hg, 100 Ca, and 3.8 HCl. Electrode spacing, 1.3 cm.

Table VII Analyses of mercury metal, parts per million

Element	McDermitt Mine	Bureau of Mines	J. T. Baker Co.
Ag.....	<2.0	<2.0	<2.0
Al.....	12.0	12.0	<9.0
As.....	62.0	67.0	62.0
Au.....	<2.0	<2.0	<2.0
Ba.....	<.4	<.4	<.4
Bi.....	<17.0	<17.0	<17.0
Ca.....	<9.0	<9.0	<9.0
Cd.....	<.3	<.3	<.3
Co.....	<.8	<.8	<.8
Cr.....	<4.0	<4.0	<4.0
Cu.....	<2.0	<2.0	<2.0
Fe.....	3.0	.9	<.6
K......	<35.0	<35.0	<35.0
La.....	<3.0	<3.0	<3.0
Li.....	<.9	<.9	<.9
Mg.....	<3.0	<3.0	<3.0
Mn.....	2.3	2.4	3.7
Mo.....	<1.0	<1.0	<1.0
Na.....	<9.0	<9.0	43.0
Ni.....	<3.0	<3.0	<3.0
Pb.....	39.0	19.0	18.0
Sb.....	11.0	14.0	10.0
Si.....	<6.0	<6.0	<6.0
Ti.....	8.7	1.6	2.3
Zn.....	22.0	12.0	16.0

Table V Removal of copper from pregnant solution containing 100 g/L Hg by precipitation with $CaCO_3$

Final pH	$CaCO_3$ added, g/L	Cu remaining in solution, g/L	Hg in Cu precipitate, pct
3.5	50	4.6	0.2
4.2	80	.37	.5
4.6	100	.026	.1
5.0	120	.001	.2

Electrolysis in 200-A Cell

The larger cell was operated continuously for 24 h to determine if sludge on the cathode or other problems would be encountered in a larger scale operation for longer periods of time. Feed electrolyte composition was 150 g/L Hg, 100 g/L Ca, 0.5 g/L Cu, and 1 g/L HCl. Electrolyte was continuously pumped from the reservoir tank to the 10-L cell at a rate of 1 L/min. Total volume of electrolyte was 95 L. Spent electrolyte was replaced with fresh electrolyte after 12 h of electrolysis. At the end of the 12 h, the spent electrolyte contained about 50 g/L Hg. Anode-cathode spacing was maintained at about 2.5 cm by siphoning 4.5 kg of mercury from the cell every 6 h. Operating data for the cell are given in Table VIII. The mercury cathode remained sludge-free during the test, and current efficiency was 98.6 pct. The high electrode current densities and the 2.5-cm electrode spacing were responsible for the high energy requirement of 2 kW·h/kg.

Mercury Removal From Wastewater

Three methods evaluated for mercury removal from wastewater were: cementation with iron powder, steel wool, or zinc; precipitation with H_2S or Na_2S, and adsorption on activated carbon. A solution containing 10 g/L Hg, 50 g/L Ca, and 1 g/L HCl was used in the experiments. When solid reagents were used, 1 L of solution was stirred with the reagent for 2 h at ambient temperature. With H_2S, the gas was slowly sparged into the solution for 2 h. The results of the experiments are shown in Table IX. Iron and zinc powder, H_2S, and activated carbon decreased the mercury concentration to 0.01 mg/L or below.

Table VIII Operating data for 200-A electrowinning cell[1]

Average cell current.........A..	200
Average voltage..............V..	7.0
Cathode current density..A/cm^2..	0.30
Anode current density....A/cm^2..	0.48
Electrolyte temperature.....°C..	60
Electricity used...........A·h..	4,800
Metal produced..............kg..	17.7
Current efficiency........pct..	98.6
Energy required.....kW·h/kg Hg..	1.9

[1]Initial electrolyte composition, in grams per liter: 150 Hg, 100 Ca, 0.5 Cu, 1 HCl. Electrode spacing, 2.5 cm.

Table IX Removal of mercury from process waste solutions[1]

Extraction reagent	Weight g/L	Hg conc, mg/L
Steel wool.........	20	1.0
Iron powder........	20	.01
Zinc powder........	20	.01
H_2S..............	5	.01
Na_2S.............	10	.05
Activated carbon[2]..	50	<.004

[1]Solution contained, in grams per liter: 10 Hg, 50 Ca, 1 HCl.
[2]CAL 12x40 from Pittsburgh Activated Carbon Co.

Precipitation by H_2S sparging is advantageous because the HgS can be returned to leaching, and no additional processing steps are necessary. A lower concentration of mercury in solution would probably be possible with any of the methods by either increasing the amount of extraction reagent or extending the treatment time.

SUMMARY AND CONCLUSIONS

Cupric chloride leaching has potential for treating mercury concentrates with minimal mercury vapor emissions. Leaching was rapid, and mercury extractions exceeded 99 pct. Chlorine sparged into the slurry during leaching decreased the copper required for leaching. Before mercury was recovered from the solution by electrolysis, copper was removed by increasing the solution pH to 4.5 with calcium carbonate. Failure to remove copper caused sludges to build up on the mercury cathode during electrolysis and decreased current efficiency to 30 pct. After copper removal, current efficiency was 99 pct and the energy requirement was as low as 0.8 kW·h/kg.

Process wastewater containing mercury was treated by cementation, sulfide precipitation, and carbon adsorption. All methods removed the mercury to low levels (0.01 mg/L or less); sulfide precipitation is advantageous because the HgS product can be returned to leaching without additional process steps.

REFERENCES

1. Murphy, J. E., Henry, H. G., and Eisele, J. A. Recovery of Mercury From Concentrates by Cupric Chloride Leaching and Aqueous Electrolysis. Bureau of Mines Report of Investigations 9140, 1987, 9 pp.

2. Atkinson, G. B., Murphy, J. E., and Eisele, J. A. Recovering Mercury From a Flotation Concentrate by Continuous Leaching-Electrolysis. Bureau of Mines Report of Investigations 8769, 1983, 9 pp.

3. Butler, J. N. Studies in the Hydrometallurgy of Mercury Sulfide Ores. Nevada Bureau of Mines, Report 5, 1963, 58 pp.

4. Scheiner, B. J., Lindstrom, R. E., and D. E. Shanks. Recovery of Mercury From Cinnabar Ores by Electrooxidation. Bureau of Mines Report of Investigations 7750, 1973, 14 pp.

5. Town, J. W., McClain, R. S., and Stickney, W. A. Flotation of Low-Grade Mercury Ores. Bureau of Mines Report of Investigations 5598, 1960, 34 pp.

6. Town, J. W., Link, R. F., and Stickney, W. A. Caustic Sulfide Leaching of Mercury Products. Bureau of Mines RI 5748, 1962, 39 pp.

7. Town, J. W., Link, R. F., and Stickney, W. A. Precipitation and Electrodeposition of Mercury in Caustic Solutions. Bureau of Mines RI 5960, 1962, 199 pp.

8. Town, J. W., and Stickney, W. A. Cost Estimates and Optimum Conditions for Continuous-Circuit Leaching and Recovery of Mercury. Bureau of Mines Report of Investigations 6459, 1964, 28 pp.

9. Town, J. W., and Stickney, W. A. Beneficiation and Hydrometallurgical Treatment of Complex Mercury Sulfide Products. Bureau of Mines Report of Investigations 6569, 1964, 35 pp.

Arsenic fixation and tailings disposal in METBA's gold project

M. Stefanakis
METBA S.A., Athens, Greece
A. Kontopoulos
National Technical University of Athens, Department of Mining and Metallurgical Engineering, Athens, Greece

ABSTRACT

Research for arsenic fixation from liquid effluents has been lately intensified worldwide due to the need to treat ores and concentrates of lower grade and higher impurity content and the necessity to comply with stringent environmental regulations for residue disposal.

The objective of this work is to highlight the stability and describe the disposal technique of arsenic containing solid effluents arising from the treatment of gold bearing Olympias pyrite – arsenopyrite concentrate using the aqueous pressure oxidation method (SHERRITT process).

The results of a miniplant campaign indicate that approximately 90% of the arsenic contained in the feed precipitates during autoclave oxidation and is fixed in a crystalline form, unidentified as yet, but resembling the structure of basic ferric sulphates. The residual arsenic solubilized during autoclave oxidation is precipitated in the form of "basic ferric arsenates" with excess limestone–lime addition. Standard EPA toxicity tests and long term stability studies of above residues indicate that these can be safely disposed without adverse environmental impact.

Adoption of sub-aerial disposal technique in place of the conventional sub-aqueous technique provides a further safeguard for environmental protection. An extensive testwork program including triaxial cyclic tests, settling, permeability and chemical stability was undertaken to design the tailings pond. The main design criteria of the tailings pond include :
- secure containment of all solids and liquids and controlled collection, removal and recycling of liquids to the plant with underseal liner permeability less than 10^{-7}cm/s.
- liquefaction resistance to 8.5 Richter magnitude earthquake.
- compatibility with 24 h heavy rainstorm and reclamation of the tailings surface with topsoiling and revegetation.

INTRODUCTION

METBA's gold plant is designed to treat 100000 tpy of the Olympias pyrite concentrate to produce 2535 kg/y gold and 175 kg/y silver. The chemical and screen analysis of the feed material is given in Table I. The design basis of the plant is given in Table II. The concentrate is produced at the Olympias mine of Hellenic Chemical and Fertilizers Products Company (HCFP S.A.) in Northern Greece. Proven reserves to date amount to 11 Mt and inferred to 4Mt with an average chemical analysis 3.61% Pb, 5.22 % Zn, 16.6% S, 3.3% As, 0.19% Cu, 0.15% Sb, 7 g/t Au and 130 g/t Ag. The run-of-mine mixed sulphide ore is treated by differential flotation technique to produce sphalerite, galena and pyrite concentrates.

Table I
Average chemical and granulometric composition of the Olympias arsenical pyrite concentrate

Chemical analysis		Screen analysis	
	wt %	Mesh Tyler	Retained, wt%
Fe	40		
S	41	+48	1.5
As	12	−48/+65	4.0
Zn	0.8	−65/+100	10.5
Cu	0.08	−100/+150	18.5
Pb	0.7	−150/+200	16.8
Sb	0.12	−200/+325	18.1
Au (g/t)	26	−325	30.6
Ag (g/t)	35		
Insolubles	4.8		
Moisture	7		

The refractory nature of the pyrite concentrate due to the fine dissemination of the contained gold, and, its high arsenic content rendered evaluation of alternative technologies for its treatment a rather difficult task. The process selection, detailed elsewhere[1], concluded with the selection of aqueous pressure oxidation process (SHERRITT TECHNOLOGY), mainly because of environmental superiority and increased gold recovery.

Table II
Design Basis of METBA's Gold Project

Capacity	100000	tpy of concentrate
Availability	85	%
Gold recovery	97.5	%
Silver recovery	5	%
Gold production	2535	kg/y
Silver production	175	kg/y

The aqueous pressure oxidation process is the only hydrometallurgical process that has found industrial application for the treatment of refractory gold sulphide ores and concentrates. It is used in Homestake's McLaughling plant in California[2-3], in Gencor's Sao-Bento plant in Brazil[4-5] and in FMG Getchel plant in Nevada[6]. A number of other plants are advancing to basic and detailed engineering stage like the Porgera Gold Project/Papua New Guinea[7] and the Lihir Project/Lihir Island[8].

In this work we shall describe the origin and stability of arsenates to be produced in METBA's gold plant, the deposition technique for disposal of arsenic containing solid effluents and the measures taken to minimize environmental impact.

PLANT PRODUCTION PROCESS

The SHERRITT aqueous pressure oxidation process has been described in detail by Weir and Berezowsky[9-10]. The modelling of the reactions involved has been presented by Demopoulos and Papangelakis[11]. Below, a short description of the plant production process is given, to provide continuity in the understanding and assessement of arsenates produced.

The block diagram of the production circuit is given in Figure 1. The production circuit includes the following sections :

Concentrate Pretreatment. The concentrate after wet grinding to 95% - 325 mesh is mixed with oxidized solids recycled from the wash circuit. This recycling serves for :
- decomposition of carbonates in the feed material with consequent decrease of autoclave venting and hence increase in oxygen utilization.

- dispersion of intermediately produced elemental sulphur during autoclave oxidation to avoid formation of agglomerates that adversely affect autoclave operability and reduce gold recovery because of incomplete oxidation.

Aqueous Pressure Oxidation. After above pretreatment the feed pulp is introduced into two horizontal, six-compartment, oxygen sparged autoclaves, where the concentrate is oxidized at a temperature of 185-190° C and 1800 kPa pressure. The oxygen required is produced on the site in a cryogenic oxygen plant with a capacity of 315 tpd. The oxidized pulp is discharged from the autoclaves in flash tanks and advanced to the washing circuit.

Washing circuit. The oxidized pulp from the autoclaves is washed countercurrently in a battery comprising three thickeners, in order to achieve solid/liquid separation and washing of the oxidised solids from acid and other cyanicide impurities, before advancement to the cyanidation circuit.

Cyanidation - Gold Recovery. Leaching of gold and silver is effected with NaCN in a battery of cascading tanks using the carbon-in-pulp technique. The gold adsorbed on activated carbon is stripped in a modified Zadra circuit and the pregnant solution is advanced to electrowinning cells. The cathodes produced with addition of fluxes are melted in an induction furnace to produce gold dore containing 93.5% Au and 6.5% Ag.

Neutralization - Tailings disposal. The waste acidic solution from the washing circuit is neutralized with limestone and lime addition to precipitate arsenic and other impurities. The precipitated solids after thickening are mixed with the cyanidation tailings pulp following destruction of residual cyanide, and then, disposed to the tailings pond.

WASTE HANDLING - FERRIC ARSENATE STABILITY

Arsenic precipitation at atmospheric pressure

During autoclave oxidation all of the arsenic contained in the feed concentrate is solubilized. Approximately 90% reprecipitates in a form to be discussed later and the rest remains in solution and eventually reports to the waste solution of the washing circuit. This acidic waste stream containing arsenic and other impurities is neutralized with excess limestone and lime addition. The solution composition before and after neutralization is shown in Table III.

The results of Table III and other data not cited otherwise originate from the testwork in a miniplant campaign simulating the production circuit. As shown in Table III the average arsenic concentration after neutralization was 0.13 mg/l and decreased even further at the end of the run to 0.02 mg/l. The high Fe/As molar ratio averaging 6.4, and as high as 28, contributed to the low arsenic levels attained.

The neutralization residue consisting of gypsum, ferric arsenates and metal hydroxides was subjected to EPA (Environmental Protection Agency) toxicity test[13] the outcome of which is shown in Table IV.

The drainage water from the sub-aerial deposition modelling of tailings produced in the miniplant campaign contained less than 0.1 mg/l arsenic as shown in Table V.

Table III[12]

Average chemical analysis of the wash circuit waste solution before and after neutralization

	Before mg/l	After mg/l
As	2140	0.13
Cd	NA	<0.05
Fe	10200	0.22
Mg	30	9.2
Mn	30	<0.01
H_2SO_4 (pH)	11400	(9.6)
Zn	280	<0.05
Fe/As [M]	6.4:1	-

NA: not analysed

Table V[12]

Median miniplant tailings drainage water analysis, mg/l

Sb	As	Ca	Cu	Fe	Pb	Mg	Mn	Zn
<0.1	<0.1	551	<0.05	<0.05	<0.05	24.1	0.64	<0.05

Despite the above encouraging results with respect to arsenic stability, METBA has undertaken an extensive literature and experimental study program, to assess the long-term stability of the ferric arsenates the results of which are described elsewhere[14-16]. The main conclusions of past work which is continued in METBA's laboratory, assessed in conjuction with literature data are :

a. Arsenic precipitation at atmospheric conditions with an Fe/As molar ratio in solution above 4 leads to formation of ferric arsenates suitable for safe disposal in a tailings pond based on stability testing of 2-year duration. These findings are in accordance with those reported earlier by Krause and Ettel[17-18]. They termed the above precipitates as "basic ferric arsenates" and postulated that these constitute discrete compounds of the form $(FeAsO_4 \cdot xFe(OH)_3)$ and not just physical mixtures of amorphous scorodite $(FeAsO_4 \cdot 2H_2O)$ and ferric hydroxide $(Fe(OH)_3)$. Kinetic study of ferric arsenate precipitation in a recent METBA's work[16] supports the theory of basic arsenate compound formation since it was observed that the Fe/As ratio in the precipitate remains identical from the onset to end of precipitation and equals the Fe/As ratio in the starting solution. Therefore there is no preferential precipitation of $FeAsO_4$ followed by $Fe(OH)_3$ as would be expected from thermodynamic analysis of the pure components but coprecipitation of $FeAsO_4$ and $Fe(OH)_3$ in the form of $FeAsO_4 \cdot xFe(OH)_3$.

The increased stability of ferric arsenates is also documented in Harris and Monette work[19-20] in which, based on 722 days stability testing, it is reported that ferric arsenates with Fe/As molar ratio greater than 3 are stable in the pH range 3-7 with As solubility less than 0.5 mg/l. Addition of base metals such as Cu, Cd, Zn extends the stability pH range to 3-10.

b. Robins who has pioneered work on the arsenates stability[21-28] does not accept the theory of basic ferric arsenates and claims instead that As is adsorbed onto $Fe(OH)_3$ which eventually will convert to a-FeOOH with release of arsenic into the enviroment. This is apparently so for

Table IV

Stability of neutralization residue according to EPA toxicity test

	Ag	Al	As	Sb	Ca	Cd	Cu	Cr	Fe	Mg	Mn	Pb	Zn
Filtrate analysis, mg/l	0.08	0.037	0.2	ND	984	0.005	<0.005	ND	<0.005	81.3	0.975	<0.05	0.21
EPA toxicity limit, mg/l	5.0	-	5.0	-	-	1.0	-	5.0	-	-	-	5	-

ND : Not detected by ICP, ICPMS methods

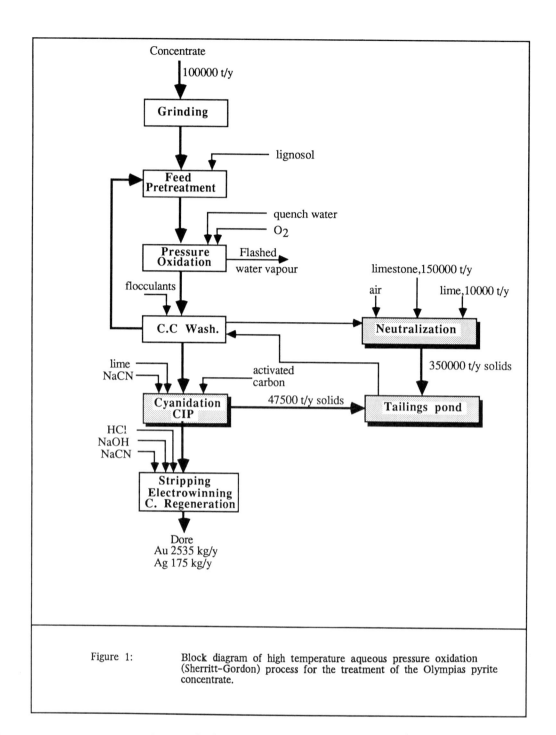

Figure 1:   Block diagram of high temperature aqueous pressure oxidation
            (Sherritt-Gordon) process for the treatment of the Olympias pyrite
            concentrate.

stoichiometric ferric arsenate (Fe/As [M]=1). The thermodynamic analysis of Robins does not cater for compounds such as basic ferric arsenates though and therefore their predominance area in an As solubility -pH diagram cannot be inferred.

In cases that the effect of high Fe/As ratios is evaluated by Robins[27] to infer removal of arsenic by adsorption, the arsenic concentration in the starting solution is low (<150 mg/l in most cases) and the pH used for adsorption calculations high (pH = 4) such as that the precipitates produced to correspond to the area where $Fe(OH)_3$ is predominant. In those experiments no release of arsenic was observed for a period of two years but it was reported that aging over longer periods has resulted in the growth of FeOOH with release of arsenic.

Slow kinetics of scorodite conversi FeOOH are reported in a recent wo Robins[28]. This retardation was attributed to incorporation into the FeOOH crystals and proposed as an ideal method of controlled arsenic release to alleviate the problems of long term storage.

c. The increased stability of non-stoichiometric ferric arsenates has been well documented before[17-20]. Krause and Ettel[17-18] reported that even 2:1 iron to arsenic precipitates are stable over a period of 905 days yielding As solubility below 0.74 mg/l. Increasing the iron to arsenic ratio from 4:1 to 17:1 results in As solubility consistently below 0.2 mg/l after 636 days at pH 5. Similar data have been reported in Harris and Monette[19-20] work where is shown that ferric arsenates with an Fe/As molar ratio above 3 are stable in the pH range 3-7 yielding As solubility less than 0.5 mg/l over a 722 days period. Addition of base metals Cu, Cd, Zn was found to extend the stability pH range to 3-10. In the studies above[17-20] no change of the XRD pattern was observed to indicate transformation of "basic ferric arsenates" to more soluble amorphous scorodite.

In an attempt to predict long term stability of ferric arsenates Krause and Ettel[18] conducted an accelerated aging test on an Fe/As precipitate with 8:1 molar ratio at 100°C, pH 5, 20% solids. It was found that the As solubility increased from 0.3 mg/l to 1.8 mg/l at 15 and 222 days of contact respectively. A partial transformation of the amorphous phase to hematite (Fe_2O_3) was observed but no scorodite was identified in the aged solids. Since the major aging product was Fe_2O_3 it was considered that the test does not simulate aging at ambient temperature where goethite formation predominates and cannot therefore be taken in favor of Robins theory[21-28].

The same authors[18] also showed that natural ore samples of varying Fe and As content (molar Fe/As ratio 2.39-4.05) exhibit lower solubility than that expected for precipitates with same Fe/As ratio. It was thus concluded that the solubility of basic ferric arsenates will actually decrease during aging and not increased as suggested by Robins[25].

The validity of any stability test to predict the long term behaviour of ferric arsenates would always be debatable and quite rightly so. Nevertheless the above experimental findings suggest that under certain conditions ferric arsenate precipitates can be extremely insoluble and environmentally safe for the disposal of arsenic.

The main factors contributing to increased stability of ferric arsenates are operating during arsenic precipitation at atmospheric pressure in METBA's plant, and include :
- the high Fe/As molar ratio averaging 6.4 during the integrated miniplant run,
- the stabilization effect induced by gypsum coprecipitation,
- the protective alkalinity provided with excess limestone-lime addition,
and therefore the anticipated stability of the resultant precipitates would be extremely high.

Arsenic precipitation in autoclaves

Under the conditions of aqueous pressure oxidation (190° C, 1800 kPa) the concentrate fed to the autoclaves is fully solubilized according to the reactions :

$$4FeAsS + 11O_2 + 2H_2O \rightarrow 4HAsO_2 + 4FeSO_4$$
$$2FeS_2 + 7O_2 + 2H_2O \rightarrow 2FeSO_4 + 2H_2SO_4$$

$$MS + 2O_2 \rightarrow MSO_4 , \quad M = Cu, Zn, Pb$$
$$4FeSO_4 + O_2 + 2H_2SO_4 \rightarrow 2Fe_2(SO_4)_3 + 2H_2O$$
$$2HAsO_2 + O_2 + 2H_2O \rightarrow 2H_3AsO_4$$

Concurrently with above dissolution the major part of arsenic (approximately 90%) precipitates in a variety of forms to be discussed later and represented for simplicity by the following reaction :

$$Fe_2(SO_4)_3 + 2H_3AsO_4 + 2H_2O \rightarrow 2FeAsO_4 \cdot 2H_2O + 3H_2SO_4$$

Part of ferric sulphate is hydrolyzed and precipitates as hematite (Fe_2O_3), basic iron sulphate ($Fe(OH)SO_4$) and hydronium, lead or argento-jarosite. Details of these reactions and others occuring to a lesser extent dependent on the conditions prevailing during autoclave oxidation are given elsewhere[9-11].

Following oxidation, the oxidized solids produced are washed, conditioned with lime and leached with sodium cyanide prior to be disposed to the tailings pond. The above operations alter only marginally the structure of the oxidized solids which remain practically identical to the autoclave discharged solids. Arsenic fixation and stability of solid effluents exiting the cyanidation circuit depends therefore merely on the reactions occuring in the autoclave.

Mineralogical analysis of the oxidised solids produced from the miniplant run confirmed the presence of hematite and ferric sulphates described above. Arsenic, however, could not be exactly tied to a tabulated arsenical mineral in the XRD files. Its diffraction pattern was found to resemble that of scorodite ($FeAsO_4 \cdot 2H_2O$) and symplesite ($Fe_3(AsO_4)_2 \cdot 8H_2O$). In several other of the miniplant campaigns at SHERRITT to treat refractory gold ores and concentrates, a complex species containing iron, arsenic and sulphur in the approximate molar ratio of 2:1:1, resembling sarmientite or bukovskyite, $Fe_2(AsO_4)(SO_4)OH \cdot nH_2O$ has been identified in the oxidation products[12].

The EPA toxicity test results on the cyanidation effluent solids produced from the miniplant run is shown in Table VI. The arsenic concentration in the filtrate is 0.4 mg/l well below the EPA toxicity concentration of 5 mg/l.

The stability of cyanidation barren solids when codisposed with the neutralization residue was tested using the following procedure. A sample of tailings material was taken from the tailings box which was used to simulate the sub-aerial deposition technique during the integrated run. The tailings material with a moisture content of 13% (two years after the end of the run) was equilibrated with water at 10g/l solids, 20-25°C, and varying pH. The pH of the solution and the As solubility over a period of one year are shown

Table VI

Stability of cyanidation residue according to EPA toxicity test

	Ag	Al	As	Sb	Ca	Cd	Cu	Cr	Fe	Mg	Mn	Pb	Zn
Filtrate analysis, mg/l	0.06	<0.02	0.4	ND	1.0	ND	<0.005	ND	<0.005	23	3.6	<0.05	0.05
EPA toxicity limit, mg/l	5.0	-	5.0	-	-	1.0	-	5.0	-	-	-	5.0	-

ND: Not detected by ICP, ICPMS methods

in Table VII. During this period the solution was replaced five times with fresh solution of the starting pH.

The As solubility measured after 305 days was ≤0.6 mg/l at relatively high pH (7.5), which is indicative of the stability of the tailings that ultimately will be disposed to the tailings pond.

Table VII

Stability of waste solids to be disposed to the tailing pond at various pHs

Initial pH	14 days		305 days	
	pH	As, mg/l	pH	As, mg/l
4	7.53	0.4	7.50	0.6
5	7.47	0.4	7.70	0.4
6	7.65	0.2	7.80	0.38
7	7.80	0.6	7.80	0.35
8	7.84	0.4	7.73	0.38
9	7.80	0.6	7.78	0.21
10	7.86	0.6	7.85	0.42

The stability of ferric arsenates produced at high temperatures in autoclaves has not been investigated to any great extent. Krause and Ettel[18] reported that the stability of crystalline scorodite produced at 160°C at CANMET is 2 orders of magnitude higher than that of amorphous scorodite and the As solubility measured was 0.6 mg/l at pH=5.4. The arsenic solubility from a natural scorodite sample was somewhat higher 0.4-3.6 mg/l at pH 5. Similar data were reported by Harris and Monette[20]. Above precipitates were produced from synthetic solutions with a molar Fe/As ratio 1.

The results of arsenic precipitation from research work undertaken by METBA aiming to study the effect of precipitation temperature, Fe/As ratio and acidity indicate so far that[29]:

- In the temperature range 140-170° C crystalline scorodite precipitates from synthetic solutions with an Fe/As molar ratio 3. Above 170° C and up to 200° C a different phase precipitates no yet identified. The X-ray diffraction pattern of this phase resembles that of the compound $MgSO_4 \cdot zMg(OH)_2 \cdot nH_2O$. Based on the chemical composition of the residue, formation of compounds resembling sarmientite and bukovksyite minerals is inferred. The above solids after washing were subjected to stability testing at 25° C, 1 g/l solids and natural pH (2.5-3.5). The As solubility measured after one month was <0.03 mg/l except at 140, 200° C for which the arsenic solubility was 0.5 mg/l. The equilibrium pH in the latter case was the lowest (pH=2.5).
- Increase of the precipitation temperature from 140° C to 200° C increases the SO_4^{2-} content in the residue from 8 to 27%. At the same time the Fe/As molar ratio in the residue increases from 1.15 to 2.4 through precipitation of ferric iron. Whether ferric iron coprecipitates as a distinct phase or as ferric-arsenate-sulphate compound or both needs to be further investigated.
- At constant temperature 180° C crystalline scorodite precipitates for Fe/As molar ratio 1-2. For Fe/As molar ratio 3 and 4 precipitation of the unidentified phase X occurs.
- X-ray diffraction analysis of the residue produced from oxidation of the Olympias pyrite concentrate at 190° C in a laboratory autoclave indicated the presence of crystalline scorodite and possibly the unidentified phase X reported above. Further work is underway to resolve the structure of the oxidation residue.

Based on the above it is reasonable to expect that the stability of ferric arsenates produced in autoclaves would be far superior to those produced under atmospheric conditions. The Fe/As molar ratio in the Olympias pyrite concentrate is 4.5. Due to faster dissolution of FeAsS in the autoclaves compared to FeS$_2$ the operational Fe/As molar ratio during precipitation of ferric arsenates would be lower than that in the feed material. Nevertheless the excess ferric iron in the system that precipitates subsequently or coprecipitates through redissolution of the crystalline scorodite formed in the first stage is expected to have a stabilizing effect on the residue similar to that reported for "basic ferric arsenates" formed under atmospheric conditions. Of course this is only a theory, the validity of which needs to be established through detailed investigation.

SOLID EFFLUENTS DISPOSAL

The advantages of sub-aerial disposal technique presented in a number of papers[30-36] prompted its adoption in METBA's Gold Project. The miniplant integrated run included disposal of tailings produced in a tailings box simulating sub-aerial deposition technique. The arrangement used was identical to that employed for the Key-Lake project development[37]. Deposition was carried out in thin layers recording quantity and quality of decant and drain solution, pore pressure and shrinkage. Infra-red lamps and fans were used to simulate climatic conditions of the actual tailings site. Samples from the tailings box and also in slurry form, were subsequently subjected by a contractor to a detailed strength, permeability and geochemical test program to establish the design basis of the tailings pond. Details of design criteria and tailings characterisitics are given below. Due to tailing site alteration during Project evolution, part of the work undertaken has to be repeated to reflect conditions at the new site. It is anticipated though that sub-aerial deposition would be favoured since climatic data are similar to the previous site, soil stability is higher and seismicity is lower.

Design criteria - objectives

The principal design objectives of the tailings pond to ensure environmental protection include:
- Permanent, secure and total confinement of all solid waste materials.
- Controlled collection, removal and recycling of all liquids associated with the disposed solids.
- Permanent arrangements to monitor all aspects of the performance of the tailings pond.
- A tailings deposition system such that at any time during, and, on completion of operation, the tailings are fully drained and consolidated in order to eliminate long term seepage and enhance seismic stability.

The above objectives can be achieved by providing a secure initial embankment and underseal, provision of drainage systems upstream of the embankment and at the base of the tailings, by using sub-aerial deposition technique for tailings disposal and surface topsoiling and vegetation to minimize infiltration after decommissioning.

Tailings characteristics

Physical characteristics

Tailings material characterisation in laboratory tests has shown that :
The material is dark red brown with approximate composition,

$FeAsO_4 \cdot 2H_2O$	9.64 %
$Fe(OH)_3$	15.04 %
$CaSO_4 \cdot 2H_2O$	54.79 %
$CaCO_3$	16.87 %

The material has a specific gravity of 3.05 and can be classified as a non-plastic silt with a trace of sand, designation ML under the Unified Soil Classification System.

The material settles sub-aqueously to relatively low densities, in the range of 0.6 t/m^3 which are only marginally increased after drainage. Air drying of the material, however, increases the density to approximately 1.0 t/m^3. At this density it is relatively incompressible but exhibits anomalously low values of c_v, coefficient of consolidation, which may be associated with time dependent creep behaviour. The coefficient of permeability at a density of 1.0 t/m^3 is in the range of 1 to 2 x 10^{-7} m/s.

A consolidated undrained triaxial shear strength test on the material indicates a high strength, with an angle of internal friction of 33° at low confining pressures (up to 200 kPa), increasing to 41° at a confining pressure of 500 kPa. The implications of this are that the tailings can be used for support of structural sections of the enbankment provided that the material is fully air-dried, drained and unsaturated.

Liquefaction resistance

The factor of safety against liquefaction was assessed by comparing the cycling resistance ratio of tailings to the cycling stress ratio caused by the earthquake as follows:

Factor of safety against liquefaction (F.O.S) = $\dfrac{\text{Cycling resistance ratio of tailings}}{\text{Cycling stress ratio caused by earthquake}}$

The measured values of cyclic resistance ratio for the resaturated samples from the tailings box are relatively high primarily because of the dendritic structure of the gypsum crystals. Consolidated samples prior to air-drying simulating near-surface tailings discharge during winter months, also exhibit cycling resistance ratio in the same order as above.

A further significant result of the cycling triaxial tests is the high values measured for the post-liquefaction undrained strength. Even after liquefaction was induced the material showed significant dilation and strength gain on shearing, indicating that a loss of strength on "flow slide" is not possible.

The worst case Factor of Safety against liquefaction from an 8.5 Richter magnitude earthquake was found from above tests to be 2.11. A Cornell type seismic risk analysis based on historical earthquake records for the previous site indicated return periods of approximately 153 years for magnitude 8.5 events. The seismicity for the new site is lower and the highest magnitude earthquake reported in the last 57 years is 5.7 Richter magnitude.

Geochemical tests

The short term stability tests and other testwork undertaken by METBA to predict the longterm stability of tailings produced in METBA's Gold Plant has been discussed in detail before. The following test results from contracted testwork were also produced :
(i) The acid generation potential test showed that the tailings are non-acid producing containing a substantial excess of reactive alkalinity equivalent to 142.3 kg H_2SO_4/tonne of tailings
(ii) Successive wetting and drying of tailings from a humidity cell test with 3 weeks duration showed arsenic dissolution of 0.3-0.4 mg/l.
(iii) ASTM solid waste extraction test to assess long term water leachable constituents showed As dissolution of 0.1 mg/l. This compares with arsenic concentration of less than 0.1 mg/l in the drainage water from the simulated tailings deposition.
(iv) Routine tests to determine naturally occuring radionuclides confirmed that no significant radionuclides are present. The results obtained are :

solids/Ra 226	0.04 Bq/g
liquids/Gross alpha	1.3 Bq/l
/Gross beta	7.0 ± 1.7 Bq/l

Tailings pond construction

Construction requirements for the new tailings site will be defined following detailed site investigation. The principal design features of the construction schedule are as follows and illustrated in figure 2.

(i) Initial embankment will be constructed from residual soil excavated from within the basin area which will be placed and compacted to form an homogeneous embankment.Upstream construction techniques will be used to raise the embankment using the tailings as the main structural material. After the overall embankment has been raised by approximately 4m, fill material will be spread and compacted to form an access bench at the new elevation, and similar material will be placed on the exposed tailings forming the dowstream face of the embankment to act as erosion protection which can be revegetated.

(ii) A low permeability underseal, constructed of in-situ soil with some bentonite addition extending over the basin area, to prevent significant vertical seepage from the tailings facility. The underseal will be designed for a premeability objective of less than 10^{-7} cm/s.
(iii) A network of strip underdrains underlying the inside perimeter of the facility to assist in draining the tailings in areas where upstream construction techniques will be used.
(iv) A staged system of decants to remove water from the tailings surface.
(v) A process water pond for recycling to the mill.
(vi) Runoff diversion ditches to route runoff water away from the facility.
(vii) A series of groundwater monitoring wells around the dowstream perimeter of the facility to monitor ground water quality downgradient of the tailings facility.

Tailings pond operation

The operation of the tailings pond is based on the following tailings material parameters:

- Tailings production rate	: 350000 t/year
- Solids content at discharge	: 32%
- Initial water recovery from tailings	: 35%
- Average dry density of tailings	: 0.95 t/m^3
- Precipitation runoff coefficient of tailings beach and underseal	: 0.80

The hydrometeorological data for the new site need to be further analysed to establish the water mass balance of the tailings pond. The mode of tailings pond operation will essentially be as described below.

Tailings slurry from the plant will be delivered to the tailings storage facility via a pipe terminating at the tailings header pipe laid along the perimeter-to be defined-of the facility. The tailings slurry will be discharged from multiple offtake points around the header pipe in thin layers of 10-15 cm thickness. Deposition will be carried out on a controlled rotational basis using 3 offtakes at approximately 20 m centres. During this process the liquid bleeding from the surface flows to the bottom end of the gently sloped beach (0.5-1.0% slope) and collects in the supernatant pond around the decant structure and directed to the process water pond for recycle to the plant. Drainage of liquid from the newly placed layer also occurs during deposition and the liquid drawn from the bottom of the layer is absorbed by the air voids in the partially saturated layer beneath. After the bleeding has ceased the surface of the layer is left exposed and additional moisture is removed by evaporation. Large negative pore pressures are induced by evaporation and drainage which cause further consolidation of the solid particles.

Important characteristics of the sub-aerial

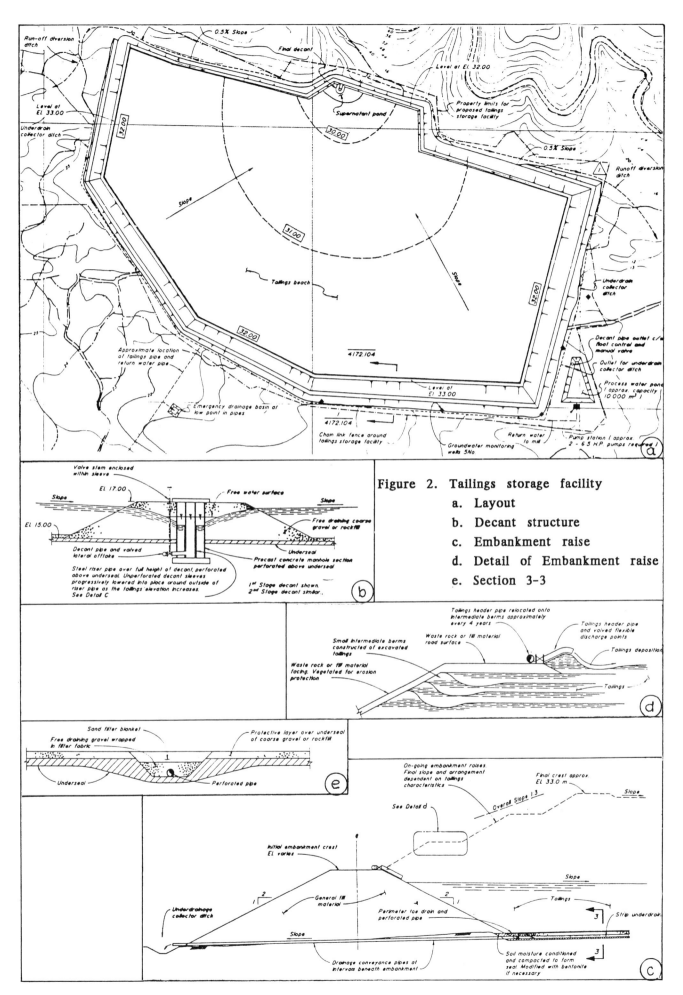

Figure 2. Tailings storage facility
   a. Layout
   b. Decant structure
   c. Embankment raise
   d. Detail of Embankment raise
   e. Section 3-3

297

deposition technique and associated environmental benefits are:
- Particle segregation during settling results in an anisotropic condition in which the vertical coefficient of permeability is by two orders of magnitude lower than the horizontal one. This condition assists in reducing the seepage through the tailings. Once the depth of tailings has exceeded approximately 1.5 m, the moisture movement through the tailings becomes extremely small and the underdrainage system serves as a long term monitoring system. The tailings material itself becomes a positive seal against downward movement of water.
- The installation of a filter blanket covering the whole basin area of the facility immediately above the underseal material allows for continuous release of all free draining liquid contained in the stored tailings and provides a system for lateral drainage of infiltrating water within the filter blanket. This drainage prevents any build up of pore pressures within the tailings and reduces the head built-up on the underseal with consequent reduction of seepage through the seal.
- At completion of plant operation the residual liquid contained in the tailings will be permanently locked in by capillary tension with minimal lateral seepage. Reclamation of the tailings facility can proceed immediately after decommissioning. Any surplus water remaining on the beach will be recirculated over the beach and wasted through evaporation. The laminated nature of the deposited layers resulting in structural integrity will allow foundation of a seal on the tailings surface topped with a rooting layer of adequate thickness for the vegetation intended.

The selection of sub-aerial deposition technique and the adoption of strict design criteria for construction and operation of the tailings pond as detailed above are rigorous enough to abide even by the stringent environmental standards worldwide for solid effluents disposal.

CONCLUSIONS

The adoption of a hydrometallurgical process to treat the Olympias pyrite-arsenopyrite concentrate, the enforcement of strict design criteria to safeguard safety and hygiene of the plant atmosphere and the incorporation of high fidelity-integrity instrumentation for process control reflect the attitude that METBA has followed for project implementation. The objective all along was to expoit the indigenous mineral resources of Greece and contribute to its industrial and technological development with the installation of a cost effective and modern plant with the least possible environmental impact. At this stage with detailed engineering almost completed it can be said that the above objectives have been met since:

- The project is quite profitable with internal rate of return (IRR) on investment and equity before taxes 25.8 and 40.3% respectively.

- The technology adopted is of the highest calibre.
- The plant is environmentally competent. No gaseous effluents exist except steam and no liquids are discharged to the environment.

Nevertheless the discharge of arsenic containing solid effluents was of concern and extensive research has been undertaken both abroad and in METBA's laboratory to investigate their stability with respect to arsenic dissolution. The outcome of this research described in this work indicates that:
- The ferric arsenate precipitates produced at atmospheric pressure during the neutralization of waste acidic solution with excess limestone and lime addition are insoluble due to the high Fe/As molar ratio in the waste solution and the stabilization effect of metal impurities and gypsum coprecipitation. In these precipitates arsenic is probably fixed in compounds such as $"FeAsO_4·xFe(OH)_3"$ distinctly different from amorphous scorodite $FeAsO_4·2H_2O$. The arsenic solubility in all stability tests was less than 0.4 mg/l.
- The ferric arsenates produced in autoclaves at 190° C concurrently with arsenopyrite-pyrite oxidation contain crystalline scorodite $(FeAsO_4·2H_2O)$ and other arsenic compounds resembling sarmientite and bukovskyite $(Fe_2(AsO_4)(SO_4)(OH)·nH_2O)$. The arsenic solubility of those precipitates was 0.4 mg/l.
- Despite the above encouraging short and long term stability tests METBA actively participates in a project pursued by MIRO (Mineral Industry Research Organization) and sponsored by 15 industries in order to resolve the structure of arsenate precipitates in correlation to arsenical minerals, and predict the long term stability in comparison to the weathering behaviour of arsenic containing ore-bodies.

The adoption of sub-aerial deposition technique for tailings disposal provides a further safeguard for minimazion of environmental impact since:
- long term seepage is eliminated through an extensive underdrainage collection system installed above the basin underseal.
- the laminated structure of deposited layers and their low vertical coefficient of permeability constitutes a positive barrier against any downward movement of liquids.
- the structural integrity of deposited tailings enhances the liquefaction resistance and seismic stability.
- decommissioning and surface reclamation of the tailings can start immediately on completion of plant operation.

Acknowledgements: METBA's Gold Project development has only been possible with the dedication of all its personnel. Special thanks are due to Dr. N. Papassiopi for her enthusiastic contribution to arsenic stabilization research. The miniplant campaign was carried out at SHERRITT's Research Centre, Fort Saskachewan, CANADA and the solids effluent disposal testwork was undertaken by KNIGHT & PIESOLD, Vancouver, CANADA. The contribution of above Companies and the high quality work produced is appreciated. The financial contribution of General Secreteriat of Research and Technology of Greece through the grants of Research programs ΠΑΕΤ 9705(EPE)901 and ΠΑΒΕ 88ΒΕ149 is gratefully acknowledged. Thanks also extend to E. Koutsoukou for typing the manuscript.

REFERENCES

1. A. Kontopoulos and M. Stefanakis. Process Selection for the Olympias Refractory Gold Concentrate. In: Precious Metals '89, Eds. M.C.Jha, S.D.Hill and M.El. Guindy, TMS, Warrendale, Pennsylvania, 1988, p. 179-209.
2. R. Guinivere. McLaughlin project : Process, project and construction development. In: Proceedings of the First International Symposium on Precious Metals Recovery, Reno, Nevada, June 10-14, 1989.
3. J. R. Turney, R.J. Smith and W.J. Janhumen Jr. The Application of Acid Pressure Oxidation to the McLaughlin Refractory Ore. In: Precious Metals '89, Eds. M.C.Jha, S.D.Hill and M.El. Guindy, TMS, Warrendale, Pennsylvania, 1988, p. 25-45.
4. T.M. Carvalho et al. Start-up of the SHERRITT Pressure Oxidation Process at Sao bento, Randol Gold Conference, Randol Perth International, Perth, Australia, 28 Oct. - 1 Nov., 1988.
5. E.J. da Silva et al. Process Selection, Design, Commissioning and Operation of the Sao-Bento Mineracao Refractory Gold Ore Treatment Complex. In: World Gold '89, Reno, Nevada, Chapter 38, p. 322-332.
6. P.G. Mason and R.F. Nanna. A New Beginning for the Getchel Mine. In: Precious Metals '89, Eds. M.C.Jha, S.D.Hill and M.El. Guindy, TMS, Warrendale, Pennsylvania, 1988, p. 3-12.
7. Engineering and Mining Journal, July 1989, p. 64.
8. M. Conway, FLUOR DANIEL, Private Communication.
   D.R. Weir and R.M.G.S. Berezowsky. The role of pressure oxidation. In: Gold 100, Proceedings of International Conference on Gold. Vol. 2, Extractive Metallurgy of Gold, SAIMM, Johannesburg 1986, p. 275-285.
10. D.R. Weir and R.M.G.S Berezowsky. Gold Extraction from Refractory Concentrates. Paper Presented at the 11th Annual Hydrometallurgical Meeting of Canadian Institute of Mining and Metallurgy, Timmins, Ontario, Oct. 1984.
11. V.G. Papangelakis and G.P. Demopoulos. Computer-aided Simulation of the Pressure Oxidation Process for Refractory Gold Recovery. In: Control 90-Mineral and Metallurgical Processing, Eds. R.K. Rajamani and J.A. Herbst, Society for Mining, Metallurgy and Exploration. Inc., Littleton, Colorado, 1990, p. 121-126.
12. M.J. Collins, R.M. Berezowsky and D.R. Weir. The Behaviour and Control of Arsenic in the Pressure Oxidation of Uranium and Gold Feedstocks.In: Arsenic Metallurgy-Fundamentals and Applications. Eds. R.G. Reddy, J.L. Hendrix and P.B. Queneau, The Metallurgical Society of AIME, Warrendale, Pennsylvania, 1988, p. 115-134.
13. U.S. Federal register, Vol. 45, No. 98, paragraph 261.2.4.
14. M. Stefanakis and A. Kontopoulos. Production of Environmentally Acceptable Arsenites - Arsenates. In: Arsenic Metallurgy-Fundamentals and Applications. Eds. R.G. Reddy, J.L. Hendrix and P.B. Queneau, The Metallurgical Society of AIME, Warrendale, Pennsylvania, 1988, p. 287-304.
15. N. Papassiopi, M. Stefanakis and A. Kontopoulos. Removal of Arsenic from Solutions by Precipitation as Ferric Arsenates. In: Arsenic Metallurgy-Fundamentals and Applications. Eds. R.G. Reddy, J.L. Hendrix and P.B. Queneau, The Metallurgical Society of AIME, Warrendale, Pennsylvania, 1988, p. 321-336.
16. A. Kontopoulos, N. Papassiopi and M. Stefanakis. Arsenic control in hydrometallurgy by precipitation as Ferric Arsenates. Proceedings of the 1st International conference on Hydrometallurgy. ICHM '88. Eds: Zheng Yulian and Xu Jiazhong, International Academy Publishers, 1988, p. 672-77.
17. E. Krause and V. A. Ettel. Ferric Arsenate Compounds: Are they Environmentally Safe? Solubilities of Basic Ferric Arsenates. In: Impurity Control and Disposal, Proceedings 15th Annual Hydrometallurgical Meeting of Canadian Institute of Mining and Metallurgy, Vancouver, Canada, Aug. 18-22, 1985, p. 5/1-5/20.
18. E. Krause and V.A. Ettel. Solubilities and Stabilities of Ferric Arsenates. In: Crystallization and Precipitation, Eds, G.L. Strathdee, M.O. Klein and L.A. Melis, Pergamon Press 1987, p. 195-210.
19. G.B. Harris and S. Monette. The stability of Arsenic-Bearing Residues. In: Arsenic Metallurgy-Fundamentals and Applications, eds. R.G. Reddy, J.L. Hendrix and P.B. Queneau, The Metallurgical Society of AIME, 1988, p. 469-489.
20. G.B. Harris and S. Monette. The Disposal of Arsenical Solid Residues. Presented at Productivity and Technology in the Metallurgical Industries, TMS-AIME/GDMB Joint Symposium, Cologne, West Germany, Sept. 17-22, 1989.
21. R.G. Robins. The Solubility of Metal Arsenates. Metallurgical Transactions B. 12(B), 1981, p. 103-109.
22. R.G. Robins and K. Tozawa. Arsenic removal from gold processing waste waters: the potential ineffectiveness of lime. Canadian Institute of Metallurgy Bulletin 75, April 1982, p. 171-74.
23. R.G. Robins. The Stabilities of Arsenic (V) and Arsenic (III) Compounds in Aqueous Metal Extraction Systems. In: Hydrometallurgy Research Development and Plant Practice, eds. K. Osseo-Asare and J.D. Miller, The Metallurgical Society of AIME, 1983, p. 291-310.
24. R.G. Robins. The Stability of Arsenic in Gold Mine Processing Wastes. In: Precious Metals, eds. V. Kudryk, D.A. Corrigan and W.W. Liang, The Metallurgical Society of AIME, 1984, p. 241-49.
25. R.G. Robins. The Aqueous Chemistry of Arsenic in Relation to Hydrometallurgical Processes. In : Impurity Control and Disposal,

Proceedings 15th Annual Hydrometallurgical Meeting of Canadian Institute of Metallurgy, Vancouver, Canada, Aug. 18-22, 1985, p. 1/1-1/26.

26. R.G. Robins. Arsenic Hydrometallurgy. In: Arsenic Metallurgy – Fundamentals and Applications, eds. R.G. Reddy, J.L. Hendrix and P.B. Queneau, The Metallurgical Society of AIME, 1988, p. 215-248.

27. R.G. Robins and J.C.Y. Haung. The Adsorption of Arsenate Ion by Ferric Hydroxide. In: Arsenic Metallurgy-Fundamentals and Applications, eds. R.G. Reddy, J.L. Hendrix and P.B. Queneau, The Metallurgical Society of AIME, 1988, p. 99-114.

28. R.G. Robins. The Stability and Solubility of Ferric Arsenate: Un Update. In: Proceeding EPD90 Symposium, TMS Annual Meeting, Anaheim, California, Feb. 19-22, 1990, p. 93-104, ed. D.R. Gaskel.

29. M. Stefanakis, N. Papassiopi. Unpublished work.

30. R.B. Knight and J.P. Haile. Sub-aerial Tailings Deposition. In: Proc. 7th Panamerican Conference on Soil Mechanics and Foundation Engineering, Vancouver, B.C., Canada, 1983, p. 627-639.

31. J.P. Haile and D.R. East. Recent Developments in the Design of Drained Tailings Impoundments. In: Proceedings 8th Symposium on Geotechnical and Geophysical Aspects of Waste management, Colorado State University, Fort Collins, Feb. 5-7, 1986, p. 301-306.

32. P.C. Lighthall. Innovative Tailings Disposal Methods in Canada. International Journal of Surface Mining, 1, 1987, p. 7-12.

33. J.P. Haile and K.J. Brower. Design and Construction of the Montana Tunnels Tailings Disposal Facility. Canadian Insitute of Mining and Metallurgy Bulletin, Volume 81, No. 919, November 1988, p. 70-74.

34. R. Dorey. Site and Method Evaluations for Tailings Disposal. In: Proc. Randol Gold Forum '88, Scottsdale, Arizona, Jan. 23-24, 1988, p. 91-94.

35. J.P. Haile and B.S. Brown. Design and Construction of Clay Liners for Gold Tailings Facilities and Heap Leach Pads. Paper presented at 2nd International Conference and Gold Mining, Vancouver, B.C., Nov. 7-9, 1988.

36. D.R. East. Recent developments in Environmentally Favorable Tailings Disposal Technologies. In: Proceedings Randol Gold Forum 88, Scottsdale, Arizona, Jan. 23-24, 1988, p. 95-98.

37. P.J. Clarke et al. The Key-Lake Uranium Process. Presented at the 19th International Conference of Metallurgists of the CIM, Halifax, N.S., August 1980.

300

Acid mine drainage from sulphide ore deposits

Fiona M. Doyle
Department of Materials Science and Mineral Engineering, University of California at Berkeley, Berkeley, California, U.S.A.

SYNOPSIS

Acid mine drainage is generally regarded as the principal environmental problem caused by the mining of sulphide ore deposits. Acid mine drainage results from the reaction of pyritic material with oxygen and water:

$$FeS_2 + 15/4\,O_2 + 7/2\,H_2O = Fe(OH)_3 + 2H_2SO_4$$

In general, oxidation proceeds by electrochemical mechanisms, with the cathodic reduction of oxygen being sufficiently strongly polarized that dissolution is slow. However, the solubility of Fe(III) increases with decreasing pH. This introduces an alternative, less strongly polarized cathodic reaction, which accelerates the oxidation of pyrite and release of acid. Oxidation may also be accelerated by naturally occurring bacteria, such as *Thiobacillus ferrooxidans*, which metabolize sulphide, sulphur and Fe(III) species. These bacteria are generally most active in acidic waters. Hence both the electrochemical and biological oxidation mechanisms are autocatalytic; once acid mine drainage develops, the problem tends to worsen and is extremely difficult to halt. Appreciable work has been done on acid mine drainage from coal mines. However sulphide ore deposits present an additional environmental threat; many sulphides are oxidized by acidic waters containing Fe(III), and release alarming amounts of heavy metals into affected drainage. Unfortunately, there is evidence that coal and ore pyrite exhibit different oxidation behaviour, thus existing preventive and mitigative technology addressing acid mine drainage from coal mines is not necessarily appropriate for sulphide ore deposits.

The West Shasta Copper-Zinc Mining District of northern California exemplifies the problems that can result from poor mining practices. This district contains a large number of abandoned and inactive mines in sulphide deposits. Large amounts of pyrite and other sulphides remain in the highly fractured ore-bodies, and in poorly situated tailings and waste piles. The local geology, the climate of mild, very wet winters and hot, dry summers, together with an extremely resilient strain of *T. ferrooxidans* all provide suitable conditions for rapid oxidation of pyrite, and release of large volumes of acid mine drainage. pH values as low as 0.6 have been measured in drainage from the Iron Mountain Mine, which is on the U.S. National Priority (Superfund) List. In addition to periodic fish kills, acid mine drainage from the district is responsible for much of the chronically high loading of heavy metals in the Sacramento River. Construction of a series of dams for irrigation of the Central Valley and hydroelectric power generation further exacerbated the problems, because this reduced the flow of clean water for dilution of the high volumes of contaminated runoff that follow heavy storms. The principal control and abatement strategies being used in the region are hydrologic controls and mine seals. The efficacy and cost of these methods are discussed critically, along with other techniques such as backfilling, water treatment, and proposed in-situ leaching. Although some of the practices that caused problems in the West Shasta Mining District, such as poor siting of waste management facilities, would not be allowed under current regulations, it is not certain that some of the problems might not arise from present and future development of sulphide ore deposits. The region provides useful insight into practices that should be avoided, and those that may be beneficial in ensuring that modern mines do not develop acid mine drainage problems.

INTRODUCTION

Acid mine drainage is widely considered to be the most serious environmental problem caused by mining. It is caused by the oxidation of pyrite, FeS_2, its dimorph marcasite, and pyrrhotite, $Fe_{1-x}S$. These are among the most common waste minerals found in sulphide ores. When exposed to oxygen and water, pyritic minerals are oxidized, releasing sulphuric acid, dissolved iron, sulphate, solid

301

iron(III) phases, and heat. The overall reactions are:

$$FeS_2 + 7/2\,O_2 + H_2O = Fe^{2+} + 2\,SO_4^{2-} + 2H^+ \quad (1)$$

$$Fe^{2+} + H^+ + 1/4\,O_2 = Fe^{3+} + 1/2\,H_2O \quad (2)$$

$$2Fe^{3+} + 3H_2O - 2Fe(OH)_3 + 6H^+ \quad (3)$$

Basic sulphates such as $Fe_4(SO_4)(OH)_{10}$ and jarosite may precipitate in place of $Fe(OH)_3$, but acid is still released[1,2] Numerous reaction paths and mechanisms have been proposed for the oxidation processes[3]. In addition, the kinetics of pyrite oxidation have been found to vary considerably, depending on the exact conditions, and the source of the pyrite. This is a fundamental obstacle to developing advanced technology for preventing and mitigating acid mine drainage. This paper discusses our current understanding of the chemistry and kinetics of pyrite oxidation. The problems of acid mine drainage are then exemplified by a description of the West Shasta Mining District, which is plagued by the worst acid mine drainage problems in California. Finally, appropriate preventive and mitigative measures are discussed in terms of fundamental and field considerations.

FORMATION OF ACID MINE DRAINAGE

It is generally accepted that when pyrite starts to oxidize in natural waters, reaction 1 is electrochemical, involving anodic dissolution of pyrite, and cathodic reduction of oxygen on the pyrite surface:

$$FeS_2 + 8H_2O = Fe^{2+} + 2SO_4^{2-} + 16H^+ + 14e \quad (4)$$

$$E = 0.367 - 0.0675\,pH + 0.0084\,\log\{SO_4^{2-}\} + 0.0042\,\log\{Fe^{2+}\} \text{ at } 25°C$$

$$O_2 + 4H^+ + 4e = 2H_2O \quad (5)$$

$$E = 1.23 - 0.0591\,pH + 0.0148\,\log\{O_2\} \text{ at } 25°C$$

where $\{SO_4^{2-}\}$, $\{Fe^{2+}\}$ and $\{O_2\}$ denote the thermodynamic activity of each species. Water is saturated with respect to atmospheric oxygen at 10 ppm, thus under ambient conditions the potential of reaction 5 is much lower than 1.23V. Moreover, the exchange current density is low for most sulphide minerals under abiotic conditions, because of the high activation energy for dissociative adsorption of O_2 molecules on the mineral surface. This gives low corrosion currents at the mixed potential, and slow initial oxidation. However, as acid is released by reactions 1 and 3, the pH falls. This first increases the potential of reaction 5, giving higher corrosion currents.

When the pH of water in an aerated waste management facility or mine falls below about 4.5, sulphide dissolution becomes markedly faster, typically by fivefold or more[4]. This is due to the action of naturally occurring, acidophilic chemolithotropic bacteria, which catalyze the oxidation of sulphides, sulphur, or Fe^{2+}. The bacteria provide biochemically mediated reaction paths for the oxidation processes, with lower activation energies than the abiotic paths. This enables them to utilize the energy released by the oxidation processes. The most common of these bacteria are *Thiobacillus ferrooxidans* and *T. thiooxidans*, although numerous other bacteria have been isolated from mine waters. Bacteria attached to sulphide surfaces give direct oxidation. The overall reaction would be similar to 1, although Fe(III) is usually produced in place of Fe(II), and pyrite oxidation may be non-stoichiometric.[5]

As the pH continues to fall, the equilibrium position of reaction 3 moves to the left, giving higher Fe(III) concentrations. Dissolved ferric ions can be reduced to ferrous:

$$Fe^{3+} + e = Fe^{2+} \quad (6)$$

$$E = 0.771 + 0.0591\,\log\{Fe^{3+}\} - 0.0591\,\log\{Fe^{2+}\} \text{ at } 25°C$$

Fe^{3+} reduction is much less strongly polarized on pyrite than is O_2 reduction, hence this cathodic reaction can give much more rapid dissolution rates. Under abiotic conditions, ferric reduction is of limited importance, because the reoxidation of ferrous ions to ferric by oxygen is about three orders of magnitude slower than the rate of ferric reduction[6]:

$$Fe^{2+} + H^+ + 1/4\,O_2 = Fe^{3+} + 1/2\,H_2O \quad (7)$$

Consequently, in sterile waters that are in contact with sulphide minerals, most of the dissolved iron is present as ferrous. However, bacteria such as *T. ferrooxidans* also accelerate reaction 7 by up to six orders of magnitude.[7,8] High ferric to ferrous ratios can then be maintained in solution, promoting even more rapid pyrite oxidation. Thus at low pH values, bacteria contribute to the release of acid mine drainage by indirect oxidation of pyritic material via Fe^{3+}, as well as by direct oxidation.

Pyritic waste generated during the utilization of sulphide orebodies contains other sulphide minerals, such as galena, chalcocite and sphalerite. The Fe^{3+} released by pyrite also accelerates the indirect bacterial oxidation of these minerals. Although these minerals do not produce acid, they release heavy metals, exacerbating the adverse environmental impact of the acid mine drainage. The dissolution of non-pyritic sulphide minerals may be accelerated by galvanic interactions with pyrite, which usually has a rest potential appreciably higher than that of many other sulphides. The pyrite becomes cathodic, while the minerals with lower rest potentials become anodic and dissolve preferentially, provided that the aqueous phase is already acidic.[9] Galvanic interactions have been reported to be enhanced by the presence of bacteria,[10,11] with the effects being largely additive for chalcopyrite/pyrite and sphalerite/pyrite mixtures.[12] However in waste management facilities it is unlikely that individual sulphide grains would remain in electrical contact for long, because oxides, sulphur and other insulating phases will form at the grain boundaries as dissolution progresses.

Pyrite oxidation has been extensively studied.[13,14,15,16,17,18] Even in well controlled laboratory experiments, there are significant differences in the behaviour of pyrite samples from different sources, particularly between coal and ore pyrite.[19] Although fine-grained coal pyrite tends to be most reactive,[20] the oxidation rate is not necessarily related to the BET surface area.[21] XPS measurement of the oxidation rates per unit area showed coal pyrite to be much less reactive after hydrothermal alteration.[22]

The electrochemical behaviour of pyrite reveals further anomalies between coal and ore pyrite. When pyrite is contacted with a sulphate solution, containing a fixed amount of oxygen, it should equilibrate with the solution. The Eh-pH diagram for the iron-sulphur-water system at 25°C[23] suggests that at low pH, the pyrite should adopt a rest potential of around 0.3V (depending on the solution composition). The measured rest potential of ore pyrite is usually significantly higher.[24,25,26] These anomalous rest potentials have been ascribed to passivation by sulphur layers,[9] metal-deficient sulphide layers[27] or iron oxide layers.[28] However, although the rest potentials of ore pyrite are high, those of coal pyrite are close to the thermodynamic value.[29] It has also been noted that there are significant differences in the charge transfer resistance of coal and ore pyrite.[30] If passivation were occurring, then pyrite from either source should behave similarly. Chander and Briceno have proposed that ore pyrite forms coherent films, whereas coal pyrite forms fractured films.[30] None of these proposals is completely satisfactory, however. Passivation would reduce the current density seen at a given potential, but would not allow ore pyrite to pass cathodic currents at potentials where coal pyrite passes anodic currents.

It is possible that the differences between coal pyrite and ore pyrite, along with the differences within each type, are due to differences in the semiconductor properties. Pure n-type and p-type semiconductors show different electrochemical behaviour.[31,32] Varying dopant concentrations and deviations from stoichiometry have been shown to affect the oxidation rates of synthetic sulphides and oxides.[33,34] Coal pyrite tends to be a p-type semiconductor, whereas ore pyrite is usually n-type. Oxidation of the sulphides proceeds by formation of sulphur-rich layers, which would be expected to be p-type. When present on n-type material, these layers would form n-p junctions that are reverse biased in the anodic direction, and hence could only pass small anodic currents. Conversely, p-type layers on p-type bulk material would not interfere with the passage of an anodic current. This could explain the gross differences in the electrochemical behaviour of ore and coal pyrite. During cyclic voltammetry, ore pyrite passes much higher anodic currents on the second and subsequent cycles than it does on the first cycle, indicating that the passage of current is altering some of the fundamental characteristics of the pyrite[35]. This suggests that electrochemical tests are not the best method for characterizing the oxidation kinetics of a given pyrite sample.

During electrochemical tests, current can be affected by junctions between bulk and surface materials because it passes through the pyrite. During chemical oxidation, however, both cathodic and anodic reactions occur at the surface, and the highest current densities will be in the surface layers, where junction effects are not important. This is the type of oxidation responsible for acid mine drainage, and needs to be better understood. The rate of charge transfer at a semiconductor-solution interface (and hence the rate of the resulting oxidation process) is determined by the extent of overlap between occupied states in the solid and solution. The energy levels of the oxidized and reduced species of a redox couple in solution follow two Gaussian distributions, above and below the standard reduction potential.[36] The surface states of pyrite are at energy levels different from the bulk states, because of both defects associated with termination of the lattice, and trace impurities. There are significant variations in the surface states of different natural pyrite samples, because of the wide range of impurities present at differing concentrations. This may account for the observed differences in reactivity exhibited by ore pyrite or coal pyrite samples from different sources.

Of particular concern here is the fact that a common method for determining the total "acid forming potential" of pyritic material involves "accelerated" oxidation with oxidants such as hydrogen peroxide, with redox potentials significantly higher than that of the ferric/ferrous couple. There is some experience that waste that was inert in this test has subsequently released appreciable acid during oxidation under milder conditions in modern waste management facilities. This could well reflect the fact that the energy levels of the strong redox couple do not overlap the surface states of a sample, whereas those of the milder ferric/ferrous couple do. A definitive understanding of the chemical oxidation of pyrite is clearly needed, to develop reliable methods for predicting the likelihood of forming acid mine drainage.

PREVENTION OF ACID MINE DRAINAGE

The insidious feature of acid mine drainage is the fact that both the chemical and bacterial mechanisms responsible for its formation are autocatalytic; once acid mine drainage develops, the problem tends to worsen, and is extremely difficult to halt. This determines the strategies that are appropriate for addressing acid mine drainage. Once pyritic waste has started to release acid mine drainage in appreciable quantities, often all that can be done is to treat the drainage, to remove acid and dissolved metals, or to separate the waste from the environment in a zero-discharge or controlled-discharge facility. Given the large volumes of waste involved, and the fact that acid mine drainage can be released for decades, even centuries from an affected site, any mitigative action is exceedingly costly, in terms of both capital and operating costs. Instead, preventive action is far preferable, especially when planning new management facilities for pyritic wastes. Because both air and water are needed to oxidize pyrite, oxidation can be prevented by completely excluding one or both from pyritic wastes. For example, if wastes or abandoned mine workings can be flooded completely, the total rate of oxidation is limited by diffusion of oxygen into the water. In principle, oxidation could also be prevented by excluding water from pyritic wastes, although this is not practicable. Nevertheless, it is advantageous to minimize the volume of water coming into contact with pyritic waste, thereby minimizing the volume of water that might need treatment.

Even if pyrite oxidation cannot be completely prevented, it can be controlled at a slow rate by maintaining moderate pH values in any waters in contact with the pyrite. It is common practice to ensure that "sufficient" basic material is present to neutralize any acid evolved. Unfortunately, because of the unpredictable oxidation kinetics of pyrite discussed earlier, it is not clear how much base is actually needed. The amount of pyrite in a waste and the acidity of drainage are only weakly correlated.[37,38,39] Balancing the total amount of acid-generating material with a stoichiometric equivalent of base, or even a reasonable excess, is not necessarily effective, because the rate at which acid is released is unlikely to be equivalent to the rate at which lime can dissolve. Furthermore, inhomogeneities in the waste can cause "hot spots" to develop, in which acid generation overwhelms the capacity of any base present, leading to more generalized formation of acid mine drainage. If the reliability of management procedures for pyritic

waste is to be improved, it will be necessary to predict the oxidation kinetics of the material, in the form in which it is to be disposed of, rather than in the unrepresentative, finely divided and homogenized forms used in typical test procedures.

WEST SHASTA COPPER-ZINC MINING DISTRICT, SHASTA COUNTY, CALIFORNIA

In California, the area most severely afflicted by acid mine drainage problems is the West Shasta Copper-Zinc Mining District, in Shasta County, at the northern end of the Sacramento River Valley. This area clearly illustrates the enormous, adverse environmental impact of acid mine drainage, along with the problems associated with alleviating the problem. The district comprises a series of north-east trending ridges, giving steep, V-shaped valleys, covering an area 2 miles wide by 8 miles long, west of the Sacramento River. Several base-metal mines produced ore in the district, mainly in the first part of this century. Most of the mineral deposits in the the district are thought to have originated as volcanogenic deposits, and contain massive pyrite, chalcopyrite, sphalerite and minor gold and silver. These deposits occur in the Balaklala Rhyolite, a Middle-Devonian volcanic sequence of pyroclastic rocks and porphyritic rhyolite divided into three units, Upper, Middle and Lower, based on the size and occurrence of phenocrysts of quartz and plagioclase in the rock.[40,41] The ore deposits typically occur near the contact between the Upper and Middle units. In addition to stratigraphic control, the orebodies were influenced by folds, faults, bedding plane foliation and fracture cleavage. Some faults acted as channels for the hydrothermal solutions, which formed ore bodies where the channels intersected zones of foliation and cleavage. The rock of the deposits is, therefore, usually highly fractured. In contrast, the country rock away from the zones of mineralization generally has a low hydraulic conductivity (overall 10^{-5} to 10^{-6} cm/sec, depending on the degree of fracturing).

The region experiences heavy precipitation, mainly in winter; the mean annual precipitation varies from about 1.23 m at Shasta Lake to 2.03 m at higher elevations, although in wet years these figures can exceed 2.28 m and 2.54 m, respectively. The average daily temperatures range from -1°C in winter to 32°C in summer. The mean evapotranspiration is about 1 m per annum. The thin soil absorbs water, which infiltrates to the top of the bedrock, giving a thin saturated zone at the soil/rock interface. Ground water flow in this zone follows the top of the rock until it discharges to the surface as a seep or spring, or encounters a fracture that channels it through the rock mass. These channels tend to intersect the orebodies, because of their higher permeability. On former mine sites, large openings such as collapsed workings provide another route for infiltration.[41]

Several mines operated in the district, from the late 1870's until the early 1960's. All are now inactive or abandoned. However there are appreciable amounts of pyritic material still underground, as well as in waste dumps that were placed poorly, often in creeks. The climate favours rapid oxidation of the pyrite, and all mines discharge drainage contaminated by copper, zinc and cadmium, with a low pH. This drainage has had severe effects on aquatic life in the region. The rugged, remote terrain hinders mitigative efforts at the mines. The largest of the mines was Iron Mountain, a 4,400 acre mine complex. The original massive sulphide deposits totalled 22.9 million tonnes, of which 12.9 million tonnes remain, along with 2.8 million tonnes of proven gossan deposits.[42,43,44,45] In 1956, an open pit was developed to replace underground mining; this now tends to retain precipitation, which infiltrates the highly permeable ore and underground workings. Water in the mine has very low pH values that reach 0.6 to 1.0 at the portals, underground temperatures are about 55°C, and there are stalactites and stalagmites of salts coming out of solution.[46] These conditions are more extreme than can be withstood by most strains of *T. ferroxidans*. Moreover, the high temperatures promote convective flow of air, ensuring continuing supplies of oxygen. The drainage is eventually discharged through the mine adits or underground seepage. Water quality data for the mine, and for Spring Creek which carries drainage from the site to the Sacramento River, are given in Table 1.

The Mammoth Mine complex consists of three mines, the largest of which produced 3.0 million tonnes of direct-smelted copper ore and 77,000 tonnes of high-grade zinc ore from 1905 to 1925. The average grade was 3.95% Cu, 4.62% Zn, 0.039 oz/t Au and 2.32 oz/t Ag. Another mine produced 32,300 tonnes of ore with an average grade of 7.44% Cu, 0.08 oz/t Au and 6.4 oz/t Ag. The principal source of acid mine drainage from the complex is the underground workings of the mines, where unmined pyrite is exposed to air and water. A secondary source is surface runoff that passes through poorly sited waste rock dumps, tailings piles and disturbed areas. Table 2 shows the total discharges of heavy metals from the Iron Mountain and Mammoth Mines. Although Mammoth is clearly a much less significant source of pollution, it is still problematic.

Balaklala Mine was worked intermittently from the 1890's until 1956, and yielded an estimated 1.1 million tonnes of ore, averaging 2.8% Cu, 1.3% Zn, 0.028 oz/t Au and 1.0 oz/t Ag. A collapsed stope has created two sink holes for infiltration above two of the three main portals. Acid mine drainage discharges from all three portals, although the most severely affected portal has now been sealed. Waste dumps near the portals contain weathered waste rock, old timbers, etc. They also produce acidic drainage and are devoid of vegetation other than algae. Two other mines near Balaklala have similar problems. The Keystone mine yielded about 112,000 tonnes of ore, averaging 6.0% Cu, 8.0% Zn, 0.06 oz/t Au and 2.7 oz/t Ag, while about 77,000 tonnes of lower grade ore were removed from the Shasta King Mine. Water quality data for the Balaklala area are given in Table 3.

Table 1   Water quality data for Iron Mountain Mine

	Richmond and Hornet Portals	Spring Creek (draining into Sacramento)
Cu (mg/l)	250	3 - 15
Zn (mg/l)	1,400	10 - 80
Cd (mg/l)	12	0.05 - 0.3
Fe (mg/l)	10,000	50 - 350
pH	≥ 0.6	3.0
Flowrate	114 - 1515 l/min	8.5 - 1700 m³/min

Table 2   Estimated total discharges of heavy metals from Iron Mountain Mine and Mammoth Mine complex

	Iron Mountain	Mammoth	
		June 1983	December 1983
Cu (kg/day)	136	23.8	68.4
Zn (kg/day)	1,066	47.8	93.9
Cd (kg/day)	23	0.35	0.24
Fe (kg/day)	-	153.0	646.9

Table 3 Water quality data for Balaklala Mine area

	Windy Camp Portals	Weil Portal (before sealing)	West Squaw Creek (draining into Lake Shasta)
Cu (mg/l)	10 - 15	170	2 - 13
Zn (mg/l)	6 - 30	180	3 - 21
Cd (mg/l)	0.1	-	0.02 - 0.12
Fe (mg/l)	50 - 240	-	9 - 140
pH	2.6 - 2.9	2.0	-
Flowrate m^3/min	0.136 - 39.37	-	-

Table 4 Water quality standards for Sacramento River Basin

	Regional Basin Maximum	California Drinking Water Limits
Cu (mg/l)	0.0056	1.0
Zn (mg/l)	0.016	5.0
Cd (mg/l)	0.010	0.22
pH	6.5 - 8.5	None

Surface runoff from the region drains into Shasta Lake, or lower reservoirs on the Sacramento River. These lakes were constructed for agricultural irrigation and hydroelectric power generation, but are also used for recreation and fishing, and support extensive wildlife. Drinking water for Redding, a city of 45,000 - 50,000, is taken from the Sacramento River a few miles below the West Shasta Region. As is common in California, the Water Quality Control Plan for the Sacramento River Basin has objectives that are significantly more stringent than drinking water requirements, because of the sensitivity of many aquatic organisms to heavy metals. These standards are shown in Table 4.[47,48] The acid mine drainage from the West Shasta Mining District clearly has compromised these standards.

Contaminated creeks are nearly devoid of aquatic life. High concentrations of heavy metal precipitates have been found in reservoir sediments, and these have significantly reduced the number of aquatic invertebrates downstream of discharges of contaminated drainage. In affected areas, fish have concentrations of copper and cadmium in liver tissues exceeding levels found to be detrimental to fish reproduction. The loss of salmonoids in the Sacramento River from copper and zinc toxicity was first noted in 1944. Periodic fish kills occur when there is inadequate dilution of contaminated drainage.[40] This problem was magnified by the completion of Shasta Dam, which reduced upstream dilution flows in the Sacramento River.

CONTROL AND ABATEMENT OF ACID MINE DRAINAGE IN WEST SHASTA

It is evident that the West Shasta Mining District is plagued by numerous abandoned mines discharging acid mine drainage. Pyrite oxidation long ago reached the "runaway" stage, and it is futile to attempt to completely prevent further oxidation. Nevertheless, there is an urgent need for corrective action. The large volumes of contaminated drainage, along with the large seasonal variations in flowrates, preclude the sole use of water treatment to remove acid and heavy metals, on both economic and technical grounds. Given the huge volumes of pyritic material in place, any control measures will be exceedingly costly. It is, therefore, highly instructive to consider which measures have been identified as most cost effective in the numerous engineering studies that have been made of the region.

Hydrologic Controls

The mine most severely affected by acid mine drainage is Iron Mountain. This was placed on the U.S. National Priority List in 1983, making the site eligible for corrective action under the U.S. Superfund program. Thus more capital is available for corrective action than at other sites, and it has been possible to consider more costly approaches, which are likely to be effective over a long time-span. The Environmental Protection Agency (EPA) is overseeing corrective action at the site, and will seek to recover costs from responsible parties, which include Iron Mountain Mines Inc., the current owner, Stauffer Chemical Co., a former owner, and Mountain Copper Ltd., which did most of the mining. EPA considers that by way of corporate acquisitions, ICI Americas holds liability for both Stauffer Chemical Co. and Mountain Copper Ltd. After extensive studies, EPA decided on a two stage approach.[49,50] The first stage, now underway, focuses on hydrologic controls.

About 2.5 acres of cracked and caved ground above one of the ore bodies has been capped, and drainage ditches have been constructed to reduce infiltration through the orebody. Creeks in the region have been diverted, to minimize the volume of pristine surface water that becomes contaminated by drainage from the mine. By reducing the flow in Spring Creek (see Table 1), the existing Spring Creek Reservoir downstream of the site is expected to have sufficient capacity to store the affected drainage. Releases can then be controlled according to water flow in the Sacramento River, to ensure adequate dilution. Surface water has also been diverted around waste rock and slide debris at the site, thereby preventing contamination.

The 1986 Record of Decision also included studies to establish the feasibility of using low-density cellular concrete for underground hydrologic control. It was hoped that placement of seals at strategic points could significantly lower infiltration through the entire ore bodies. It was subsequently found that there has been appreciable caving underground. Given the high underground temperatures and acidic environment, it is now thought that it would be too costly to place concrete underground.[46] However similar measures implemented at the time of mine closure may well prove to be worthwhile. The cost of all measures, both interim and projected final actions, listed in the 1986 Record of Decision was estimated at $72.2 million. The cost of the first stage work is about $10 - 12 million.[46]

Infrared photography of the site has revealed that there are still steam vents above the mineralized zone. This indicates that although capping has been partially effective, there is still significant infiltration. This will be addressed in a second stage of work. It is also likely that some water treatment will be needed.

Hydrologic controls were also considered at the Mammoth Mine complex, and dismissed because of lack of funds. The cost of diverting groundwater from the mineralized zones of each mine, by driving a diversion adit and using long-hole drilling to intercept incoming freshwater, was estimated at $8.3 to $15.7 million.[40] There is significant subsidence in a 16 acre region over the Main Mammoth orebody; however, it was estimated that the acid mine drainage from the mine could be reduced by up to 50% by using impermeable clay material and diversion ditches to reduce infiltration from this.

In June 1980, the owners of Balaklala Mine attempted to divert runoff water around the sink holes that serve as a major source of inflow. Their numerous problems with the diversion pipe, which is now severely corroded and ineffective,[41] demonstrate the need for good engineering when installing hydrologic controls.

Mine Seals

Seals placed judiciously in mine adits, stopes, and occasionally shafts, can prevent the outflow of contaminated water from the mine. In addition, if the seals allow the mine to flood, oxygen is excluded, halting the formation of acid mine drainage. Flooding is probably not a realistic goal in the West Shasta district, because of the high permeability of the ore, and because the seasonal rainfall would cause the water-table to rise and fall appreciably. Nevertheless, seals have proved to be a highly cost effective means of reducing contaminated drainage at some sites. ICI Americas has proposed to install seals at Iron Mountain, but the risks at this site are probably too great.[46]

One of the problems that can be encountered in retrofitting seals in fractured ore deposits is demonstrated by the experience at Mammoth. In 1981, a concrete plug was installed in the Main Mammoth portal, which was the principal source of acid mine drainage from the Mammoth Complex. This reduced metal loadings in the receiving waters by 80-90%. About 3 months later, flow and metals concentrations increased from another portal. A few weeks after this was plugged, flows increased by a factor of 3 from another portal, about 65 meters higher. The valve at one of the lower plugs was opened in spring 1984, at the request of the Regional Water Quality Control Board, which was concerned about fish kills. The Mammoth Mine complex was studied in 1985, to evaluate the effectiveness of the portal plugging program, and assess the efficacy of installing additional plugs.[40] Although it was felt that up to 29 additional plugs might be needed to seal the mine effectively, at $5.5 million, this was still the most cost-effective approach.

In December 1981, a steel-reinforced concrete plug was installed in the Balaklala Weil adit, 235 feet from the portal entrance. Although there have been problems with leakage, the seal has controlled a major percentage of the acid mine drainage to West Squaw Creek; 1983 data indicated that the total copper outflow was reduced by 90% from the levels before sealing (see Table 3). The cost of installing similar seals at the other portals was estimated at $65,000.

The cost of installing seals, and their subsequent performance, is strongly dependent on the rock quality. This problem can be minimized if the mine is designed to facilitate sealing on closure.

Given the prevalence of acid mine drainage, this should be done on all new underground mines in sulphide deposits.

Backfilling

Backfilling was considered for the Mammoth mine. This would be technically difficult, and the most expensive of the alternatives ($31 to 59 million); attempts at backfilling the mine workings at Mammoth in the 1940's were unsuccessful, and geophysical techniques would be needed now to locate the old workings. Moreover, new fill material would have to be quarried. Nevertheless, backfilling with waste rock or tailings during mining or at the time of closure may well be appropriate for modern underground mines.

Water Treatment

Cementation plants were operated intermittently at Iron Mountain from the 1960's, to remove copper from drainage collected from mine portals and conveyed by a system of flumes. However cementation does not remove acid or other heavy metals. For these, lime/limestone neutralization is needed. The cost of installing a lime treatment plant for the drainage from the main portals at Iron Mountain was estimated at $10 million for construction of the treatment plant and $1 million annually for operating and maintenance. In addition to high costs, it is difficult to ensure that all contaminated drainage from an extensive site is diverted to a treatment plant. Moreover, lime neutralization creates large volumes of sludge, contaminated by heavy metals, requiring disposal.

Other treatment methods, such as ion exchange, solvent extraction, electroflocculation and reverse osmosis have been evaluated for use at Iron Mountain, in the hope of producing a marketable product (i.e. Cu, Zn, Au/Ag, etc.). The feasibility of these processes appears to be highly doubtful. Again, however, these techniques should be considered for water treatment at new facilities.

In-situ leaching

Iron Mountain Mines Inc., which owns the Iron Mountain site, has proposed in-situ leaching of the orebody to extract copper, zinc, iron and precious metals, and recovery of the base metals as industrial and agricultural chemicals.[43,45]

The in-situ operation would require sealing the main portals, and recirculating acid mine drainage back into the mountain to build up reasonable concentrations of metal ions. Concentrated leach solution would be treated in a copper solvent extraction plant, to produce 6 tpd cathode copper and 24 tpd copper sulphate. A bleed from the copper solvent extraction would pass through carbon columns (or ion exchange columns) for precious metals recovery. The solution would then go for precipitation of jarosite with ammonia. Some of the jarosite would be "acidulated" (presumably to desorb heavy metals) for sale, the rest would be dissolved, and used to produce ferrous sulphate. The solution would then go to cadmium cementation with zinc, followed by zinc solvent extraction and production of 12 tpd zinc sulphate or hydroxide. The bleed solution would finally go to alum and gypsum recovery, followed by lime neutralization and discharge.

EPA chose not to accept this proposal, because the published proposal was not sufficiently detailed to assess its technical feasibility, or economic viability. Moreover, Iron Mountain Mines Inc. did not have adequate financial backing for the venture. Of particular concern was the iron and sulphate balance; unlike copper in-situ operations being tested and developed in chalcopyrite or bornite deposits in other parts of the United States, Iron Mountain is primarily a pyrite deposit. The pyrite generates acid, which would require neutralization before solvent extraction.

CONCLUSIONS

Acid mine drainage is probably the greatest environmental problem that can arise from the development of sulphide ore deposits. The oxidation behaviour of pyritic material is autocatalytic; once it starts, it is almost impossible to halt. Hence every effort must be made to prevent oxidation from occurring in waste management facilities. The kinetics of pyrite oxidation vary considerably with the source, and a much better understanding is needed before the behaviour of waste can be predicted reliably. The West Shasta Mining District in northern California demonstrates the widespread environmental damage that is caused by acid mine drainage. The mitigative efforts in this district are extremely costly, and difficult to maintain. Experience suggests that new underground mines should be designed to facilitate backfilling on closure, with suitable hydrologic controls.

ACKNOWLEDGEMENTS

This work has been sponsored by the United States Department of the Interior, Bureau of Mines, under Grant Number G1175132, and by the Legislature of the State of California. All opinions stated in this publication, however, are those of the author. The author wishes to thank Rick Sugarek, Remedial Project Manager - Iron Mountain Mine, U.S. Environmental Protection Agency, for reviewing the manuscript.

References

1. McAndrew, R.T., Wang, S.S., and Brown, W.R. Precipitation of iron compounds from sulphuric acid leach solutions. CIM Bulletin, vol. 68, no. 753, 1975, p. 101-110.
2. Haigh, C.J. The hydrolysis of iron in acid solutions. Proceedings of the Australian Institute of Mining and Metallurgy, Sept. 1967, p. 49-56.
3. Onysko, S.J. Chemical abatement of acid mine drainage formation. Ph.D. Dissertation, University of California at Berkeley, 1985.
4. Robertson, A.M. Alternative acid mine drainage abatement measures. Report, Steffen Robertson and Kirsten (B.C.) Inc., Vancouver, B.C., Canada, 1987.
5. Andrews, G. An examination of the kinetics of bacterial pyrite decomposition. In: Biotechnology in minerals and metal processing. Edited by B.J. Scheiner, F.M. Doyle and S.K. Kawatra. Littleton, CO: Society of Mining Engineers, Inc., 1989. p. 87-93.
6. Singer, P.C. and Stumm, W. Acid mine drainage: the rate determining step, Science, vol. 167, 1970, p. 1121-1123.
7. Lacey, D.T. and Lawson, F. Kinetics of the liquid-phase oxidation of acid ferrous sulfate by the bacterium Thiobacillus ferrooxidans. Biotechnology and Bioengineering, vol. 12, 1970, p. 29-50.
8. Nordstrom, D.K. The rate of ferrous iron oxidation in a stream receiving acid mine effluent. In: Selected papers in the hydrological sciences. U.S. Geological Survey Water-Supply Paper 2270, 1985, p. 113-119.
9. Peters, E. Leaching of sulfides. In: Advances in mineral processing. Edited by P. Somasundaran. Littleton, CO: Society of Mining Engineers, Inc., 1986, p. 445-462.
10. Berry, V.K., Murr, L.E. and Hiskey, J.B. Galvanic interaction between chalcopyrite and pyrite during bacterial leeching of low-grade waste. Hydrometallurgy, vol. 3, 1978, p. 309-326.
11. Murr, L.E. and Mehta, A.P. The role of iron in metal sulfide leaching by galvanic interaction. Biotechnology and Bioengineering, vol. 25, 1983, p. 1175-1180.
12. Mehta, A.P. and Murr, L.E. Fundamental studies of the contribution of galvanic interaction to acid-bacterial leaching of mixed metal sulfides. Hydrometallurgy, vol. 9, 1983, p. 235-256.
13. Frost, D.C. et al. An XPS study of oxidation of pyrite and pyrites in coal. Fuel, vol. 56, 1977, p. 277-280.
14. Michell, D. and Woods, R. Analysis of oxidized layers on pyrite surfaces by X-ray emission spectroscopy and cyclic voltammetry. Australian Journal of Chemistry, vol. 31, 1978, p. 27-34.
15. Mishra, K.K. and Osseo-Asare, K. Photodissolution of coal pyrite. Fuel, vol. 66, 1987, p. 1161-1162.
16. Mishra, K.K. and Osseo-Asare, K. Electrodeposition of H^+ on oxide layers at pyrite (FeS$_2$) surfaces. Journal of the Electrochemical Society, vol. 135, 1988, p. 1898-1901.
17. Kawakami, K. et al. Kinetic study of oxidation of pyrite slurry by ferric chloride. Industrial Engineering and Chemical Research, vol. 27, 1988, p. 571-576.
18. Mishra, K.K. and Osseo-Asare, K. Aspects of the interfacial electrochemistry of semiconductor pyrite (FeS$_2$). Journal of the Electrochemical Society, vol. 135, 1988, p. 2502-2509.
19. Briceno, A. and Chander, S. Kinetics of pyrite oxidation. Minerals and Metallurgical Processing, Aug. 1977, p. 171-176.
20. Caruccio, F.T. The quantification of reactive pyrite by grain size distribution. In Preprints: Third symposium on coal mine drainage research. Pittsburgh, PA, 1970. p. 123-131.
21. Esposito, M.C., Chander, S. and Aplan, F.F. Characterization of pyrite from coal sources. AIME Annual Meeting, February 1987, Denver, CO.
22. Hammack, R.W., Lai, R.W. and Diehl, J.R. Methods for determining fundamental chemical differences between iron disulfides from different geologic provenances. In: Mine drainage and surface mine reclamation, Vol. 1. U.S. Department of the Interior,Bureau of Mines Information Circular 9183, 1988, p. 136-146.
23. Ferreira, R.C.H. High temperature E-pH diagrams for the systems S-H$_2$O, Cu-S-H$_2$O and Fe-S-H$_2$O. In: Leaching and reduction in hydrometallurgy. Edited by A.R. Burkin. London: Institution of Mining and Metallurgy, 1975, p. 67-83.
24. Sato, M. Oxidation of sulfide ore bodies, II. Oxidation mechanisms of sulfide minerals at 25°C. Economic Geology, vol. 55, 1960, p. 1202-1231.
25. Peters, E. The electrochemistry of sulphide minerals. In: Trends in electrochemistry. Edited by J. O'M. Brockris, D.A.J. Rand and B.J. Welch. New York: Plenum Publishing Corp., 1977. p. 267-290.
26. Hiskey, J.B. and Pritzker, M.D. Electrochemical behavior of pyrite in sulfuric acid solutions containing silver ions. Journal of Applied Electrochemistry, vol. 18, 1988, p. 484-490.
27. Buckley, A.N., Hamilton, I.C. and Woods, R. In: Proceedings of the international symposium on electrochemistry in mineral and metal processing. Edited by P.E. Richardson, S. Srinivasan and R. Woods. The Electrochemical Society, 1984. p. 234-246.
28. Ennaoui, A., Fiechter, S., Jaegermann, W. and Tributsch, H. Photoelectrochemistry of highly quantum efficient single-crystalline n-FeS$_2$ (pyrite). Journal of the Electrochemical Society, vol. 133, 1986, p. 97-106.
29. Ogunsola, O.M. and Osseo-Asare, K. The electrochemical behaviour of coal pyrite, 1. Effects of mineral source and composition. Fuel, vol. 65, 1986, p. 811-815.
30. Chander, S. and Briceno, A. The rate of oxidation of pyrites from coal and ore sources - an AC impedance study. In: Mine drainage and surface mine reclamation, vol. 1. U.S. Department of the Interior, Bureau of Mines Information Circular 9183, 1988, p. 164-169.

31. Gerischer, H. Semiconductor electrode reactions. In: _Advances in electrochemistry and electrochemical engineering._ Edited by P. Delahay and C.W. Tobias. Interscience Publishers, Inc., 1961. p. 139-232.

32. Myamlin, V.A. and Pleskov, Y.V. _Electrochemistry of semiconductors._ New York: Plenum Press, 1967. p. 450.

33. Simkovich, G. and Wagner, J.B., Jr., The influence of point defects on the kinetics of dissolution of semiconductors. _Journal of the Electrochemical Society,_ vol. 110, 1963, p. 513-516.

34. Eadington, P. and Prosser, A.P. Oxidation of lead sulfide in aqueous suspensions. _Transactions of IMM,_ vol. 78, 1969, p. C74-82.

35. Doyle, F.M. and Mirza, A.H. Understanding the mechanisms and kinetics of acid and heavy metals release from pyritic wastes. In: _Mining and mineral processing wastes._ Edited by F.M. Doyle. Littleton, CO: SME, 1990. p. 43-51.

36. Crundwell, F.K. The influence of the electronic structure of solids on the anodic dissolution and leaching of semiconducting minerals. _Hydrometallurgy,_ vol. 21, 1988, p. 155-190.

37. diPretoro, R.S. and Rauch, H.W. The use of acid-base accounts in premining prediction of acid drainage potential: A new approach for northern West Virginia. In: _Mining drainage and surface mine reclamation, vol. 1._ U.S. Department of the Interior, Bureau of Mines Information Circular 9183, 1988. p. 2-10.

38. Erickson, P.M. and Hedin, R.S. Evaluation of overburden analytical methods as means to predict post-mining coal drainage quality. In: _Mining drainage and surface mine reclamation, vol. 1._ U.S. Department of the Interior, Bureau of Mines Information Circular 9183, 1988. p. 11-19.

39. Miller, S.D. and Murray, G.S. Application of acid-base analysis to wastes from base metal and precious metal mines. In: _Mining drainage and surface mine reclamation, vol. 1._ U.S. Department of the Interior, Bureau of Mines Information Circular 9183, 1988. p. 29-32.

40. Mammoth Mine water quality management planning study. CH2M HILL. Prepared for the California Regional Water Quality Control Board, Central Valley Region, August 1985.

41. Control of acid and heavy metal discharge from Balaklala, Keystone and Shasta King Mine sites. Advanced Environmental Consultants, Inc. Pittsburgh, PA. Prepared for the California Regional Water Quality Control Board, Central Valley Region, June 1983.

42. Alternatives for control of toxic metal discharges from representative surface areas at Iron Mountain Mine, near Redding, California. Ott Water Engineers. Prepared for the California Regional Water Quality Control Board, Central Valley Region, October 1982.

43. Iron Mountain Mines, Inc., Alternative Proposal. In-situ leaching and hydrometallurgical recovery of metals and metal salts for agriculture and industry. Iron Mountain Mines, Shasta County, Davy McKee, July 1985.

44. Summary of remedial alternative selection. Iron Mountain Mine. Redding, CA: U.S. Environmental Protection Agency, September 19, 1986.

45. Turk, T.W., McLean, D.C. and Arman, T.W. In-situ mining of massive sulfides at Iron Mountain Mines, California. _SME Fall Meeting,_ Littleton, CO: Society of Mining Engineers, Inc., September 1986, preprint 86-343.

46. Sugarek, R., Remedial Project Manager, Iron Mountain Mine (U.S. Environmental Protection Agency). Private communication, June 12, 1990.

47. California Regional Water Quality Control Board, Central Valley Region. Resolution No. 84-054, April 1984.

48. State Water Resources Control Board. Resolution No. 84-55, August 1984.

49. Record of Decision, Iron Mountain Mine, Redding, CA, U.S. Environmental Protection Agency, October 3, 1986.

50. Iron Mountain Mine Superfund Site. Fact sheet, Interim Remedial Action Program and Source Control Pilot Study. U.S. Environmental Protection Agency, July 1987.